I0882557

Advanced Material Innovations and Industrial Applications

Mohamed El Khaili
https://orcid.org/0000-0001-8730-2981
Hassan II University of Casablanca, Morocco

Naoual Belouaggadia
Hassan II University of Casablanca, Morocco

Abdellah Zamma
Hassan II University of Casablanca, Morocco

IGI Global
Scientific Publishing
Publishing Tomorrow's Research Today

Vice President of Editorial Melissa Wagner
Director of Acquisitions Mikaela Felty
Director of Book Development Jocelynn Hessler
Production Manager Mike Brehm
Cover Design Jose Rosado

Published in the United States of America by
IGI Global Scientific Publishing
701 East Chocolate Avenue
Hershey, PA, 17033, USA
Tel: 717-533-8845 | Fax: 717-533-7115
Website: https://www.igi-global.com E-mail: cust@igi-global.com

Copyright © 2027 by IGI Global Scientific Publishing. All rights reserved. No part of this publication may be reproduced, stored or distributed in any form or by any means, electronic or mechanical, including photocopying, without written permission from the publisher. Use of this publication to train generative artificial intelligence (AI) technologies is expressly prohibited. The publisher reserves all rights to license its use for generative AI training and machine learning model development.

Product or company names used in this set are for identification purposes only. Inclusion of the names of the products or companies does not indicate a claim of ownership by IGI Global Scientific Publishing of the trademark or registered trademark.

Library of Congress Cataloging-in-Publication Data

Names: El Khaili, Mohamed, 1969- editor. | Belouaggadia, Naoual, 1973- editor. | Zamma, Abdellah, 1974- editor.
Title: Advanced material innovations and industrial applications / edited by Mohamed El Khaili, Naoual Belouaggadia, Abdellah Zamma.
Description: Hershey, PA : Engineering Science Reference, [2027] | Includes bibliographical references and index. | Summary: "The book "Advanced Material Innovations and Industrial Applications" amounts to an optimal platform for researchers, engineers, academicians as well as industrial professionals from all over the world. The authors will have the opportunity to publish their research results, development activities, and cases of study in advanced materials science, aiming to promote the sustainable developments and innovation. This book provides opportunities for the chapter authors to exchange new ideas and experiences to establish research networking and to find global partners for future collaboration"-- Provided by publisher.
Identifiers: LCCN 2024042529 (print) | LCCN 2024042530 (ebook) | ISBN 9798369366271 (hardcover) | ISBN 9798369366288 (paperback) | ISBN 9798369366295 (ebook)
Subjects: LCSH: Materials--Technological innovations.
Classification: LCC TA403.6 .A334 2025 (print) | LCC TA403.6 (ebook) | DDC 620.1/1--dc23/eng/20241223
LC record available at https://lccn.loc.gov/2024042529
LC ebook record available at https://lccn.loc.gov/2024042530

British Cataloguing in Publication Data
A Cataloguing in Publication record for this book is available from the British Library.

All work contributed to this book is new, previously-unpublished material.
The views expressed in this book are those of the authors, but not necessarily of the publisher.
This book contains information sourced from authentic and highly regarded references, with reasonable efforts made to ensure the reliability of the data and information presented. The authors, editors, and publisher believe the information in this book to be accurate and true as of the date of publication. Every effort has been made to trace and credit the copyright holders of all materials included. However, the authors, editors, and publisher cannot assume responsibility for the validity of all materials or the consequences of their use. Should any copyright material be found unacknowledged, please inform the publisher so that corrections may be made in future reprints.

Table of Contents

Detailed Table of Contents

This chapter offers a thorough analysis of the state-of-the-art discoveries and wide range of industrial uses for graphene and carbon nanotubes (CNTs). The remarkable mechanical, electrical, thermal, and chemical capabilities of these nanomaterials are well known, and they have sparked important developments in a number of different fields. The chapter starts with a brief history of the discovery of graphene and carbon nanotubes (CNTs), and then explores their distinct structural properties and the different synthetic techniques used to create them. Their remarkable tensile strength and elasticity are highlighted in-depth examinations of their mechanical characteristics, which make them perfect for high-performance composite materials. The exceptional electrical qualities of these nanomaterials, such as their high conductivity and electron mobility, which have transformed the electronics and optoelectronics industries, are also covered in this chapter. Examined are thermal attributes such strong thermal conductivity.

Fly ash, a waste product from coal power plants, emerges as a viable and eco-friendly building material. This chapter explores the advantages of fly ash bricks, including cost-efficiency, durability, and thermal insulation. Fly ash's incorporation in construction reduces environmental impact by lowering greenhouse gas emissions and conserving landfill space. While challenges like slower curing and freeze-thaw susceptibility exist, fly ash bricks offer superior strength, fire resistance, and lower maintenance needs. The chapter compares fly ash bricks to alternatives and explores the projected market growth driven by sustainability concerns. Research highlights fly ash's role in pollution reduction and its suitability for various construction applications. Market trends indicate robust growth, with the fly ash brick sector projected to expand at a CAGR of 6.34% to exceed 1090 million tonnes by 2032, led by Asia-Pacific demand and supported by global industry players. Overall, fly ash bricks represent a promising path towards a more sustainable construction industry.

The objective of this research is to present a new approach to physical desalination using Lorentz forces, electrostatic forces and flow velocity. This process involves a combination of magnetic and electric fields to separate the charged ions from the soluble salts in seawater. The added value of this desalination process lies in the downstream coupling of Magnetohydrodynamics (MHD) and Capacitive Deionization (CDI) under an external magnetic field, followed by centrifugation and crystallization for efficient brine valorization and Zero Liquid Discharge (ZLD). Our contribution is aimed at developing an environmentally sustainable process that not only produces fresh water, but also solves the problem of brine discharge and its environmental impact on the ecosystem. The recovery of residual brine as a by-product of the seawater desalination process allows valuable materials to be extracted, by recovering ions of heavy metals such as lithium, bromine, phosphorus, arsenic, zinc, copper, nickel, silver, iron and manganese through a mechanism of concentration and precipitation in the solid state. This green technology is chemical-free and sustains the environment.

Advanced optical materials have driven innovations in telecommunications, biomedicine, and energy by manipulating light at the nanoscale. These materials enhance absorption, emission, and nonlinear optical responses. Quantum dots and photonic crystals boost data transmission and device efficiency, while quantum dots and plasmonic nanoparticles advance imaging and diagnostics in biomedicine. In energy, quantum dot solar cells and plasmonic nanostructures improve solar energy harvesting. Additionally, these materials are utilized in environmental monitoring, high-resolution displays, and security systems, aiding in pollution detection, energy efficiency, and national security. Despite progress, challenges in stability, scalability, and cost-effectiveness persist. Ongoing interdisciplinary research and collaboration are essential for overcoming these challenges and expanding applications. Advanced optical materials promise to address societal issues and drive technological progress.

While procurement can be considered a critical process in supply chain management, since it can provide new opportunities for supply chains to improve their efficiency and effectiveness. Most of companies have been thinking about how logistics 4.0 and blockchain could help them be more efficient and minimizing risks of error and disputes. Therefore, the applications of Industry 4.0 technologies such as blockchain, the Internet of Things and Artificial Intelligence in procurement processes have assisted organizations to improve their performance and to deal with SCM's environmental uncertainties. Implementing a materials supply policy via logistics 4.0 involves leveraging advanced and cutting-edge technologies to optimize the sourcing process and to revolutionize SC management and the way materials are sourced, managed, and distributed. What can offer several advantages such as enhanced transparency, traceability and efficiency in the SC. The adoption of blockchain technology into the materials procurement process has huddled a crucial consideration within industry in recent years.

Magnetic materials have revolutionized modern industry with their versatile applications across various sectors. This chapter explores the comprehensive landscape of these materials, starting with their historical evolution and fundamental principles. It delves into traditional and modern synthesis methods, highlighting techniques like sol-gel, sputtering, and vapor deposition. Detailed characterization methods such as X-ray diffraction, SEM, and VSM are discussed to elucidate the properties of magnetic materials. The chapter also examines their pivotal role in energy applications, including renewable energy systems and superconducting magnets, and their integration into electronics and sensing technologies. Biomedical applications, such as drug delivery and MRI contrast agents, are explored with a focus on biocompatibility. Emerging trends like quantum materials and spintronics are considered, providing a forward-looking perspective. The concluding part summarizes key advancements, addresses industry challenges, and predicts future research and application trends.

This study conducts a comprehensive bibliometric analysis of machine learning (ML) applications in Structural Health Monitoring (SHM) from 2003 to early 2024. Utilizing the OpenAlex database, we analyzed 806 publications using keywords related to ML and SHM. The data was processed using VOSviewer software to examine co-citation patterns, keyword frequency, author/institution collaboration networks, and citation analysis. Our findings reveal an exponential growth in research output, with 80% of publications occurring in the last four years. The United States, China, and the United Kingdom emerge as leading contributors. Keyword analysis indicates a shift towards advanced ML techniques like convolutional neural networks. The study identifies key research clusters, influential publications, and emerging trends in the field. This analysis provides valuable insights for researchers and practitioners, highlighting potential areas for future investigation in ML-based SHM.

Preface

INTRODUCTION

Advances in materials science continue to shape the future of engineering, manufacturing, energy, infrastructure, healthcare, and environmental sustainability. As industries face increasing demands for high-performance materials, resource efficiency, and environmentally responsible technologies, the development of innovative materials and their practical applications have become a cornerstone of scientific and industrial progress. *Advanced Material Innovations and Industrial Applications* bring together cutting-edge research that highlights these advancements while emphasizing their role in addressing contemporary global challenges.

This volume serves as a platform for researchers, engineers, academicians, and industrial professionals to share novel discoveries, technological developments, and practical case studies spanning a wide range of materials science disciplines. The contributions reflect the interdisciplinary nature of modern materials research, illustrating how innovations in material design, characterization, manufacturing processes, and digital technologies are transforming industrial practices across sectors including construction, renewable energy, transportation, logistics, environmental engineering, electronics, and healthcare.

The primary objective of this book is to make recent advances in materials science accessible to a broad audience while presenting state-of-the-art research applicable to diverse industrial contexts. It promotes collaboration among the global scientific community by encouraging the exchange of ideas, methodologies, and emerging technologies that foster future research partnerships. In addition, the volume highlights innovative analytical tools, computational models, and next-generation approaches that support the development and optimization of advanced materials, with particular attention to sustainable manufacturing, energy efficiency, recycling, waste management, and environmentally conscious industrial processes. Through the inclusion of both theoretical discussions and practical case studies, the book enriches the growing body of literature dedicated to advanced materials and their real-world implementation.

Designed primarily for scientific researchers, doctoral candidates, and graduate students specializing in materials science and engineering, this book also provides valuable insights for practicing engineers, industrial decision-makers, and professionals seeking to implement advanced material technologies within their organizations. By bridging fundamental research with industrial applications, the volume aims to support innovation, stimulate interdisciplinary collaboration, and contribute to the sustainable advancement of materials science worldwide.

CHAPTER OVERVIEW

Chapter 1 examines the remarkable properties of graphene and carbon nanotubes, exploring their synthesis, mechanical, electrical, thermal, and chemical characteristics alongside their rapidly expanding applications in high-performance composites, electronics, and optoelectronic systems.

Chapter 2 reviews the use of fly ash as a sustainable construction material, discussing its environmental benefits, structural performance, economic advantages, market trends, and its growing role in promoting greener construction practices.

Chapter 3 presents an innovative electromagnetic desalination process that combines magnetic, electric, centrifugation, and crystallization mechanisms to produce freshwater while recovering valuable minerals from brine in an environmentally sustainable, chemical-free approach.

Chapter 4 explores recent advances in optical materials and nanostructures, highlighting their transformative applications in telecommunications, biomedical imaging, renewable energy, environmental monitoring, security systems, and advanced display technologies while discussing ongoing challenges and future research directions.

Chapter 5 investigates the integration of blockchain technology, Logistics 4.0, artificial intelligence, and the Internet of Things into materials procurement, demonstrating how digital technologies improve transparency, traceability, efficiency, and resilience within modern supply chains.

Chapter 6 provides a comprehensive overview of magnetic materials, covering their synthesis, characterization, and applications in renewable energy, electronics, sensing technologies, biomedical engineering, superconductivity, quantum materials, and spintronics while outlining future industrial opportunities.

Chapter 7 presents a bibliometric analysis of machine learning applications in structural health monitoring, identifying major research trends, collaboration networks, influential publications, and the rapid growth of artificial intelligence techniques in infrastructure monitoring.

Chapter 8 examines piezoelectric energy harvesting technologies using conventional and nanocomposite materials, comparing their performance under linear vibrations and demonstrating how advanced materials can significantly improve energy conversion efficiency for sustainable power generation.

Chapter 9 explores the application of machine learning models to predict Mohs hardness, evaluating multiple algorithms while demonstrating how data-driven approaches enhance material characterization and support industrial and scientific decision-making.

Chapter 10 concludes the volume by examining policy and strategic frameworks supporting green hydrogen and renewable energy deployment in emerging economies through the case of Morocco, offering practical insights into regulatory development, infrastructure planning, investment strategies, and sustainable energy transitions.

CONCLUSION

The chapters collected in this volume demonstrate the breadth and depth of contemporary materials science, showcasing innovations that span nanomaterials, smart materials, digital manufacturing technologies, sustainable construction, renewable energy, and data-driven materials engineering. Collectively, these contributions illustrate how interdisciplinary research is accelerating technological progress while addressing critical global priorities related to sustainability, resource efficiency, and industrial competitiveness.

It is our hope that this book will serve as a valuable reference for researchers, educators, students, and industry professionals alike, inspiring further scientific inquiry, technological innovation, and collaborative efforts that advance the development and responsible application of advanced materials for the benefit of society.

Mohamed El Khaili
Hassan II University of Casablanca, Morocco

Naoual Belouaggadia
Hassan II University of Casablanca, Morocco

Abdellah Zamma
Hassan II University of Casablanca, Morocco

Chapter 1 Carbon Nanotube and Graphene: Advancements and Industrial Applications

Niharika Das
http://orcid.org/0009-0008-4524-5198
Centurion University of Technology and Management, India

Siddharth Parthasarathy
http://orcid.org/0000-0001-8319-3239
Centurion University of Technology and Management, India

Subhraraj Panda
http://orcid.org/0000-0003-4224-4663
Centurion University of Technology and Management, India

Santanu Kumar Nayak
http://orcid.org/0009-0004-2143-7655
Centurion University of Technology and Management, India

Dipan Kumar Das
http://orcid.org/0000-0003-0224-3295
Centurion University of Technology and Management, India

Padmaja Patnaik
http://orcid.org/0000-0003-4468-7871
Centurion University of Technology and Management, India

ABSTRACT

This chapter offers a thorough analysis of the state-of-the-art discoveries and wide range of industrial uses for graphene and carbon nanotubes (CNTs). The remarkable mechanical, electrical, thermal, and chemical capabilities of these nanomaterials are well known, and they have sparked important developments in a number of different fields. The chapter starts with a brief history of the discovery of graphene and carbon nanotubes (CNTs), and then explores their distinct structural properties and the different synthetic techniques used to create them. Their remarkable tensile strength and elasticity are highlighted in-depth examinations of their mechanical characteristics, which make them perfect for high-performance composite materi-

DOI: 10.4018/979-8-3693-6627-1.ch001

2027, IGI Global Scientific Publishing. Copying or distributing in print or electronic forms without written permission of IGI Global Scientific Publishing is prohibited. Use of this chapter to rative artificial intelligence (AI) technologies is expressly prohibited. The publisher reserves all rights to license its use for generative AI training and machine learning model development.

als. The exceptional electrical qualities of these nanomaterials, such as their high conductivity and electron mobility, which have transformed the electronics and optoelectronics industries, are also covered in this chapter. Examined are thermal attributes such strong thermal conductivity.

1. INTRODUCTION

1.1 Overview of Carbon Nanotubes and Graphene

Carbon nanotubes (CNTs) and graphene are advanced carbon-based materials with unique structures and properties. CNTs are cylindrical tubes composed of rolled-up graphene sheets, available in single-walled (SWCNTs) or multi-walled (MWCNTs) forms, known for their exceptional strength, electrical conductivity, and lightweight nature. They find applications in electronics, materials science, and biomedicine, owing to their robust mechanical properties and conductivity. Graphene, on the other hand, consists of a single layer of carbon atoms arranged in a hexagonal lattice, offering extraordinary strength, flexibility, and high electrical and thermal conductivity. It is utilized in diverse fields such as electronics (for ultrafast transistors and transparent conductive films), energy storage (batteries and supercapacitors), and biomedical applications (drug delivery and biosensors), thanks to its unique 2D structure and remarkable properties. Together, CNTs and graphene represent cutting-edge materials with immense potential for technological advancement across various industries.

Carbon nanotubes (CNTs) and graphene are remarkable carbon-based materials with diverse applications in various fields. CNTs, synthesized through processes like laser ablation and chemical vapor deposition, possess high thermal and electronic conductivity, making them ideal for drug delivery systems and as antioxidants in medicine and pharmacy. On the other hand, graphene, along with CNTs, is crucial in energy storage, generation, and conversion applications like batteries and photovoltaics. Both materials exhibit exceptional mechanical properties, optical and electronic characteristics, and are utilized in photonics, optoelectronics, and nanoelectronics (Subhedar, K. M., & Singh, B. P.,2023). Moreover, CNTs and graphene are promising for terahertz detection due to their outstanding electron mobility and spectral absorption properties (Nimanpure et al., 2023). Additionally, these materials show great potential in flexible thermoelectric devices, offering high performance and cost-efficiency (Tyagi et al., 2023). Overall, carbon nanotubes and graphene represent cutting-edge materials with a wide range of applications in modern technology and science.

The author does an exhaustive comparison between carbon nanotubes (CNTs) and graphene, considering them together in terms of their structure, properties, and use. There are references here to the method of preparation (CVD, arc discharge, laser ablation, epitaxy, reduction of graphene oxide) and other properties such as the amount of defects and chirality. There is also an important part of the text dedicated to the combination of CNTs and graphene, referring to the binding process and the transport of charges. Aspects such as regulations, the environment, and security are also included, as well as the prospects of use.

1.2 Historical Background and Discovery

Carbon nanotubes (CNTs) and graphene, discovered in 1991 and 2004, respectively, have revolutionized nanotechnology and materials science. CNTs, first observed by Sumio Iijima, were discovered due to their unique properties, leading to further research into mechanical strength, electrical conductivity, and potential applications in various industries, including electronics, energy, and medicine.

The historical background and discovery of carbon nanotubes (CNTs) and graphene showcase their evolution from basic carbon materials to advanced nanoscale structures with exceptional properties. Initially used as current conductors for light emission, carbon's low intensity and durability led to the development of CNTs and graphene, offering lightweight, efficient, and stable alternatives with superior electrical and mechanical performance (Dariyal et al., 2023). These materials have become integral in various fields, including energy harvesting devices like photovoltaics, where SWCNTs demonstrate ideal characteristics for photovoltaic applications due to their wide band gaps, high charge carrier mobility, and light absorption capabilities (Kumar, P.,2023). Furthermore, CNTs and graphene have found applications in terahertz electronics and photonics, showcasing outstanding electron mobility and spectral absorption properties suitable for terahertz detection systems. Overall, the historical journey of CNTs and graphene highlights their transformation into key materials for diverse technological advancements and future applications (Subhedar, K. M., & Singh, B. P.,2023).

Table 1. properties of CNTS and graphenes

Property	Carbon Nanotubes (CNTs)	Graphene
Structure	Cylindrical structure formed by rolled-up graphene sheets	Single layer of carbon atoms arranged in a hexagonal lattice
Dimensionality	One-dimensional (1D)	Two-dimensional (2D)
Strength	Extremely high tensile strength and stiffness	Strongest known material

continued on following page

Table 1. Continued

Property	Carbon Nanotubes (CNTs)	Graphene
Conductivity	Excellent electrical and thermal conductivity	Exceptional electrical and thermal conductivity
Flexibility	Can be bent and twisted without breaking	Highly flexible and elastic
Surface Area	High surface area due to its tubular structure	Extremely high surface area
Optical Properties	Can exhibit unique optical properties, such as photoluminescence	Transparent and highly absorbent of light
Applications	Electronics, composites, energy storage, sensors	Electronics, composites, energy storage, filtration, sensors

2. SYNTHESIS METHODS

2.1 Chemical Vapor Deposition (CVD)

Chemical Vapor Deposition (CVD) is a versatile technique widely used for the synthesis of carbon nanotubes (CNTs) and graphene. In CNT growth, a hydrocarbon precursor gas (such as methane or ethylene) is introduced into a high-temperature reactor along with a catalyst material, typically transition metals like iron, nickel, or cobalt. Under controlled conditions, the hydrocarbon decomposes on the catalyst surface, forming carbon atoms that nucleate and grow into nanotubes. The catalyst particles play a crucial role in determining the diameter and chirality of the nanotubes produced. Alternatively, for graphene synthesis via CVD, a carbon-containing gas (often methane or ethylene) is introduced onto a heated metal substrate, such as copper or nickel. The carbon atoms from the gas adsorb onto the substrate surface, where they self-assemble into a single layer of graphene. CVD offers precise control over the growth parameters, enabling the production of high-quality CNTs and large-area graphene films suitable for various applications in electronics, energy storage, and composite materials.

Chemical Vapor Deposition (CVD) is a pivotal technique for synthesizing high-quality graphene and carbon nanotubes due to its controllability and efficiency. The research highlights the importance of gas-phase reaction dynamics in CVD processes, showcasing their impact on the growth rates of graphene under different conditions, with surface kinetic reaction rates and species transport playing crucial roles (Li et al., 2024). Additionally, the application of CVD-grown carbon nanotubes in IC interconnects is explored, emphasizing the advantages of CVD in achieving compatibility with integrated circuit manufacturing processes and addressing challenges like low-temperature growth and contact resistance (Chu et al., 2023). Furthermore, studies on the effects of electric fields during CVD synthesis demonstrate significant impacts on graphene growth and the chirality of carbon nanotubes, suggesting new avenues for controlling the synthesis of low-dimensional materials (Kong,2023). Overall, CVD stands out as a versatile and promising method for producing large-area graphene films and high-quality carbon nanotubes, offering a range of synthesis possibilities and optimization strategies for future research endeavors(Yang et al., 2023) (Hernández et al., 2022).

2.2 Arc Discharge Method

The arc discharge method is a pioneering technique for the production of carbon nanotubes (CNTs) and has also been employed for graphene synthesis. In this method, a high electrical current (~100 A) is passed between two carbon electrodes in an inert atmosphere (typically helium). The intense heat generated vaporizes the carbon electrode tips, creating a plasma arc. Within this arc, carbon atoms recombine and condense to form carbon nanotubes, which are collected on the cooler parts of the reactor walls or on substrates placed nearby. The arc discharge method can produce both single-walled and multi-walled carbon nanotubes, depending on the experimental conditions. Similarly, graphene can be synthesized using a modified arc discharge setup, where the carbon deposition occurs on a substrate placed in the path of the plasma arc. This method is known for its simplicity and scalability, although it requires careful control of parameters such as temperature, pressure, and electrode composition to achieve desired material quality and yield for both CNTs and graphene.

The arc discharge method is a widely used technique for synthesizing carbon nanotubes (CNTs) and graphene due to its simplicity and effectiveness. This method involves generating an electric arc between graphite electrodes to produce soot containing multiwall carbon nanotubes (MWCNTs) along with impurities, which are then purified through various methods like thermal and chemical treatments to yield pure CNTs (Shifa et al., 2024). The arc discharge method has been instrumental in producing both finite and infinite linear carbon chains, with the ability to control parameters like discharge frequency and onset to influence the nucleation and growth of carbon nanotubes (Zhang,2022) (Saucedo-Jimenez et al.,2023). Characterization techniques such as scanning electron microscopy (SEM), transmission electron microscopy (TEM), and X-ray diffraction (XRD) are employed to evaluate the quality and purity of the synthesized carbon nanotubes . Despite its long-standing use, challenges remain in fully understanding the growth mechanisms and optimizing synthesis parameters for enhanced CNT production (M Stetsenko et al.,2023) (Adeel et al., 2022).

2.3 Laser Ablation

Laser ablation is a prominent method for synthesizing carbon nanotubes (CNTs) and graphene. In this technique, a high-power laser beam is focused onto a carbon target in the presence of a reactive gas (typically methane or carbon monoxide) at elevated temperatures. The intense laser pulse vaporizes the carbon target, creating a plasma plume composed of carbon atoms and clusters. Under controlled conditions, these carbon species reassemble and nucleate to form carbon nanotubes or graphene flakes, which are collected on substrates or within a deposition chamber. Laser ablation offers advantages such as precise control over the growth parameters, which can influence the diameter and chirality of the CNTs or the size and quality of graphene produced. This method is particularly favored for its ability to produce high-purity carbon nanostructures with well-defined characteristics, making it suitable for applications in electronics, sensors, and composite materials where material quality Laser ablation in liquids and gases has emerged as a versatile and environmentally friendly method for synthesizing carbon nanostructures, including carbon nanotubes (CNTs) and graphene. By utilizing high-power lasers, researchers have been able to precisely control the growth parameters, influencing the characteristics of the produced nanostructures (Shavelkina et al., 2024). Laser-induced nanomaterials have shown promise in various applications, such as electronic devices, batteries, supercapacitors, and sensors, due to their high purity and well-defined properties (Alex et al.,). Additionally, laser ablation techniques have been employed to fabricate graphene-silicone polymer composites, showcasing the method's ability to induce chemical bonding and modify material structures effectively (Tham et al., 2024). Laser-assisted carbonization, a subset of laser ablation, offers a pathway for the selective synthesis of functional carbon-based materials with tunable properties, making it suitable for flexible electronic applications like sensors, energy storage devices, and electrodes (Devi et al.,2023). and uniformity are critical.

2.4 Epitaxial Growth on Silicon Carbide

Epitaxial growth of carbon nanotubes (CNTs) and graphene on silicon carbide (SiC) substrates is a significant method for producing high-quality carbon-based nanostructures. In this process, SiC substrates are heated to high temperatures (typically above 1300°C) in an inert atmosphere, causing silicon atoms to evaporate from the surface and leaving behind a carbon-rich layer. Under controlled conditions, carbon atoms self-assemble into either graphene sheets or nucleate into carbon nanotubes on the SiC surface. The epitaxial relationship between the SiC substrate and the carbon nanostructures ensures excellent structural alignment and minimal defects, resulting in high-quality materials suitable for various applications. Epitaxial growth on SiC is advantageous for its scalability, potential integration with existing semiconductor technologies, and the ability to produce large-area graphene films or aligned arrays of carbon nanotubes with controlled properties, making it promising for next-generation electronics, sensors, and energy storage devices.

Epitaxial growth on silicon carbide (SiC) has been extensively explored for the synthesis of carbon-based materials like graphene and carbon nanotubes. Research has shown that using a Ni/Cu catalyst can facilitate the growth of graphene on SiC, leading to consistent coverage over large scales (Aiswarya et al., 2024). Additionally, a method involving pre-patterning a catalytic alloy has been developed for the direct synthesis of nanopatterned epitaxial graphene on SiC substrates, enabling the creation of graphene structures at the wafer scale (Katzmarek et al., 2023). Epitaxial graphene on SiC has been identified as a promising platform for the deposition of metals and insulators without the need for additional surface pre-treatment steps, showcasing its versatility in device fabrication (Ivan, 2021). Furthermore, the tribological properties of epitaxial graphene grown on SiC have been investigated, with the number of defects playing a crucial role in determining the frictional behavior of the material (Ünverdi,2018). Moreover, innovative approaches involving microstructures on SiC layers have been proposed to reduce dislocation defects during epitaxial growth, enhancing the quality of the resulting material.

2.5 Reduction of Graphene Oxide

The synthesis of carbon nanotubes (CNTs) and graphene from graphene oxide (GO) involves a reduction process aimed at restoring the sp^2 hybridized carbon network typical of pristine graphene. Initially, graphene oxide, derived from graphite through oxidation processes, contains oxygen functional groups (such as hydroxyl and epoxy groups) that disrupt the π-conjugated structure. Reduction methods, such as thermal reduction, chemical reduction (using agents like hydrazine or sodium borohydride), or electrochemical reduction, are employed to remove these oxygen groups and restore the graphene-like structure. During reduction, oxygen-containing functional groups are cleaved off, and the resulting reduced graphene oxide (rGO) sheets can subsequently nucleate into graphene domains or provide nucleation sites for the growth of carbon nanotubes under appropriate conditions. This approach leverages the versatile properties of graphene oxide as a precursor, allowing for controlled synthesis and functionalization of carbon-based nanostructures suitable for applications in electronics, energy storage, and biomedical devices.

The synthesis of Carbon Nanotubes (CNTs) and Graphene through the reduction of Graphene Oxide (GO) involves various methodologies and reducing agents to achieve the desired properties. Different studies have explored the reduction process of GO to produce reduced Graphene Oxide (rGO) with varying carbon-to-oxygen ratios (Khan, M. U., & Shaida, M. A.,2023). For instance, the reduction of GO using hydrazine hydrate has shown a higher carbon to oxygen ratio in rGO compared to other reducers like sodium borohydride and ascorbic acid, indicating the superior reducing capability of hydrazine hydrate (Bansal et al.,2022). Additionally, utilizing natural substances like tea powder for reduction has demonstrated promising results in enhancing the electrical conductivity of rGO, showcasing the potential of eco-friendly reduction routes for efficient synthesis processes (Khan, J., & Jaafar, M.,2021). These studies emphasize the importance of the reduction methodology, choice of reducing agents, and the resulting properties of CNTs and Graphene synthesized from GO, paving the way for advancements in large-scale production of graphene-based materials.

3. STRUCTURAL PROPERTIES

3.1 Atomic Structure and Bonding in CNTs and Graphene

Carbon nanotubes (CNTs) exist in two primary types: single-walled (SWCNTs) and multi-walled (MWCNTs), each distinguished by their structural complexity and properties. SWCNTs consist of a single cylindrical layer of graphene rolled into a seamless tube, typically with diameters ranging from about 0.4 to 2 nanometers. They exhibit exceptional mechanical strength and unique electronic properties, varying based on their chirality—how the graphene sheet is rolled up. SWCNTs are highly sought after for applications requiring high conductivity, such as in nanoelectronics and conductive films. In contrast, MWCNTs comprise multiple concentric layers of graphene tubes nested within each other, resembling a coaxial cable structure. MWCNTs have larger diameters (typically 2 to 100 nanometers) and are known for their superior mechanical strength and thermal conductivity compared to SWCNTs. They find applications in structural reinforcement, composite materials, and energy storage devices due to their robustness and scalability in production. Both SWCNTs and MWCNTs continue to be extensively researched for their unique properties and potential applications across various industries.

The atomic structure and bonding in carbon nanotubes (CNTs) and graphene are crucial for understanding their properties. Research utilizing self-consistent density functional tight binding (SCC-DFTB) methods has revealed that the connection between CNTs and graphene, especially at defect sites, significantly influences the configurations and electrical properties of the coupled systems (Wei, L., & Zhang, L.,2023) (Wei, L., & Zhang, L.,2022). Additionally, studies have shown that CNTs can be directly grown on graphene, forming a CNT-graphene heterostructure with C–C bonding, enabling both lateral and vertical conduction paths within the structure (Zhou et al., 2016). Furthermore, investigations into co-doped CNT/graphene systems connected by oxygen molecules have demonstrated the transformation of semiconducting materials into semimetallic composites, indicating the potential for self-assembled connections between nanotubes and graphene (Wei et al.,2016). These findings collectively contribute to a deeper understanding of the atomic interactions and bonding mechanisms in CNTs and graphene.

The arc discharge technique is generally preferred for the synthesis of highly crystalline carbon nanotubes (CNTs) with few defects, resulting in high electrical conductivity, mechanical strength, and heat stability. This technique is versatile enough to synthesize both single-walled (SWCNTs) and multi-walled CNTs. The CNTs produced from the given technique are highly used in electronics and nano-electronic applications because of the high conductivity and low defects, in the case of field-emission devices, because of the beneficial geometry. Furthermore, the arc discharge technique is very essential in the research study that requires materials with near-perfect crystallinity.

Figure 1. Types of bond in CNTS and Graphene

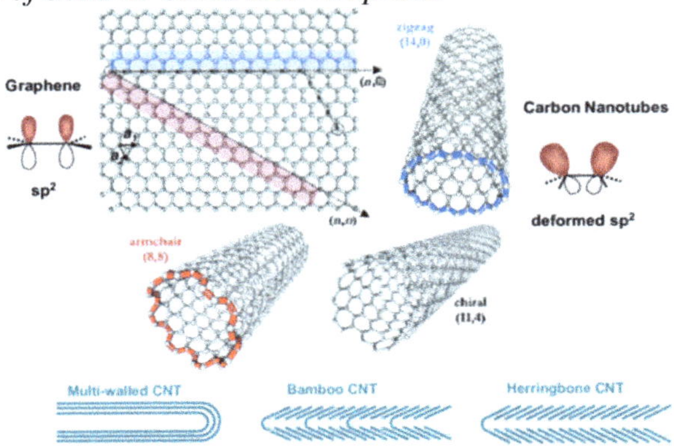

(Filchakova & Saik, 2021)

3.2 Types of CNTs: Single-Walled and Multi-Walled

Carbon nanotubes (CNTs) are classified into single-walled carbon nanotubes (SWCNTs) and multi-walled carbon nanotubes (MWCNTs). SWCNTs consist of a single graphene sheet rolled into a cylindrical shape with a diameter of about 1 nanometer, exhibiting unique properties such as high electrical conductivity, exceptional tensile strength, flexibility, and either metallic or semiconducting behavior depending on their chirality. In contrast, MWCNTs comprise multiple concentric graphene cylinders, resembling a series of nested tubes, which provide additional structural stability and higher tensile strength compared to SWCNTs, though they typically have lower electrical conductivity. Both types of CNTs are synthesized using methods like arc discharge, laser ablation, and chemical vapor deposition (CVD), and they find applications in various fields including electronics, materials science, and nanotechnology due to their remarkable mechanical, electrical, and thermal properties.

Carbon nanotubes (CNTs) are categorized into two main types: single-walled and multi-walled structures. Single-walled CNTs consist of a single layer of graphene wrapped into a cylindrical tube, while multi-walled CNTs are composed of multiple layers of graphene sheets concentrically arranged. Both types of CNTs exhibit unique physical and chemical properties that make them valuable in various fields such as genetic engineering, drug delivery, and water treatment (Pawar et al., 2023). Multi-walled CNTs, in particular, have been shown to be effective vectors for delivering DNA into cells without causing damage, making them crucial in gene delivery applications (Alomar et al.,2023). Additionally, the functionalization of CNTs, especially when combined with other molecules, can help reduce their toxic effects, enhancing their applicability in different industries (Cid-Samamed et al.,2023)

Figure 2. Types of CNTS in single walled and multi-walled

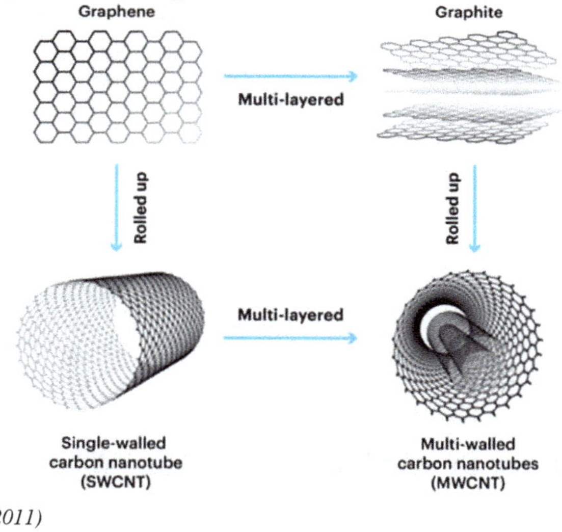

(Esplandiu et al.,2011)

3.3 Graphene Layers and Their Stacking

Graphene layers are single sheets of carbon atoms arranged in a two-dimensional honeycomb lattice, renowned for their exceptional electrical, thermal, and mechanical properties. When multiple graphene layers stack, they form graphite, where the stacking arrangement significantly influences the material's properties. The most common stacking pattern is the AB (Bernal) stacking, where every other layer is aligned, while ABC (rhombohedral) stacking is less common but can affect the electronic properties differently. These stacking configurations impact the electronic band structure, leading to variations in electrical conductivity and other characteristics. Understanding and controlling the stacking of graphene layers is crucial for tailoring the properties of graphene-based materials for specific applications in electronics, materials science, and nanotechnology.

Graphene layers can exhibit different stacking arrangements, influencing their properties. Studies have shown that the stacking type, such as AA- and AB-stacking, significantly impacts the thermal conductivity of multilayer graphene nanoribbons (GNRs). While AB-stacked GNRs display weaker phonon coupling due to a phonon mismatch between layers, AA-stacked GNRs exhibit higher thermal conductivity due to their more symmetrical lattice structure (Zanane et al.,2022). Moreover, introducing nanospacers like cellulose nanofibers or nano-diamonds during the formation of porous graphene sponges can reduce the stacking order of multilayer graphene, promoting a randomly stacked structure and preserving the superior properties of single-layer graphene in bulk scale (Xu et al.,2022). Additionally, doping graphene layers with elements like boron, nitrogen, oxygen, and sulfur can enhance quantum capacitance, with sulfur dopants inducing surface distortions that enhance stability by avoiding steric clashes with stacking layers (Prasert et al.,2024). These findings highlight the importance of understanding and manipulating graphene layer stacking for various applications.

4. MECHANICAL PROPERTIES

4.1 Tensile Strength and Elasticity

The tensile strength and elasticity of carbon nanotubes (CNTs) and graphene play crucial roles in determining the mechanical properties of composite materials. Research indicates that CNTs possess ultimate intrinsic tensile strengths in the 100–200 GPa range, with experimental values varying due to structural factors like chiral structure and diameter (Takakura et al.,2019). Additionally, the inclusion of CNTs in ceramics can enhance toughness by affecting grain boundaries and dislocations, ultimately influencing the strength of the composites (Xie etal.,2023). Moreover, studies on elastic properties reveal that increasing the volume fraction of CNTs and graphene significantly enhances the Young's modulus and shear modulus of reinforced composite materials, showcasing the potential for improved mechanical performance in these materials (Zheng et al.,2022). These findings collectively highlight the importance of understanding the structural characteristics and volume fractions of CNTs and graphene in optimizing the tensile strength and elasticity of composite materials.

4.2 Comparison with Traditional Materials

Carbon nanotubes (CNTs) and graphene exhibit superior mechanical, electrical, and thermal properties compared to traditional materials, making them highly attractive for various applications. CNTs, classified as single-walled (SWCNTs) or multiwalled (MWCNTs), possess exceptional tensile strength surpassing steel, ideal for structural materials (Kumaravel, A., & Sathyamoorthi, S.,2024). Both CNTs and graphene are promising for terahertz detection due to their outstanding electron mobility and spectral absorption properties, crucial for next-generation devices (44). Furthermore, CNTs and graphene are valued for their excellent thermoelectric performance, offering high power factors and flexibility for thermoelectric devices (Tyagi et al., 2023). In the realm of energy storage, generation, and conversion, CNTs and graphene find applications in batteries, supercapacitors, fuel cells, and photovoltaics, thanks to their unique mechanical, optical, and electronic properties (Subhedar, K. M., & Singh, B. P.,2023). Additionally, CNTs and graphene are utilized in biosensors for their exceptional chemical, electrical, and structural characteristics, enabling high sensitivity and stability in detection applications (Das, A., & Wazeer, A.,2022).

4.3 Applications in Composite Materials

Carbon nanotubes (CNTs) and graphene have revolutionized the field of composite materials with their exceptional properties and diverse applications. CNT-reinforced polymer composites (CNT-RPCs) have gained significant recognition in biomedicine, offering benefits like biocompatibility, electrical conductivity, and efficient drug delivery (Mhd Razali et al.,2023). Graphene, a two-dimensional nanomaterial, possesses impressive characteristics such as high electron mobility, large specific surface area, and high thermal conductivity, making it ideal for applications in energy conversion, sensors, catalysis, and biomedicines (Mahalingam et al.,2023). Additionally, the incorporation of CNTs and graphene in conductive composites has shown remarkable improvements in conductive and mechanical properties, leading to enhanced performance in various scientific fields. Furthermore, vegetable oil-based polymer nanocomposites utilizing graphene and CNTs as reinforcement materials have been developed, showcasing the eco-friendliness and versatility of these materials in biomedical and industrial applications (Hasnat et al.,2024). The utilization of these advanced materials in diverse sectors like automotive, aerospace, marine, military, and bio-medical areas highlights their growing importance and potential for future breakthroughs.

5. ELECTRICAL PROPERTIES

5.1 Conductivity and Electron Mobility

Carbon nanotubes (CNTs) and graphene are promising materials for high electrical conductivity and electron mobility in nanoelectronics. The exceptional electrical properties of these carbon-based materials stem from their high carrier mobility and low impurity scattering, enabling efficient charge transport. Models have been developed to predict the conductivity of graphene-based heterostructures, considering factors like charged impurity scattering and optical phonons (Gosling etal.,2021) (Gosling et al.,2020). Additionally, composite models have been proposed to study the electrical conductivity of polymer nanocomposites containing carbon fillers, accounting for filler loading, filler agglomeration, and electron tunneling effects (Wang, Y., & Weng, G. J.,2018). These materials offer a wide range of applications in electronics, such as field-effect transistors, batteries, supercapacitors, and materials for electromagnetic interference shielding, showcasing their potential for future technological advancements.

5.2 Quantum Hall Effect in Graphene

The quantum Hall effect in graphene exhibits unique characteristics due to its two-dimensional structure and relativistic-like charge carriers. Unlike conventional 2D electron systems, graphene displays a half-integer or anomalous quantum Hall effect, visible even at room temperature (Bolotin et al.,2009). This stems from the linear dispersion relation of electrons in graphene, which behave as massless Dirac fermions. Despite the fundamental differences between graphene and conventional 2D systems, the Hall resistance quantization remains universal, expressed in terms of h/e^2 and an integer. Recent experiments have observed the fractional quantum Hall effect in suspended graphene sheets, along with a magnetic-field-induced insulating state at low carrier density. The fractional quantum Hall effect in graphene exhibits a unique sequence of states, different from that of conventional 2D electron gases, due to the relativistic nature of electron interactions (Sahoo, S., & Das, S.K.,2009).

5.3 Field-Effect Transistors (FETs)

Carbon nanotubes (CNTs) and graphene are promising materials for next-generation electronic devices, particularly field-effect transistors (FETs). CNTFETs demonstrate superior performance compared to silicon-based devices, operating faster and at lower voltages (Qiu et al.,2017). Researchers have successfully fabricated CNTFETs with gate lengths as small as 5 nm, approaching the quantum limit of one electron per switching operation. Graphene nanoribbons have also been used to create FETs with performance similar to single-walled CNT devices (Thornhill et al.,2008). Both CNTs and graphene exhibit exceptional electronic properties, making them suitable for high-frequency electronics and VLSI circuit design (Farhana, S., & Noordin, M.F.,2016). When compared, CNTs and graphene show advantages in various applications, including transparent conducting films and FETs, with each material offering unique characteristics in terms of mobility, ambipolarity, and doping strategies (Biswas, C., & Lee, Y.H. (2011). These carbon-based materials hold great promise for advancing electronic device performance beyond current silicon technology limitations.

6. THERMAL PROPERTIES

6.1 Thermal Conductivity and Stability

Carbon nanotubes (CNTs) and graphene exhibit remarkable thermal properties, making them promising materials for heat transfer applications. CNT-based nano-fluids show enhanced thermal conductivity, with improvements of up to 22.2% for water containing 0.5wt% MWCNT and 0.01wt% PVP surfactant (Fadhillahanafi et al.,2014). Gum arabic can stabilize CNT nanofluids, leading to thermal conductivity enhancements of 100-250% (Rashmi et al.,2011). Graphene nanoflakes demonstrate thermal stability and do not spontaneously transform into CNTs when heated (Barnard, A.S., & Snook, I.K et al.,2008). The thermal conductivity of graphene nanoribbons (GNRs) varies depending on their structure, with zigzag GNRs showing higher thermal conductivity than armchair CNTs, while armchair GNRs have lower thermal conductivity than zigzag CNTs (Zhang et al.,2014). These findings highlight the potential of CNTs and graphene in thermal management applications, with their stability and conductivity characteristics dependent on factors such as structure, surfactants, and preparation methods.

6.2 Applications in Heat Dissipation and Thermal Management

Carbon nanotubes (CNTs) and graphene show great potential for thermal management in electronics due to their exceptional thermal conductivity. Three-dimensional composite structures combining graphene sheets and CNTs can enhance heat transfer characteristics (Yin et al.,2020). Graphene fillers in phase change materials significantly improve thermal management of lithium-ion batteries, while CNT incorporation in battery electrodes enhances their thermal conductivity (Goli,2014). Graphene-CNT hybrid materials have been explored for heat dissipation in electronics (Darmawan et al.,2017). CNT-based thermal greases, nanoliquids, and lubricating oils have demonstrated improved thermal conductivity and heat dissipation in various applications, including CPUs, LEDs, and internal combustion engines. For instance, CNT-containing thermal greases reduced CPU temperatures by 5°C, while nanoliquids decreased temperatures in processors and LEDs by 6°C and 3.5°C, respectively (Nguyen et al.,2016). These studies highlight the significant potential of CNTs and graphene in thermal management applications.

7. CHEMICAL PROPERTIES

7.1 Functionalization of CNTs and Graphene

Functionalization of carbon nanotubes (CNTs) and graphene has gained significant attention due to their exceptional properties and potential applications in energy storage and other fields (Wang et al.,2016). Various strategies for functionalization exist, including both covalent and non-covalent methods (Ferreira et al.,2016). One common approach involves treating CNTs with an acid mixture to introduce carboxylic groups, which serve as precursors for further functionalization (Le et al.,2013). Organic compounds such as dodecylamine and 3-aminopropyl triethoxysilane can be covalently attached to these modified CNTs, enhancing their versatility for applications in biology, nanocomposites, and solar energy. Functionalization is broadly defined as the modification of CNTs by introducing desired functional groups or drug molecules onto their walls (Hasnain, S., & Nayak, A.K.,2019). This process can significantly improve the performance of CNTs and graphene-based materials in various applications, particularly in supercapacitors and batteries.

7.2 Chemical Reactivity and Stability

Carbon nanotubes (CNTs) and graphene exhibit varying chemical reactivity and stability depending on their structure and environment. A model predicting CNT reactivity based on curvature and mechanical strain has been developed (Park et al.,2003). For graphene, van der Waals interactions with substrates significantly influence its chemical stability, with 2D crystal substrates enhancing resistance to functionalization compared to SiO2 (Lee et al.,2015). Thermal stability of covalent functionalization differs between isolated and paired radicals, with paired phenyls showing higher stability on graphene and CNTs (Margine et al.,2008). Chemical modifications can improve graphene's hydrophilicity for water-based nanofluid applications, with gentle oxidation treatments preserving stability under flowing conditions (Kim et al.,2017). These findings highlight the complex interplay between structure, environment, and chemical modifications in determining the reactivity and stability of CNTs and graphene, offering insights for tailoring their properties for various applications.

7.3 Sensing Applications

Graphene and carbon nanotubes (CNTs) have emerged as promising materials for sensing applications due to their unique properties. These nanomaterials exhibit high sensitivity, selectivity, and fast response times in detecting various chemicals, biomolecules, and gases (Thangamani et al.,2017). Graphene-based sensors have been developed for a wide range of applications, including gas sensing, heavy metal ion detection, and biosensing of molecules such as DNA, glucose, and disease biomarkers (Atta et al.,2015). The versatility of graphene and its derivatives allows for the creation of diverse sensor architectures, from 1D fibers to 3D foams and aerogels (Tung et al.,2017). Recent advances in synthesis techniques and functionalization approaches have further improved the performance of graphene-based sensors, enhancing their sensitivity, stability, and selectivity (Menaa et al.,2021). These developments have positioned graphene and CNTs as key materials for next-generation sensing devices in various fields, including healthcare, environmental monitoring, and security.

8. ADVANCEMENTS IN APPLICATIONS

8.1 Electronics and Optoelectronics

Carbon nanotubes (CNTs) and graphene are promising materials for soft electronics and optoelectronic devices due to their exceptional elastic, electronic, and thermal properties (Chae, S.H., & Lee, Y.H.,2014). These nano-carbon materials have shown potential in various applications, including transparent conductive films for organic photovoltaic cells, organic light-emitting diodes, and touch panels.

Research has demonstrated their use in light-emitting, nonlinear, photovoltaic, and modulating devices, as well as short-pulse fiber lasers (Yamashita,2012). Studies on individual CNT and graphene devices have revealed novel phenomena, such as photocurrent gain in CNT p-n junctions attributed to multiple electron-hole pair generation, and pronounced photocurrent generation at interfaces between single and multi-layer graphene or suspended and unsuspended graphene sheets (Gabor et al.,2009). Despite challenges, CNTs and graphene have already found commercial applications in touch panels for smartphones, indicating significant opportunities for industrial use.

8.2 Energy Storage and Conversion

Carbon nanotubes (CNTs) and graphene have emerged as promising materials for energy storage and conversion due to their unique properties, including high conductivity, large surface area, and mechanical strength (Wang et al.,2016). These nanomaterials have been extensively studied for applications in supercapacitors, batteries, solar cells, and fuel cells (Dai et al.,2012). Functionalization of CNTs and graphene has further enhanced their performance in energy storage devices. Flexible electrochemical energy storage (FEES) devices utilizing CNTs and graphene have gained attention for wearable electronics (Wen et al.,2016). CNTs are also being explored for renewable energy applications, such as solar cells and hydrogen storage (Kachhi et al.,2021). Recent advancements include the development of CNT hybrid nanocomposites, binder-free electrodes, and microscaled current collectors for improved energy storage systems. Despite significant progress, challenges remain in synthesizing and integrating these materials into practical devices.

8.3 Biomedical Applications

Carbon nanotubes (CNTs) and graphene-based materials have shown significant potential in biomedical applications due to their unique properties. These nanomaterials have been extensively studied for drug delivery, biosensing, imaging, and cancer treatment. CNTs can enhance drug lifespan and facilitate targeted delivery to cells, while also serving as efficient biosensors and bioimaging agents (Murjani et al.,2022). Their high aspect ratio and nanometer size contribute to diverse biomedical applications (Sharma et al.,2016). Functionalization of CNTs has been crucial in modifying their properties, expanding applications, and reducing toxicity. However, concerns regarding cytotoxicity and biosafety persist, hindering their clinical use for in vivo applications (Menezes et al.,2019). Recent studies have also explored graphene oxide and graphene quantum dots for bio-sensing, drug delivery, and cancer therapy (Maiti et al.,2019). Despite promising results, further research is needed to address safety concerns and standardize toxicological tests.

8.4 Environmental Applications

Carbon nanotubes (CNTs) and graphene-based materials have shown significant potential for environmental applications, particularly in water purification and air filtration. These nanomaterials exhibit excellent adsorption capabilities for removing various pollutants, including heavy metals and organic compounds from water (Fan et al.,2017) (Machado et al.,2015). Functionalized CNTs and graphene derivatives have been developed to enhance their adsorption performance and selectivity towards specific contaminants (Xu et al.,2018). The high surface area and numerous adsorption sites of these materials contribute to their effectiveness in environmental remediation (Wang et al.,2019). Additionally, CNTs and graphene can serve as platforms for sensing and monitoring environmental pollutants. Research has also explored their potential in seawater desalination and as catalysts or catalyst supports for environmental applications. Overall, these carbon nanomaterials offer promising solutions for addressing pressing environmental challenges and improving water quality.

8.5 Structural Materials

Carbon nanotubes (CNTs) and graphene have shown promise as structural materials for aerospace and automotive applications due to their superior mechanical properties. However, challenges remain in translating nanoscale properties to macroscale structures (Siochi, E.J., & Harrison, J.S.,2015). Key issues include poor load transfer, interfacial engineering, dispersion, and processing difficulties (Kinloch et al.,2018). Despite these obstacles, progress has been made in developing CNT-enhanced composites with improved damage tolerance and self-sensing capabilities. Researchers have explored various dispersion methods, chemical functionalization, and stress-induced changes in CNT composites to address these challenges (Paipetis, A.S., & Kostopoulos, V.,2013). The integration of CNTs into aerospace structures faces technical barriers, but approaches to reduce technology maturation timeframes have been proposed (Siochi, E.J., & Harrison, J.S.,2015). While the potential for CNTs and graphene in composites remains high, further research is needed to overcome existing limitations and develop practical applications in aerospace and automotive industries (Nayak et al.,2024;Das et al.,2024).

CNT networks and graphene films can replace indium tin oxide (ITO), which is brittle and indium-limited, Flexible, lightweight, mechanically robust and compatible with roll-to-roll processing. Most promising prospects for large-scale use of CNTs and graphene include Touch panels and displays, flexible OLEDs, wearable electronics

9. INDUSTRIAL SCALE PRODUCTION

Carbon nanotubes (CNTs) and graphene are promising nanomaterials with exceptional properties and diverse applications (Zarbin, A.J., & Oliveira, M.R.,2013). However, challenges persist in their large-scale synthesis, standardization, and commercialization. Mass production of various CNT types has been achieved, with global capacity exceeding 10,000 tons annually (Qiang et al.,2013). Recent breakthroughs in synthesis methods, particularly for single-walled CNTs, have advanced the field, but ongoing research is needed to address remaining challenges (Rao et al.,2018). Cost considerations and market potential are driving the exploration of sustainable, low-cost carbon sources, with biomass-derived graphene and CNTs emerging as promising alternatives. These materials have found applications in energy storage, composites, catalysis, and environmental protection. Despite progress, further efforts are required to improve production methods, ensure quality control, and expand commercial applications to fully realize the potential of CNTs and graphene in various industries (Zhou et al.,2022).

Carbon nanotubes (CNTs) and graphene are being considered for their superior properties in water treatment and purity. The properties of CNTs and graphene make them excellent adsorbents for heavy metals and organic compounds due to electrostatic attraction and surface complexation. Moreover, an enhancement in filtration and desalination can be observed with the aid of CNTs and graphene due to improved water permeability and stability. Specific surface area and surface chemistry can be controlled with the help of CNTs and graphene to selectively react with pollutants. Moreover, their catalytic performance would be improved. The best use of CNTs and graphene could be in an immobilized form to avoid dispersion in the environment. In addition, CNTs and graphene are being used for environmental sensing to detect environmental pollutants at low concentrations. This helps in monitoring environmental health.

10. CASE STUDIES

10.1 Successful Implementation in Commercial Products

Carbon nanotubes (CNTs) and graphene have shown promising commercial applications due to their exceptional properties. CNT production capacity has exceeded several thousand tons per year, with bulk powders incorporated into various products like batteries, automotive parts, and sporting goods (107). Mass production strategies for different CNT types have been developed, with production capacity reaching over 10,000 tons annually (108). CNTs and graphene-based transparent conductive films (TCFs) have emerged as potential alternatives to indium tin oxide in optoelectronic devices, with successful integration into touch panels for smartphones (109). However, challenges in large-scale production of pure nanotubes have limited widespread use. Lessons from CNT commercialization, particularly in electronics applications, can potentially accelerate graphene commercialization (110). Despite obstacles, CNTs and graphene continue to show great potential for industrial applications across various sectors.

10.2 Ongoing Research and Development Projects and Collaboration Between Academia and Industry

Recent research highlights the ongoing collaboration between academia and industry in developing carbon nanotubes (CNTs) and graphene for various applications. These carbon nanomaterials exhibit unique properties, making them suitable for use in electronics, energy capture, and biosensors (111). Researchers are exploring sustainable and cost-effective production methods, including the use of biomass-derived sources for manufacturing high-quality graphene and CNTs (112). The integration of these nanomaterials with biomaterials shows promise in overcoming limitations of current biomedical materials, such as corrosion and inadequate mechanical properties. Functionalized graphene and CNTs demonstrate biocompatibility for applications in drug delivery, biosensing, and tissue regeneration (113). The rapid development of graphene research has benefited from existing knowledge of CNTs, with researchers adapting strategies for preparation, functionalization, and applications (114). This ongoing research and development continues to drive innovation in various fields.

11. FUTURE DIRECTIONS

11.1 Emerging Trends in CNT and Graphene Research

Recent research highlights emerging trends in carbon nanotubes (CNTs) and graphene applications. Graphene-based polymer nanocomposites show promise in electromagnetic shielding, sensors, and energy storage (Idumah & Hassan, 2016). Graphene's unique properties, including high conductivity and mechanical strength, have led to its widespread use in various fields (Vashist, 2014). Advancements in large-scale synthesis techniques for both graphene and CNTs have been reported, with applications in high-speed electronics and energy harvesting (Rajavel et al., 2012). Flexible electrochemical energy storage (FEES) devices utilizing CNTs and graphene are gaining attention for wearable electronics (Wen et al., 2016). The development of chemical modification and biomolecular immobilization strategies for graphene has further expanded its potential applications (Vashist, 2014). As research progresses, sustainability and commercial viability of these materials are becoming increasingly important considerations (Idumah & Hassan, 2016).

11.2 Potential Breakthroughs and Innovations, Regulatory and Environmental Considerations

Carbon nanotubes (CNTs) and graphene offer significant potential for technological and therapeutic applications, but also raise concerns about their environmental and health impacts (Saquib Hasnain & A. Nayak, 2019). Research suggests that the surface area of carbon nanomaterials is a more appropriate metric for assessing ecotoxicological effects than weight (Flahaut et al., 2020). To facilitate regulatory decision-making, tools for rapid categorization of nanomaterials based on risk potential are needed, allowing for targeted scrutiny of high-concern materials and expedited review of lower-risk categories (Godwin et al., 2015). Despite potential hazards, CNTs show promise in various environmental applications, including water treatment, air pollution monitoring, and renewable energy technologies (Ong et al., 2010). Strategies to mitigate risks, such as modifying surface chemistry, are being explored to develop safer nanomaterials (Flahaut et al., 2020). Balancing the economic potential of CNTs with environmental and health considerations remains a key challenge for regulators and researchers.

12. CONCLUSION

12.1 Impact of CNTs and Graphene on Modern Technology

Carbon nanotubes (CNTs) and graphene have emerged as revolutionary materials with extraordinary properties, poised to transform various technological fields. In electronics, they are potential replacements for conventional materials in integrated circuits, despite challenges in processing and dispersion. These nanomaterials show great promise in implantable sensors for medical monitoring and diagnostics due to their unique properties (Wujcik & Monty, 2013). CNT- and graphene-based transparent conductive films are viable alternatives to indium tin oxide in flexible optoelectronic devices, with applications in organic photovoltaic cells, OLEDs, and touch panels (Du et al., 2014). In the field of terahertz science and technology, CNTs and graphene offer novel optical and electronic properties, enabling advancements in THz generation, manipulation, and detection. Their potential applications include THz sources, detectors, modulators, antennas, and polarizers. These materials are driving significant progress across multiple technological domains.

12.2 Outlook for Future Advancements and Applications

Carbon nanotubes (CNTs) and graphene show promise for future electronics and composite materials, but face challenges in practical implementation. For CNTs, issues include separating semiconducting from metallic specimens, alignment, and contact reproducibility. Graphene appears more promising for integrated circuits and transparent conducting films. In composites, both materials struggle with poor load transfer, interfacial engineering, dispersion, and processing difficulties (Kinloch et al., 2018). Future advancements may involve combining CNTs and graphene to enhance device properties, using CNTs as precursors for graphene nanoribbons (Pribat, 2011), and developing selection rules for specific applications (Kinloch et al., 2018). Ongoing research focuses on growth and selection, chemical modifications, substitutional doping, and engineering 3D structures (Ma et al., 2014). Despite challenges, these nanomaterials hold potential for next-generation carbon technologies across various applications

REFERENCES

Adeel, Aabir., Muhammad, Yasin, Naz., Shazia, Shukrullah. (2022). Synthesis of Carbon Nanotubes via Plasma Arc Discharge Method. Advances in bioinformatics and biomedical engineering book series, DOI: 10.4018/978-1-7998-8398-2.ch005

Aiswarya, P., David, Cortie., Erin, Smyth., Anton, P., Le, Brun., Francesca, Iacopi. (2024). Epitaxial graphene growth on cubic silicon carbide on silicon with high temperature neutron reflectometry: An operando study. *RSC Advances*, *14*(5), 3232–3240. Advance online publication. DOI: 10.1039/D3RA08289J

Alex, A., Sugunan, S. K., & George, G. 1 Synthesis of Micro and Nanostructured Materials by Laser-Assisted Methods Towards Energy Production, Conversion, and Storage Applications. In Materials for Energy Production, Conversion, and Storage (pp. 1-20). CRC Press.

Alomar, M., Arshad, N., Irshad, M. S., Abdelmohsen, S. A., Ahmed, I., Alhoshani, N., & Alqarni, A. S. (2023). Multi-Walled CNTs/BiVO4 Heterostructures for Solar-Driven Evaporation System and Efficient Photocatalytic Activity against Oxytetracycline. *Nanomaterials (Basel, Switzerland)*, *13*(20), 2764. DOI: 10.3390/nano13202764 PMID: 37887915

Atta, N.F., Galal, A., & El-Ads, E.H. (2015). Graphene — A Platform for Sensor and Biosensor Applications.

Bansal, K., Singh, J., & Dhaliwal, A. S. (2022, February). Synthesis and characterization of Graphene Oxide and its reduction with different reducing agents. []. IOP Publishing.]. *IOP Conference Series. Materials Science and Engineering*, *1225*(1), 012050. DOI: 10.1088/1757-899X/1225/1/012050

Barnard, A.S., & Snook, I.K. (2008). Thermal stability of graphene edge structure and graphene nanoflakes. *The Journal of chemical physics, 128 9*, 094707 .

Biswas, C., & Lee, Y. H. (2011). Graphene Versus Carbon Nanotubes in Electronic Devices. *Advanced Functional Materials*, *21*(20), 21. DOI: 10.1002/adfm.201101241

Bolotin, K. I., Ghahari, F., Shulman, M. D., Stormer, H. L., & Kim, P. (2009). Observation of the fractional quantum Hall effect in graphene. *Nature*, *462*(7270), 196–199. DOI: 10.1038/nature08582 PMID: 19881489

Chae, S. H., & Lee, Y. H. (2014). Carbon nanotubes and graphene towards soft electronics. *Nano Convergence*, *1*(1), 1. DOI: 10.1186/s40580-014-0015-5 PMID: 28936384

Chu, Z., Xu, B., & Liang, J. (2023). Direct Application of Carbon Nanotubes (CNTs) Grown by Chemical Vapor Deposition (CVD) for Integrated Circuits (ICs) Interconnection: Challenges and Developments. *Nanomaterials (Basel, Switzerland), 13*(20), 2791. DOI: 10.3390/nano13202791 PMID: 37887942

Cid-Samamed, A., Correa-Duarte, M. Á., Mariño-López, A., & Diniz, M. S. (2023). Exposure to oxidized multi-walled CNTs can lead to oxidative stress in the asian freshwater clam Corbicula fluminea (Müller, 1774). *International Journal of Molecular Sciences, 24*(22), 16122. DOI: 10.3390/ijms242216122 PMID: 38003314

Dai, L., Chang, D. W., Baek, J., & Lu, W. (2012). Carbon nanomaterials for advanced energy conversion and storage. *Small, 8*(8), 1130–1166. DOI: 10.1002/smll.201101594 PMID: 22383334

Dariyal, P., Sehrawat, M., Dhakate, S. R., & Singh, B. P. (2023). Carbon nanotubes and graphene for conducting wires. In *Emerging applications of carbon nanotubes and graphene* (pp. 105–129). CRC Press. DOI: 10.1201/9781003231943-6

Darmawan, C. C., Ye, L., Samani, M. K., Fu, Y., & Liu, J. (2017). Graphene-CNT hybrid material as potential thermal solution in electronics applications. *2017 IMAPS Nordic Conference on Microelectronics Packaging (NordPac)*, 190-193. DOI: 10.1109/NORDPAC.2017.7993191

Das, A., & Wazeer, A. (2022). Graphene and carbon nanotubes (CNTs)-based biosensor for life sciences applications. In *Next Generation Smart Nano-Bio-Devices* (pp. 61–79). Springer Nature Singapore.

Das, D. K., Patnaik, P., Nayak, S. K., & Barala, M. (2024). Electronic and half-metallic behavior of copper-doped silicon carbide armchair single-walled nanotubes using density functional theory. *International Journal of Computational Materials Science and Engineering*, •••, 2450011.

Devi, M., Wang, H., Moon, S., Sharma, S., & Strauss, V. (2023). Laser-Carbonization–A powerful tool for micro-fabrication of patterned electronic carbons. *Advanced Materials, 35*(38), 2211054. DOI: 10.1002/adma.202211054 PMID: 36841955

Esplandiu, M. J., Baeza, M., Olivé-Monllau, R., Céspedes, F., & Bartrolí, J. (2011). *Development of tunable nanocomposites made from carbon nanotubes for electrochemical applications*. InTech.

Fadhillahanafi, N.M., Leong, K.Y., & Risby, M.S. (2014). STABILITY AND THERMAL CONDUCTIVITY CHARACTERISTICS OF CARBON NANOTUBE BASED NANOFLUIDS.

Fan, W., Zhang, L., & Liu, T. (2017). Graphene-CNT Hybrids for Environmental Applications.

Farhana, S., & Noordin, M.F. (2016). Technology and performance: Carbon nanotube (CNT) field effect transistor (FET) in VLSI circuit design. *2016 IEEE 7th Annual Ubiquitous Computing, Electronics & Mobile Communication Conference (UEMCON)*, 1-4.

Ferreira, F.V., Cividanes, L.D., Brito, F.S., Menezes, B.R., Franceschi, W., Nunes Simonetti, E.A., & Thim, G.P. (2016). Functionalizing Graphene and Carbon Nanotubes: A Review. *Functionalizing Graphene and Carbon Nanotubes*.

Filchakova, M., & Saik, V. (2021). *Single-walled carbon nanotubes: structure, properties, applications, and health & safety*. Tuball.

Flahaut, E., Evariste, L., Liné, C., Mottier, A., Mouchet, F., Barret, M., Pinelli, E., Galibert, A. M., Soula, B., Larue, C., & Gauthier, L. (2020). (Invited) Environmental Impact of Carbon Nanotubes & Graphene and Related Materials (GRMs). *ECS Meeting Abstracts, MA2021-01*, 591-591.

Gabor, N.M., Xu, X., & Alden, J.S. (2009). Optoelectronics in Carbon Nanotube Photodiodes and Graphene Hetero-Interface Devices.

Godwin, H. A., Nameth, C., Avery, D., Bergeson, L. L., Bernard, D., Beryt, E., Boyes, W. K., Brown, S. C., Clippinger, A. J., Cohen, Y., Doa, M. J., Hendren, C. O., Holden, P. A., Houck, K. A., Kane, A. B., Klaessig, F. C., Kodas, T. T., Landsiedel, R., Lynch, I., & Nel, A. E. (2015). Nanomaterial categorization for assessing risk potential to facilitate regulatory decision-making. *ACS Nano*, *9*(4), 3409–3417. DOI: 10.1021/acsnano.5b00941 PMID: 25791861

Goli, P. (2014). Thermal Management Using Graphene and Carbon-Nanotubes.

Gosling, J. H., Makarovsky, O., Wang, F., Cottam, N. D., Greenaway, M. T., Patanè, A., . . . Fromhold, T. M. (2020). Universal mobility characteristics of graphene originating from electron/hole scattering by ionised impurities. arXiv preprint arXiv:2005.07961.

Gosling, J. H., Makarovsky, O., Wang, F., Cottam, N. D., Greenaway, M. T., Patanè, A., Wildman, R. D., Tuck, C. J., Turyanska, L., & Fromhold, T. M. (2021). Universal mobility characteristics of graphene originating from charge scattering by ionised impurities. *Communications on Physics*, *4*(1), 30. DOI: 10.1038/s42005-021-00518-2

Hasnain, S., & Nayak, A. K. (2019). *Functionalization of Carbon Nanotubes*. Carbon Nanotubes for Targeted Drug Delivery. DOI: 10.1007/978-981-15-0910-0_4

Hasnain, S., & Nayak, A. K. (2019). *Regulatory Considerations of Carbon Nanotubes*. Carbon Nanotubes for Targeted Drug Delivery. DOI: 10.1007/978-981-15-0910-0_15

Hasnat, A., Moheman, A., Usmani, M. A., Bhawani, S. A., & Alotaibi, K. M. (2024). Carbon Nanotube and Graphene-Reinforced Vegetable Oil-Based Nanocomposites. In *Vegetable Oil-Based Composites: Processing, Properties and Applications* (pp. 179–190). Springer Nature Singapore. DOI: 10.1007/978-981-99-9959-0_10

Hernández, A. R. R., Cruz, A. G., & Campos-Delgado, J. (2022). Chemical Vapor Deposition Synthesis of Graphene on Copper Foils. In Graphene-A Wonder Material for Scientists and Engineers. IntechOpen

Idumah, C. I., & Hassan, A. (2016). Emerging trends in graphene carbon based polymer nanocomposites and applications. *Reviews in Chemical Engineering*, *32*(2), 223–264. DOI: 10.1515/revce-2015-0038

Ivan, S. (2021, June 22). Filippo, Giannazzo., Rositsa, Yakimova. (2021). Epitaxial Graphene on 4H-SiC (0001) as a Versatile Platform for Materials Growth: Mini-Review. *Applied Sciences (Basel, Switzerland)*, *11*(13), 5784. Advance online publication. DOI: 10.3390/app11135784

Kachhi, B., Singh, K.K., Singh, A.K., & Sharma, D.K. (2021). Role of Carbon Nanotubes as Energy Storage Materials.

Katzmarek, D. A., Mancini, A., Maier, S. A., & Iacopi, F. (2023). Direct synthesis of nanopatterned epitaxial graphene on silicon carbide. *Nanotechnology*, *34*(40), 405302. DOI: 10.1088/1361-6528/ace369 PMID: 37399800

Khan, J., & Jaafar, M. (2021). Reduction efficiencies of natural substances for reduced graphene oxide synthesis. *Journal of Materials Science*, *56*(33), 18477–18492. DOI: 10.1007/s10853-021-06492-y

Khan, M. U., & Shaida, M. A. (2023). Reduction mechanism of graphene oxide including various parameters affecting the C/O ratio. *Materials Today. Communications*, *36*, 106577. DOI: 10.1016/j.mtcomm.2023.106577

Kim, E. J., Desforges, A., Speyer, L., Ghanbaja, J., Gleize, J., Estellé, P., & Vigolo, B. (2017). Graphene for Water-Based Nanofluid Preparation: Effect of Chemical Modifications on Dispersion and Stability. *Journal of Nanofluids*, *6*(3), 603–613. DOI: 10.1166/jon.2017.1353

Kinloch, I. A., Suhr, J., Lou, J., Young, R. J., & Ajayan, P. M. (2018). Composites with carbon nanotubes and graphene: An outlook. *Science*, *362*(6414), 547–553. DOI: 10.1126/science.aat7439 PMID: 30385571

Kong, J. (2023, August). Electrical Control in the Chemical Vapor Deposition Synthesis of Graphene and Related Materials. In Electrochemical Society Meeting Abstracts 243 (No. 13, pp. 1302-1302). The Electrochemical Society, Inc. DOI: 10.1149/MA2023-01131302mtgabs

Kumar, P. (2023). Carbon Nanotubes and Graphene in Photovoltaics. In Emerging Applications of Carbon Nanotubes and Graphene (pp. 131-159). CRC Press. DOI: 10.1201/9781003231943-7

Kumaravel, A., & Sathyamoorthi, S. (2024). Carbon nanotubes-based sensors to detect environmental contaminants. In *Nanotechnology-Based Sensors for Detection of Environmental Pollution* (pp. 349–370). Elsevier. DOI: 10.1016/B978-0-443-14118-8.00018-8

Le, V.T., Ngo, C.L., Le, Q.T., Ngo, T.T., Nguyen, D.N., & Vu, M.T. (2013). Surface modification and functionalization of carbon nanotube with some organic compounds. *Advances in Natural Sciences: Nanoscience and Nanotechnology, 4.*

Lee, J. H., Avsar, A., Jung, J., Tan, J. Y., Watanabe, K., Taniguchi, T., Natarajan, S., Eda, G., Adam, S., Castro Neto, A. H., & Özyilmaz, B. (2015). Van der Waals force: A dominant factor for reactivity of graphene. *Nano Letters*, *15*(1), 319–325. DOI: 10.1021/nl5036012 PMID: 25493357

Li, Q., Luo, J., Li, Z., Rummeli, M. H., & Liu, L. (2024). Numerical investigation on the effect of gas-phase dynamics on graphene growth in chemical vapor deposition. *Journal of Applied Physics*, *135*(12), 125302. DOI: 10.1063/5.0199471

M., Stetsenko., Rashad, Abaszade. (2023). X-ray phase analysis of carbon nanotubes obtained by the arc discharge method. DOI: 10.61640/ujeas.2023.0503

Machado, F.M., Lima, E.C., Jauris, I.M., & Adebayo, M.A. (2015). Carbon Nanomaterials for Environmental Applications.

Mahalingam, S., Omar, A., Manap, A., & Abd Rahim, N. (2023). Synthesis and applications of carbon-polymer composites and nanocomposite functional materials. In Functional Materials from Carbon, Inorganic, and Organic Sources (pp. 71-105). Woodhead Publishing. DOI: 10.1016/B978-0-323-85788-8.00020-3

Maiti, D., Tong, X., Mou, X., & Yang, K. (2019). Carbon-Based Nanomaterials for Biomedical Applications: A Recent Study. *Frontiers in Pharmacology, 9.* manpure, S., Singh, G., & Jewariya, M. (2023). Carbon Nanotubes and Graphene: The Novel Materials for Terahertz Detection. In Emerging Applications of Carbon Nanotubes and Graphene (pp. 275-288). CRC Press.

Margine, E. R., Bocquet, M., & Blase, X. (2008). Thermal stability of graphene and nanotube covalent functionalization. *Nano Letters, 8*(10), 3315–3319. DOI: 10.1021/nl801718f PMID: 18767882

Menaa, F., Fatemeh, Y., Vashist, S. K., Iqbal, H., Sharts, O. N., & Menaa, B. (2021). Graphene, an Interesting Nanocarbon Allotrope for Biosensing Applications: Advances, Insights, and Prospects. *Biomedical Engineering and Computational Biology, 12*, 12. DOI: 10.1177/1179597220983821 PMID: 33716517

Menezes, B. R., Rodrigues, K. F., Fonseca, B. C., Ribas, R. G., Montanheiro, T. L., & Thim, G. P. (2019). Recent advances in the use of carbon nanotubes as smart biomaterials. *Journal of Materials Chemistry. B, 7*(9), 1343–1360. DOI: 10.1039/C8TB02419G PMID: 32255006

Mhd Razali, M. N., Ruzlan, N. N., & Sufian, A. H. (2023, August). Carbon Nanotube-Reinforced Polymer Composites for Biomedical Applications. In *Innovative Manufacturing, Mechatronics & Materials Forum* (pp. 255–264). Springer Nature Singapore.

Murjani, B. O., Kadu, P. S., Bansod, M., Vaidya, S. S., & Yadav, M. D. (2022). Carbon nanotubes in biomedical applications: Current status, promises, and challenges. *Carbon Letters, 32*(5), 1207–1226. DOI: 10.1007/s42823-022-00364-4 PMID: 40477687

Nayak, S. K., Patnaik, P., Das, D. K., & Barala, M. (2024). Electronic and half-metallic behaviour of manganese doped SiC Zigzag nanotubes using density functional theory. *Chemical Physics Impact, 8*, 100417. DOI: 10.1016/j.chphi.2023.100417

Nguyen, M.H., Bui, H.T., Pham, V.T., Phan, N.H., Nguyen, T.H., Nguyen, V.C., Le, D.Q., Phan, H.K., & Phan, N.M. (2016). Thermo-mechanical properties of carbon nanotubes and applications in thermal management. *Advances in Natural Sciences: Nanoscience and Nanotechnology, 7*.

Nimanpure, S., Singh, G., & Jewariya, M. (2023). Carbon Nanotubes and Graphene: The Novel Materials for Terahertz Detection. In Emerging Applications of Carbon Nanotubes and Graphene (pp. 275-288). CRC Press.

Paipetis, A.S., & Kostopoulos, V. (2013). Carbon nanotube enhanced aerospace composite materials: a new generation of multifunctional hybrid structural composites.

Park, S., Srivastava, D., & Cho, K. (2003). Generalized Chemical Reactivity of Curved Surfaces: Carbon Nanotubes. *Nano Letters, 3*(9), 1273–1277. DOI: 10.1021/nl0342747

Pawar, P., Anumalla, S., & Sharma, S. (2023). Role of carbon nanotubes (CNTs) in transgenic plant development. *Biotechnology and Bioengineering*, *120*(12), 3493–3500. DOI: 10.1002/bit.28550 PMID: 37691181

Prasert, K., Sanglaow, T., Liangruksa, M., & Sutthibutpong, T. (2024). Effects of stacking layers and different doping elements on the electronic structures and quantum capacitance of graphene: A DFT study. *Journal of Physics and Chemistry of Solids*, *185*, 111758. DOI: 10.1016/j.jpcs.2023.111758

Qiang, Z., Jiaqi, H., Mengqiang, Z., Weizhong, Q., & Fei, W. (2013). Review on mass production and industrialization of carbon nanotubes.

Qiu, C., Zhang, Z., Xiao, M., Yang, Y., Zhong, D., & Peng, L. (2017). Scaling carbon nanotube complementary transistors to 5-nm gate lengths. *Science*, *355*(6322), 271–276. DOI: 10.1126/science.aaj1628 PMID: 28104886

Rajavel, K., Minitha, C. R., Ranjith, K. S., & Rajendra Kumar, R. T. (2012). Recent progress on the synthesis and applications of carbon based nanostructures. *Recent Patents on Nanotechnology*, *6*(2), 99–104. DOI: 10.2174/187221012800270199 PMID: 22292609

Rao, R., Pint, C. L., Islam, A., Weatherup, R. S., Hofmann, S., Meshot, E. R., Wu, F., Zhou, C., Dee, N. T., Amama, P. B., Carpena-Núñez, J., Shi, W., Plata, D. L., Penev, E. S., Yakobson, B. I., Balbuena, P., Bichara, C., Futaba, D. N., Noda, S., & Hart, A. (2018). Carbon Nanotubes and Related Nanomaterials: Critical Advances and Challenges for Synthesis toward Mainstream Commercial Applications. *ACS Nano*, *12*(12), 11756–11784. DOI: 10.1021/acsnano.8b06511 PMID: 30516055

Rashmi, W., Ismail, A. F., Sopyan, I., Jameel, A. T., Yusof, F., Khalid, M., & Mubarak, N. M. (2011). Stability and thermal conductivity enhancement of carbon nanotube nanofluid using gum arabic. *Journal of Experimental Nanoscience*, *6*(6), 567–579. DOI: 10.1080/17458080.2010.487229

Sahoo, S., & Das, S. K. (2009). Fractional quantum Hall effect in graphene. *Indian Journal of Pure and Applied Physics*, *47*, 658–662.

Saucedo-Jimenez, D., Medina-Sánchez, I., Padilla-Pérez, D. A., Couder-Castañeda, C., & Sanchez, J. L. L. (2023). Double-walled carbon nanotubes synthesis by pulsed electric arc discharge with spinning anode.

Sharma, P. R., Mehra, N. K., Jain, K., & Jain, N. K. (2016). Biomedical Applications of Carbon Nanotubes: A Critical Review. *Current Drug Delivery*, *13*(6), 796–817. DOI: 10.2174/1567201813666160623091814 PMID: 27339036

Shavelkina, M. B., Malikov, M. M., Ivanov, P. P., Borodina, T. I., & Valyano, G. E. (2024). Nanosecond pulsed laser ablation of synthetic graphite in liquids for the synthesis of spherical graphene. *Journal of Laser Applications, 36*(2), 022021. DOI: 10.2351/7.0001329

Shifa, M., Toor, Z. S., & Tariq, F. (2024). Arc discharge synthesis and multistep purification of multiwall carbon nanotubes. *Nano, 19*(02), 2450007. DOI: 10.1142/S1793292024500073

Siochi, E. J., & Harrison, J. S. (2015). Structural nanocomposites for aerospace applications. *MRS Bulletin, 40*(10), 829–835. DOI: 10.1557/mrs.2015.228

Subhedar, K. M., & Singh, B. P. (2023). Introduction to Carbon Nanotubes and Graphene and their Emerging Applications. In *Emerging Applications of Carbon Nanotubes and Graphene* (pp. 1–22). CRC Press. DOI: 10.1201/9781003231943-1

Subhedar, K. M., & Singh, B. P. (2023). Introduction to Carbon Nanotubes and Graphene and their Emerging Applications. In *Emerging Applications of Carbon Nanotubes and Graphene* (pp. 1–22). CRC Press. DOI: 10.1201/9781003231943-1

Takakura, A., Beppu, K., Nishihara, T., Fukui, A., Kozeki, T., Namazu, T., Miyauchi, Y., & Itami, K. (2019). Strength of carbon nanotubes depends on their chemical structures. *Nature Communications, 10*(1), 3040. DOI: 10.1038/s41467-019-10959-7 PMID: 31292451

Tham, A., Siah, K. X., Noor, I. M., Ghazali, I., Tan, S. T., & Tamchek, N. (2024). Fabrication and structural modification of graphene oxide-tetraethyl orthosilicate solution via liquid-phase pulsed laser ablation. *Journal of Laser Applications, 36*(1), 012023. DOI: 10.2351/7.0001227

Thangamani, G. J., Deshmukh, K., Sadasivuni, K. K., Chidambaram, K., Ahamed, M. B., Ponnamma, D., AlMaadeed, M. A., & Pasha, S. K. (2017). Recent advances in electrochemical biosensor and gas sensors based on graphene and carbon nanotubes (CNT) - A review. *Advanced Materials Letters, 8*, 196–205. DOI: 10.5185/amlett.2017.7042

Thornhill, S., Wu, N., Wang, Z., Shi, Q., & Chen, J. (2008). Graphene nanoribbon field-effect transistors. *2008 IEEE International Symposium on Circuits and Systems*, 169-172. DOI: 10.1109/ISCAS.2008.4541381

Tung, T. T., Nine, M. J., Krebsz, M., Pasinszki, T., Coghlan, C. J., Tran, D. N., & Losic, D. (2017). Recent Advances in Sensing Applications of Graphene Assemblies and Their Composites. *Advanced Functional Materials, 27*(46), 27. DOI: 10.1002/adfm.201702891

Tyagi, K., Verma, A. K., Gahtori, B., & Dhakate, S. R. (2023). Carbon Nanotubes and Graphene-Based Thermoelectric Materials: A Futuristic Approach for Energy Harvesting. In Emerging Applications of Carbon Nanotubes and Graphene (pp. 205-226). CRC Press.

Ünverdi, Ö. (2018). Nanotribological Properties Of Epitaxial Graphene Grown On C-Terminated Face Of Silicon Carbide Semiconductor. *Eskişehir Teknik Üniversitesi Bilim ve Teknoloji Dergisi B-Teorik Bilimler*, *6*(2), 228–234. DOI: 10.20290/aubtdb.464691

Vashist, E.K. (2014). Graphene: Emerging Trends, Prospects and Challenges. *Journal of Nanomaterials & Molecular Nanotechnology, 2014.*

Wang, B., Hu, C., & Dai, L. (2016). Functionalized carbon nanotubes and graphene-based materials for energy storage. *Chemical Communications (Cambridge)*, *52*(100), 14350–14360. DOI: 10.1039/C6CC05581H PMID: 27709194

Wang, Y., Pan, C., Chu, W., Vipin, A. K., & Sun, L. (2019). Environmental Remediation Applications of Carbon Nanotubes and Graphene Oxide: Adsorption and Catalysis. *Nanomaterials (Basel, Switzerland)*, *9*(3), 9. DOI: 10.3390/nano9030439 PMID: 30875970

Wang, Y., & Weng, G. J. (2018). Electrical conductivity of carbon nanotube-and graphene-based nanocomposites. Micromechanics and nanomechanics of composite solids, 123-156.

Wei, J., Wei, Q., Ma, Z., & Zeng, H. (2016). Atomic and electronic structures of carbon nanotube covalent connecting with graphene by oxygen molecular. *The European Physical Journal Applied Physics*, *73*(2), 20401. DOI: 10.1051/epjap/2016150444

Wei, L., & Zhang, L. (2022). Atomic Simulations of (8, 0) CNT-Graphene by SCC-DFTB Algorithm. *Nanomaterials (Basel, Switzerland)*, *12*(8), 1361. DOI: 10.3390/nano12081361 PMID: 35458069

Wei, L., & Zhang, L. (2023). Structures and electron states of (4, 4) CNT-graphene hybrid system through DFTB investigation. *Materials Today. Communications*, *35*, 105576. DOI: 10.1016/j.mtcomm.2023.105576

Wen, L., Li, F., & Cheng, H. (2016). Carbon Nanotubes and Graphene for Flexible Electrochemical Energy Storage: From Materials to Devices. *Advanced Materials*, *28*(22), 28. DOI: 10.1002/adma.201504225 PMID: 26748581

Wujcik, E. K., & Monty, C. N. (2013). Nanotechnology for implantable sensors: Carbon nanotubes and graphene in medicine. *Wiley Interdisciplinary Reviews. Nanomedicine and Nanobiotechnology*, *5*(3), 233–249. DOI: 10.1002/wnan.1213 PMID: 23450525

Xie, Y., Shibata, K., & Mizoguchi, T. (2023). Tensile strength of a transverse grain boundary in a single-walled carbon nanotube. *Journal of the Ceramic Society of Japan*, *131*(10), 621–627. DOI: 10.2109/jcersj2.23062

Xu, J., Cao, Z., Zhang, Y., Yuan, Z., Lou, Z., Xu, X., & Wang, X. (2018). A review of functionalized carbon nanotubes and graphene for heavy metal adsorption from water: Preparation, application, and mechanism. *Chemosphere*, *195*, 351–364. DOI: 10.1016/j.chemosphere.2017.12.061 PMID: 29272803

Xu, Z., Inoue, T., Nishina, Y., & Kobayashi, Y. (2022). Stacking order reduction in multilayer graphene by inserting nanospacers. *Journal of Applied Physics*, *132*(17), 174305. DOI: 10.1063/5.0103826

Yamashita, S. (2012). Carbon nanotube and graphene for photonic and optoelectronic applications. *2012 Photonics Global Conference (PGC)*, 1-3.

Yang, B., Yang, N., Zhao, D., Chen, F., Yuan, X., Kou, B., Hou, Y., & Xie, G. (2023). Numerical Simulation of Graphene Growth by Chemical Vapor Deposition Based on Tesla Valve Structure. *Coatings*, *13*(3), 564. DOI: 10.3390/coatings13030564

Yin, H., Zhang, Y., Lu, P., Zhang, Y., & Liu, J. (2020). Analysis of heat dissipation characteristics of three-dimensional graphene-carbon nanotube composite structures. *2020 21st International Conference on Electronic Packaging Technology (ICEPT)*, 1-5.

Zanane, F. Z., Sadki, K., Drissi, L. B., Saidi, E. H., & Bousmina, M. (2022). Graphene multilayers nanoribbons with chirality from molecular dynamics. *Materials Science and Engineering B*, *286*, 115982. DOI: 10.1016/j.mseb.2022.115982

Zarbin, A. J., & Oliveira, M. R. (2013). Nanoestruturas de carbono (nanotubos, grafeno): Quo Vadis? *Química Nova*, *36*(10), 1533–1539. DOI: 10.1590/S0100-40422013001000009

Zhang, H., Zhou, T., Xie, G., Cao, J., & Yang, Z. (2014). Thermal transport in folded zigzag and armchair graphene nanoribbons. *Applied Physics Letters*, *104*(24), 241908. DOI: 10.1063/1.4884278

Zhang, Y. F. (2022). A review of arc-discharge method towards large-scale preparation of long linear carbon chains. *Chinese Physics B*, *31*(12), 125201. DOI: 10.1088/1674-1056/ac833f

Zheng, W., Guo, Z., Wang, Z., Boay, C. G., Wang, X., & Li, Y. (2022). Modeling analysis of elastic properties of graphene-carbon nanotube (G-C) reinforced composites. *Polymer Composites*, *43*(5), 3136–3149. DOI: 10.1002/pc.26606

Zhou, C., Senegor, R., Baron, Z., Chen, Y., Raju, S., Vyas, A. A., Chan, M., Chai, Y., & Yang, C. Y. (2016). Synthesis and interface characterization of CNTs on graphene. *Nanotechnology*, *28*(5), 054007. DOI: 10.1088/1361-6528/28/5/054007 PMID: 28029110

Zhou, Y., He, J., Chen, R., & Li, X. (2022). *Recent Advances in Biomass-derived Graphene and Carbon Nanotubes*. Materials Today Sustainability. DOI: 10.1016/j.mtsust.2022.100138

Chapter 2
Fly Ash in Advanced Construction Materials:
A Brief Review

Jagdeep Walia
http://orcid.org/0009-0006-4614-3420
Rayat Bahra University, Punjab, India

Rita Sharma
http://orcid.org/0009-0003-5119-8731
Rayat Bahra University, Punjab, India

Shalom Akhai
http://orcid.org/0000-0002-7533-457X
MM Engineering College, MM(DU), Mullana, India

ABSTRACT

Fly ash, a waste product from coal power plants, emerges as a viable and eco-friendly building material. This chapter explores the advantages of fly ash bricks, including cost-efficiency, durability, and thermal insulation. Fly ash's incorporation in construction reduces environmental impact by lowering greenhouse gas emissions and conserving landfill space. While challenges like slower curing and freeze-thaw susceptibility exist, fly ash bricks offer superior strength, fire resistance, and lower maintenance needs. The chapter compares fly ash bricks to alternatives and explores the projected market growth driven by sustainability concerns. Research highlights fly ash's role in pollution reduction and its suitability for various construction applications. Market trends indicate robust growth, with the fly ash brick sector projected to expand at a CAGR of 6.34% to exceed 1090 million tonnes by 2032, led by Asia-Pacific demand and supported by global industry players. Overall, fly ash bricks represent a promising path towards a more sustainable construction industry.

DOI: 10.4018/979-8-3693-6627-1.ch002

© 2027, IGI Global Scientific Publishing. Copying or distributing in print or electronic forms without written permission of IGI Global Scientific Publishing is prohibited. Use of this chapter to [generative] artificial intelligence (AI) technologies is expressly prohibited. The publisher reserves all rights to license its use for generative AI training and machine learning model development.

1. INTRODUCTION

Figure 1. Graphical Abstract

A sustainable alternative to conventional building materials such as clay bricks and cement is fly ash, a by-product of coal-fired power plants (Dan et al., 2021; Yadav et al., 2022; Bhagat et al., 2023). Fly ash is increasingly being used in construction projects due to its cost-effectiveness, durability, and thermal insulation properties as shown in Figure 1. (Aşıkoğlu, 2023; Sen et al., 2022; Akhai et al., 2020). In India, the government has encouraged the use of fly ash bricks because they are environmentally beneficial, cost-effective, and long-lasting (Pásztory, 2021). Fly ash can also partially replace cement, resulting in stronger and more economical building materials (Dhiman et al., 2025; Sharma & Akhai, 2019). Further, incorporating wood ash into concrete reduces landfill disposal costs, providing significant environmental advantages (da Costa et al., 2022; Wang & Haller, 2022; Akhai, 2024). Fly ash concrete consumes about 10% less water than conventional concrete. Kongkajun et al. (2020) reported that fibre cement, which combines Portland cement, water, fly ash, and wood pulp, offers longevity and resistance to cracking (Abbass & Akhai 2025). Recycled steel, glass and several waste materials can provides superior compressive strength and carbon neutrality (Dhiman et al., 2025). Similarly, bamboo concrete is recommended for humid and earthquake-prone regions (Li et al., 2023). Beyond construction, fly ash is also used as an additive in metals such as aluminium to enhance properties (Thareja & Akhai, 2016; 2017; Sharma & Akhai, 2019; Akhai et al., 2021). Fly ash particles, which are lighter in colour than the cement matrix, can replace a portion of Portland cement in concrete, reducing costs and improving durability. Hydration of fly ash produces denser calcium-aluminosilicate hydrate gel, reducing cement demand and greenhouse gas emissions while conserving landfill space.

2. PROCESS FOR MAKING FLY ASH BRICKS

The production of fly ash bricks involves several steps (Bai et al., 2022; Ndukwu et al., 2021; Rahmat & Saleh, 2023):

1. Mixing – Fly ash, sand, lime, gypsum, grit, crusher dust, cement, and water are combined in specific proportions.
2. Moulding – The mix is compressed in moulds using hydraulic systems.
3. Curing – Bricks are cured for 7–14 days under wet conditions for durability.
4. Drying – Cured bricks are dried in chambers at 60–70°C to remove excess moisture.
5. Sorting and Stacking – Bricks are inspected, sorted, and stacked for use.

This process produces high-quality, cost-efficient, and environmentally friendly bricks, serving as a sustainable alternative to clay bricks.

3. ADVANTAGES OF FLY ASH BRICKS OVER CLAY BRICKS

Fly ash bricks offer several advantages over conventional clay bricks (Singh et al., 2024; Jonnala et al., 2024, Verma & Malvi, 2021). They possess higher compressive strength, which allows for greater load-bearing capacity, and exhibit reduced water absorption, enhancing durability against dampness, sulphate, and mild acids (Rathee & Singh, 2022). Their superior thermal insulation and fire resistance make them highly suitable for disaster-prone regions, while also reducing electrical energy consumption in applications such as air conditioning by limiting external heat inflow (Kumar & Akhai, 2022; Akhai et al., 2016). In addition, fly ash bricks deliver significant economic and environmental benefits by utilizing industrial waste, lowering energy requirements, and reducing the overall carbon footprint (Amran et al., 2021; Evans & Guthrie, 2023). They are also resistant to termites due to their inorganic nature and demand less maintenance, as lower porosity decreases water retention and minimizes long-term repair costs. Collectively, these attributes make fly ash bricks both cost-efficient and eco-friendly, presenting them as stronger, more durable, and sustainable alternatives to traditional clay bricks (Akhai, 2023; Kumar et al., 2024).

4. DISADVANTAGES OF FLY ASH BRICKS

Despite these notable benefits, fly ash bricks are not without drawbacks (Aldawsari et al., 2022; Quiatchon et al., 2022). Their strength development is slower compared to clay bricks, often prolonging construction timelines (Nedunuri & Muhammad, 2021). They also exhibit extended setting times, which can delay projects unless suitable admixtures are incorporated (Mahadevan et al., 2021). In colder climates, the absence of proper air entrainment can make them susceptible to freeze–thaw damage, compromising long-term performance. Additionally, issues such as surface smoothness and colour inconsistency can reduce their aesthetic appeal (Li & Li, 2020; Minnu, 2021). Seasonal production constraints and limited size availability further restrict their application and design flexibility.

Table 1. Fly Ash Bricks vs. Clay Bricks

Feature	Fly Ash Bricks	Clay Bricks
Advantages		
Strength	Higher compressive strength, leading to increased load-bearing capacity and potentially taller structures	Can be strong, but strength varies depending on quality. May not be suitable for all structural applications.
Durability	Less water absorption, reducing dampness and improving resistance to water damage, sulfate attack, and mild acids. Lower maintenance costs over time.	Can be durable, but susceptible to water damage and cracking if exposed to harsh weather or poor construction practices. May require more frequent maintenance.
Sustainability	Made from waste product (fly ash), reducing reliance on landfills and environmental impact associated with extracting new resources. Lower carbon footprint compared to clay bricks.	Uses natural resources that need to be extracted, impacting the environment. Higher carbon footprint due to production process.

continued on following page

Table 1. Continued

Feature	Fly Ash Bricks	Clay Bricks
Cost	Lower production cost due to use of waste materials and less energy-intensive manufacturing process. Often cheaper than clay bricks.	Can be more expensive depending on location, quality, and availability.
Thermal Insulation	Provides superior thermal insulation, potentially reducing heating and cooling costs.	Generally less effective as thermal insulation.
Fire Resistance	Good fire resistance, can withstand high temperatures for extended periods.	Can also be fire-resistant, but specific properties depend on the type of clay brick.
Pest Resistance	Resistant to termites due to inorganic composition.	Susceptible to termite infestation.
Disadvantages		
Strength Development	Slower rate of strength gain compared to clay bricks, potentially extending construction timelines.	Generally faster strength gain, allowing for quicker construction progress.
Setting Time	Extended setting time may delay construction. Additional admixtures might be needed to improve setting speed.	Generally faster setting time compared to fly ash bricks.
Consistency	Color and surface uniformity can be difficult to control, potentially requiring additional finishing work.	Generally more consistent color and surface texture, reducing the need for additional finishing.
Seasonal Limitations	Production may be affected by weather conditions, limiting availability in certain seasons.	Less impacted by weather during production, offering more consistent availability.
Size Limitations	May have limitations on available sizes, restricting design flexibility.	Wider variety of sizes available, offering more design options.
Aesthetics	May have a less aesthetically pleasing appearance due to potential color variations and smooth surface.	Can offer a wider range of aesthetics depending on the type and color of clay brick.
Applications	Well-suited for load-bearing walls, foundations, and fire-resistant applications. May not be ideal for architectural elements requiring specific aesthetics.	Suitable for a wider range of applications including load-bearing walls, non-load bearing walls, paving, and architectural features.

5. ALTERNATIVES TO FLY ASH BRICKS IN CONSTRUCTION

The durability of fly ash bricks is influenced by several factors, including raw material quality, manufacturing techniques, and environmental exposure. Issues such as inconsistency in raw materials, inadequate air content control, and seasonal limitations can compromise their strength and longevity. Additionally, the smooth texture of fly ash bricks sometimes creates challenges in achieving uniform colour and surface finish, leading to concerns about aesthetics and long-term performance (Waheed et al., 2022).

Despite these drawbacks, fly ash bricks still outperform conventional clay bricks in many aspects, such as compressive strength, reduced water absorption, and enhanced resistance to termites, sulphate, and mild acids (Amran et al., 2021). Nevertheless, several alternative construction materials have been developed (Vijayan et al., 2021; Hany et al., 2021; Li et al., 2021; Thakur & Tiwary, 2021):

- Concrete blocks – Known for high compressive strength, durability, and fire resistance; available in various sizes and shapes.
- Hollow blocks – Lightweight with good insulation, suitable for walls and partitions.
- Cellular blocks – Contain smaller cells than hollow blocks, offering improved insulation and sound absorption.
- Lintel blocks – Specially designed to carry masonry loads above windows and doors.
- Thermal blocks – Provide high thermal insulation, ideal for regions with extreme climates.
- Compressed Stabilised Earth Blocks (CSEB) – Eco-friendly and economical, made from soil, stabilisers, and water.
- Sand lime bricks – Composed of sand, lime, and cement, offering strength and durability.
- Autoclaved Aerated Concrete (AAC) – Lightweight, highly insulating, and widely used for walls and roofs.

These alternatives provide advantages such as better insulation, reduced structural weight, and improved durability, making them versatile choices for sustainable construction.

6. INNOVATIVE SUSTAINABLE ALTERNATIVES

In recent years, more sustainable and creative materials have emerged (Hartman & Williams, 2024; Phonphuak & Thiansem, 2012; Yilmaz & Dal, 2015; Morel et al., 2001):

1. K-Briq – Produced with 90% recycled construction waste, requires 90% less production energy, and emits substantially less CO_2.
2. Cork blocks – Interlocking expanded cork blocks that are easy to dismantle, recyclable, and potentially carbon-negative.

3. Green Charcoal Bricks – Made with soil, charcoal, and loofah fibres; up to 20 times more porous than standard bricks, supporting urban biodiversity.
4. Gent Waste Brick – Features one-third the embodied carbon of conventional Belgian clay bricks, manufactured through a simple process that encourages community participation.
5. Building the Local – Utilises local waste resources (e.g., horse manure, wool) to reinforce unfired clay bricks, fostering circular waste management.

These innovative alternatives are gaining recognition for their eco-friendly properties, reduced carbon footprint, and creative use of recycled materials, making them promising solutions for the future of sustainable construction.

7. MARKET SIZE OF FLY ASH BRICKS IN THE NEXT 5 YEARS

The fly ash brick market is projected to grow at a CAGR of 6.34% between 2022 and 2032, reaching over 1090 million tonnes by 2032 (Chandrashekhar, 2013; Sangwan et al., 2020). This growth is driven by the expanding use of fly ash in cement, tiles, highways, landfilling, and mine filling applications (Dwivedi & Jain, 2014). The Asia-Pacific region dominates the market, with China emerging as the leading consumer due to the rapid growth of its cement industry. Key market players include Boral, Cemex S.A.B. de C.V., Separation Technologies LLC, Charah Solutions Inc., and Cement Australia (Tkachenko et al., 2023). These companies are actively pursuing product innovations, strategic acquisitions, and capacity expansions to cater to the rising global demand.

8. KEY FINDINGS OF THE RESEARCH ON FLY ASH BRICKS

1. Fly ash bricks are an eco-friendly substitute for clay bricks, reducing air, water, and soil pollution (Singh et al., 2024; Jonnala et al., 2024).
2. They have higher compressive strength and lower water absorption, ensuring structural durability.
3. They meet EPA-mandated TCLP standards, confirming safety for landfill disposal.
4. Long-term carbonation enhances strength while capturing CO_2, contributing to climate mitigation (Samal et al., 2023).
5. India consumes ~180 billion bricks annually, and fly ash bricks help reduce the ~5000 acres of topsoil lost each year due to clay brick manufacturing.

6. Applications include infrastructure, pavements, dams, canals, irrigation, and underwater projects.
7. Although COVID-19 disrupted supply chains, demand is expected to rise significantly in the next five years.

9. CONCLUSIONS

1. Sustainability: Fly ash bricks reduce pollution, conserve topsoil, and enhance public health near power plants.
2. Performance: They offer higher compressive strength, improved resistance to water, sulphate, and mild acids, and reduced water absorption compared to clay bricks.
3. Challenges: Issues such as slower strength gain, fly ash procurement, and pandemic-related supply chain disruptions remain.
4. Future Market: Despite challenges, growing demand across industries will continue driving adoption.
5. Alternatives: AAC, CSEB, sand lime bricks, and other sustainable options complement fly ash bricks for broader construction applications.

Overall, fly ash bricks are a promising sustainable replacement for clay bricks, combining strength, cost-effectiveness, and environmental benefits. The market outlook remains strong, offering significant opportunities for sustainable construction practices.

10. FUTURE SCOPE

Integration of artificial intelligence (AI) and digital technologies offers new opportunities to advance the performance and adoption of fly ash bricks. AI-based optimization can refine mix designs for higher strength and durability, while predictive models and real-time monitoring can enhance quality control in large-scale production. Furthermore, coupling AI with life cycle assessment frameworks will enable more accurate evaluation of environmental benefits, carbon savings, and long-term sustainability. In the context of smart cities, IoT-enabled systems may also be used alongside fly ash bricks for monitoring structural health and energy performance. These future directions highlight how traditional sustainable materials can be strengthened by modern technological innovations, ensuring wider acceptance and long-term impact. Figure 2 highlights fly ash bricks' advantages, sustainability benefits, market trends, and AI-driven future scope, while Table 3 provides a summary of the chapter. The infographic summary of the chapter on fly ash in advanced construction materials is presented in Figure 3.

Figure 2. Graphical Highlights of fly ash bricks

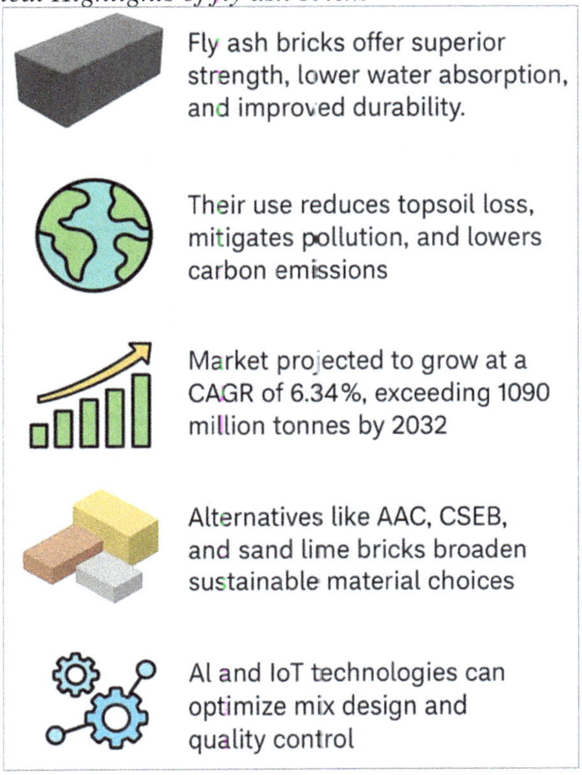

Table 3. Chapter Summary

Milestone / Section	Content	Key Findings / Outcomes
Introduction	Presents fly ash as a by-product of coal-fired power plants and a sustainable substitute for clay bricks and cement. Discusses applications in construction, composites, and metallurgy.	Fly ash reduces pollution, conserves topsoil, enhances strength and insulation, and lowers carbon emissions, aligning with eco-friendly construction goals.
Process for Making Fly Ash Bricks	Steps: mixing fly ash with lime, sand, gypsum, and water → hydraulic moulding → curing for 7–14 days → drying at 60–70°C → sorting & stacking.	Produces uniform, strong, cost-effective bricks with lower embodied energy compared to clay bricks.
Advantages	Highlights superior properties of fly ash bricks over clay bricks.	Higher compressive strength, durability against sulphates and mild acids, lower porosity, termite resistance, better thermal/fire insulation, reduced energy use, and long-term cost savings.
Disadvantages	Discusses constraints in strength development, setting time, and aesthetics.	Slower strength gain, extended setting times, freeze–thaw vulnerability, colour inconsistency, seasonal/size limitations, and reduced architectural appeal.
Alternatives	Reviews other conventional sustainable materials such as AAC, CSEB, concrete blocks, hollow blocks, and sand lime bricks.	These provide lightweight structures, improved insulation, and varied design flexibility, complementing or substituting fly ash bricks in certain contexts.
Innovative Sustainable Alternatives	Examines newer eco-materials such as K-Briq, Cork blocks, Green Charcoal Bricks, Gent Waste Bricks, and Building the Local.	Materials are low-carbon, recyclable, porous, biodiversity-supportive, and community-driven, marking new pathways in circular construction.
Market Size & Growth	Global market trends for fly ash bricks and composites.	Market projected to grow at CAGR 6.34% (2022–2032) to exceed 1090 Mt; Asia-Pacific dominates; leading companies include Boral, Cemex, Charah, Separation Technologies, and Cement Australia.
Key Research Findings	Summarizes core insights from reviewed studies.	Fly ash bricks meet TCLP safety standards, capture CO_2 through carbonation, and save ~5000 acres of Indian topsoil annually; ensure structural strength and durability in diverse applications.
Conclusions	Wraps up strengths, limitations, and outlook of fly ash bricks.	Fly ash bricks are eco-friendly, cost-efficient, and high-performing alternatives to clay bricks; challenges exist, but demand is rising steadily.
Future Scope	Explores the role of AI, IoT, and digital tools in advancing material use.	AI can optimize mix design and predict durability, IoT can monitor building health, and AI–LCA integration can quantify environmental impacts. These tools support adoption in smart cities and Industry 4.0.

Figure 3. Infographic Summary of the Chapter on Fly Ash in Advanced Construction Materials

Miistonne / Section	Content	Key Findings / Outcomes
Introduction	Presents fly ash as epbace a s ub-product of clay bricks over clay sustainable subatifs for clay bricks a-cements, Applications in constnlicrdy.	Fly ash reducec pollution, conserves surapoli, enhances strenggthth and insulation, and lowers carbon omiss ajons, atigning with eco-friendly consr-ruction goals, allsing with ecu-rnently construction goals. ighned.
Advantages	Steps; miging fly ash with lime, sand, goosum, crmu water → hydrailıd moulding → curing for 7–14 days, → dilying of 50-70 °CC → sorting & Sstacking	Higher compressive strength, durab-llity against sulphates and mild acids, lawer perssity, termite resistance; better insemalifire insulation. reduced energy use, and long-term cost savings:
Disadvantages	Highustus recpr orepar is constraints in strength development; setting timime and aestnetics.	Slower strength gain, extended setting times, frecze-thow vulnerability. colout inconsistency, sessonalisze lunitations, and reduced archifectu rat appeal.
Alternatives	Reviews, other conventional sustainable matenals such K-bıta, CSBB. concrete blocks, hollow blocks, and sand lime bricks.	Materiale are low-carbon, recyclable, porous, blodiversiry supportive, and communify driven, marking new pathways in circular construction.
Innovative Sustaindble Alternatives	Examines riewer eco-materals such as K-Brio, Cork blocks Green Charceal Bricks; Gent Waste Bricks, and Building the Local,	Market size & Growth CAGR &.34 % (2032-2083) to exceed 1090 Mt: A · Asta-Pacific Jominanies: espec: Oiorai, Cemen, Charah; Separation Technologies and Cement Australla.
Matket Size & Growth	Corciusb. isFraps up ts urowth. limitations, and outlook of fly ash bricks,	Fly ash bricks meet TCLP safety standarde, capture CO-through. carbanation, save ± 3000 acres of = indre: topsoil arinually. ~snsure structural strength and durability n diverse applications.
Key Research Findings	Explore; the rople of A:/lot and digital tool.'s in advancing material use.	Fly ash bricks are aco-friendly. cost-efticient; and high performing alterna-tives to blay bricks. Challengee existo; but demand is rising steadily.
	Wraps up s rengths. limitations, and outtook of fly ash bricka	Fly ash bricks are aco–efficient, cost-efficient, and high performing alternat-was to clay bricks.

REFERENCES

Abbass, M., & Akhai, S. (2025). Eco-friendly geopolymer bricks synthesized from alkali-activated sugarcane bagasse ash and rice husk ash. *Industrial Crops and Products, 233*, 121410. DOI: 10.1016/j.indcrop.2025.121410

Abbass, M., & Akhai, S. (2025). Exploring the sustainability potential of geopolymer concrete with coal bottom ash and basalt fibres. *Multiscale and Multidisciplinary Modeling. Experiments and Design, 8*(3), 199.

Akhai, S. (2023). Navigating the Potential Applications and Challenges of Intelligent and Sustainable Manufacturing for a Greener Future. *Evergreen, 10*(4), 2237–2243. DOI: 10.5109/7160899

Akhai, S. (2024). Trends and Environmental Impact of Paper Consumption: A Prognostic Scenario for the Indian Market by 2030-A Case Study. In *International Conference on Interdisciplinary Approaches in Civil Engineering for Sustainable Development* (Vol. 464, pp. 11-18). Springer Nature Singapore. DOI: 10.1007/978-981-97-0910-6_2

Akhai, S., Bansal, S. A., & Singh, S. (2020). A critical review of thermal insulators from natural materials for energy saving in buildings. *Journal of Critical Reviews, 7*(19), 278–283.

Akhai, S., Singh, V. P., & John, S. (2016). Human performance in industrial design centers with small unit air conditioning systems. *Journal of Advanced Research in Production Industrial Engineering, 3*(2), 5–11.

Akhai, S., Srivastava, P., Sharma, V., & Bhatia, A. (2021). Investigating weld strength of AA8011-6062 alloys joined via friction-stir welding using the RSM approach. []. IOP Publishing.]. *Journal of Physics: Conference Series, 1950*(1), 012016. DOI: 10.1088/1742-6596/1950/1/012016

Amran, M., Debbarma, S., & Ozbakkaloglu, T. (2021). Fly ash-based eco-friendly geopolymer concrete: A critical review of the long-term durability properties. *Construction & Building Materials, 270*, 121857. DOI: 10.1016/j.conbuildmat.2020.121857

Aşıkoğlu, A. (2023). Cost analysis of insulation materials used to increase energy performance in buildings with Net Present Value method. *Journal of Sustainable Construction Materials and Technologies, 8*(2), 134–145. DOI: 10.47481/jscmt.1270831

Bai, Y., Guo, W., Wang, J., Xu, Z., Wang, S , Zhao, Q., & Zhou, J. (2022). Geopolymer bricks prepared by MSWI fly ash and other solid wastes: Moulding pressure and curing method optimisation. *Chemosphere*, *307*, 135987. DOI: 10.1016/j.chemosphere.2022.135987 PMID: 35970219

Bhagat, M., Singh, S., Suresh, S., Arisutha, S., Verma, S., & Kansal, S. K. (2023). Utilization of value-added products from fly ash: An industrial waste. In *Advanced Materials from Recycled Waste* (pp. 239–254). Elsevier. DOI: 10.1016/B978-0-323-85604-1.00003-2

Chandrashekhar, A. U. (2013). Studies on potential usage of fly ash in steel fiber reinforced concrete for rigid pavements (Doctoral dissertation, National Institute of Technology Karnataka, Surathkal).

da Costa, T. P., Quinteiro, P., Arroja, L., & Dias, A. C. (2022). Environmental performance of different end-of-life alternatives of wood fly ash by a consequential perspective. *Sustainable Materials and Technologies*, *32*, e00411. DOI: 10.1016/j.susmat.2022.e00411

Dan, A. K., Bhattacharjee, D., Ghosh, S., Behera, S. S., Bindhani, B. K., Das, D., & Parhi, P. K. (2021). Prospective utilization of coal fly ash for making advanced materials. In *Clean Coal Technologies: Beneficiation, Utilization, Transport Phenomena and Prospective* (pp. 511–531). Springer International Publishing. DOI: 10.1007/978-3-030-68502-7_20

Dhiman, S., Seema, S., & Akhai, S. (2025). Performance evaluation of OPC 43 concrete incorporating micro silica and recycled ceramic tile waste as partial cement and coarse aggregate replacements. *Asian Journal of Civil Engineering*, 1-19.

Dwivedi, A., & Jain, M. K. (2014). Fly ash–waste management and overview: A review. *Recent Research in Science and Technology, 6*(1).

Evans, M., & Guthrie, W. S. (2023, May). Reducing the Carbon Footprint of Concrete by Expanding the Use of Supplementary Cementitious Materials. In *2023 Intermountain Engineering, Technology and Computing (IETC)* (pp. 253-256). IEEE. DOI: 10.1109/IETC57902.2023.10152065

Hany, E., Fouad, N., Abdel-Wahab, M., & Sadek, E. (2021). Investigating the mechanical and thermal properties of compressed earth bricks made by eco-friendly stabilization materials as partial or full replacement of cement. *Construction & Building Materials*, *281*, 122535. DOI: 10.1016/j.conbuildmat.2021.122535

Hartman, H., & Williams, J. J. (Eds.). (2024). *Materials: An environmental primer.* Routledge. DOI: 10.4324/9781003486923

Jonnala, S. N., Gogoi, D., Devi, S., Kumar, M., & Kumar, C. (2024). A comprehensive study of building materials and bricks for residential construction. *Construction & Building Materials*, *425*, 135931. DOI: 10.1016/j.conbuildmat.2024.135931

Kongkajun, N., Laitila, E. A., Ineure, P., Prakaypan, W., Cherdhirunkorn, B., & Chakartnarodom, P. (2020). Soil-cement bricks produced from local clay brick waste and soft sludge from fiber cement production. *Case Studies in Construction Materials*, *13*, e00448. DOI: 10.1016/j.cscm.2020.e00448

Kumar, H., Wadhwa, A. S., Akhai, S., & Kaushik, A. (2024). Parametric optimization of the machining performance of Al-SiCp composite using combination of response surface methodology and desirability function. *Engineering Research Express*, *6*(2), 025505. DOI: 10.1088/2631-8695/ad38ff

Kumar, H., Wadhwa, A. S., Akhai, S., & Kaushik, A. (2024). Parametric analysis, modeling and optimization of the process parameters in electric discharge machining of aluminium metal matrix composite. *Engineering Research Express*, *6*(2), 025542. DOI: 10.1088/2631-8695/ad4ba9

Kumar, P., & Akhai, S. (2022). Effective energy management in smart buildings using VRV/VRF systems. In *Additive Manufacturing in Industry 4.0* (pp. 27–35). CRC Press. DOI: 10.1201/9781003360001-2

Li, H., Yang, S., Zha, Z., Fei, B., & Wang, X. (2023). Hygrothermal properties analysis of bamboo building envelope with different insulation systems in five climate zones. *Buildings (Basel, Switzerland)*, *13*(5), 1214. DOI: 10.3390/buildings13051214

Li, J., Li, X., Liang, S., Zhang, Y., Ye, Q., Zhang, L., Pan, C., Lv, R., Li, Q., Xiao, K., Yan, Y., Chen, Y., Hu, J., Hou, H., & Yang, J. (2021). Preparation of water-permeable bricks derived from fly ash by autoclaving. *Construction & Building Materials*, *271*, 121556. DOI: 10.1016/j.conbuildmat.2020.121556

Morel, J. C., Mesbah, A., Oggero, M., & Walker, P. (2001). Building houses with local materials: Means to drastically reduce the environmental impact of construction. *Building and Environment*, *36*(10), 1119–1126. DOI: 10.1016/S0360-1323(00)00054-8

Ndukwu, M. C., Bennamoun, L., Simo-Tagne, M., Ibeh, M. I., Abada, U. C., & Ekop, I. E. (2021). Influence of drying applications on wood, brick and concrete used as building materials: A review. *Journal of Building Pathology and Rehabilitation*, *6*(1), 1–19. DOI: 10.1007/s41024-021-00119-0

Pásztory, Z. (2021). An overview of factors influencing thermal conductivity of building insulation materials. *Journal of Building Engineering*, *44*, 102604. DOI: 10.1016/j.jobe.2021.102604

Phonphuak, N., & Thiansem, S. (2012). Using charcoal to increase properties and durability of fired test briquettes. *Construction & Building Materials*, *29*, 612–618. DOI: 10.1016/j.conbuildmat.2011.11.018

Rahmat, M. N., & Saleh, A. M. (2023). Strength and environmental evaluation of building bricks using industrial waste for livable environments. *Construction & Building Materials*, *403*, 132864. DOI: 10.1016/j.conbuildmat.2023.132864

Rathee, M., & Singh, N. (2022). Durability properties of copper slag and coal bottom ash based I-shaped geopolymer paver blocks. *Construction & Building Materials*, *347*, 128461. DOI: 10.1016/j.conbuildmat.2022.128461

Samal, A. K., Panigrahi, M., Ganguly, R. I., & Dash, R. R. (2023). Historical development of construction materials–From Stone Age to Modern Age. In *Development of Geopolymer from Pond Ash-Thermal Power Plant Waste: Novel Constructional Materials for Civil Engineers* (pp. 1-70). Elsevier. DOI: 10.1002/9781394167975.ch1

Sangwan, K. S., Choudhary, K., & Agarwal, S. (2020). Sustainability assessment of sanitary ware supply chain using life cycle assessment framework—a case study. In *Enhancing Future Skills and Entrepreneurship: 3rd Indo-German Conference on Sustainability in Engineering* (pp. 167-179). Springer International Publishing.

Sen, S., Singh, A., Bera, C., Roy, S., & Kailasam, K. (2022). Recent developments in biomass derived cellulose aerogel materials for thermal insulation application: A review. *Cellulose (London, England)*, *29*(9), 4805–4833. DOI: 10.1007/s10570-022-04586-7

Sharma, V., & Akhai, S. (2019). Mechanical Behaviour of Fly Ash Reinforced Aluminum Composite Prepared by Casting. *Journal of Advanced Research in Mechanical Engineering and Technology*, *6*(1&2), 23–26. DOI: 10.24321/2454.8650.201904

Sharma, V., & Akhai, S. (2019). Trends in Utilization of Coal Fly Ash in India: A Review. *Journal of Engineering Design and Analysis*, *2*(1), 12–16.

Singh, S., Maiti, S., Bisht, R. S., Panigrahi, S. K., & Yadav, S. (2024). Large CO2 reduction and enhanced thermal performance of agro-forestry, construction and demolition waste based fly ash bricks for sustainable construction. *Scientific Reports*, *14*(1), 8368. DOI: 10.1038/s41598-024-59012-8 PMID: 38600139

Thakur, A., & Tiwary, A. K. (2021, November). A comparative study on conventional clay bricks and autoclaved aerated concrete blocks. []). IOP Publishing.]. *IOP Conference Series. Earth and Environmental Science*, *889*(1), 012061. DOI: 10.1088/1755-1315/889/1/012061

Thareja, P., & Akhai, S. (2016). Processing Aluminum Fly Ash Composites via Parametric Analysis of Stir Casting. *Journal of Advanced Research in Manufacturing. Material Science & Metallurgical Engineering*, *3*(3&4), 21–28.

Thareja, P., & Akhai, S. (2017). Processing parameters of powder aluminium-fly ash P/M composites. *Journal of advanced research in manufacturing, material science & metallurgical engineering, 4*(3&4), 24-35.

Tkachenko, N., Tang, K., McCarten, M., Reece, S., Kampmann, D., Hickey, C., ... & Caldecott, B. (2023). Global database of cement production assets and upstream

Verma, M., & Malvi, L. (2021). Comparative analysis between properties of fly ash bricks and normal clay bricks. *International Journal of Scientific Modern Research and Technology*, *4*(3), 26–33.

Vijayan, D. S., Mohan, A., Revathy, J., Parthiban, D., & Varatharajan, R. (2021). Evaluation of the impact of thermal performance on various building bricks and blocks: A review. *Environmental Technology & Innovation*, *23*, 101577. DOI: 10.1016/j.eti.2021.101577

Wang, R., & Haller, P. (2022). Applications of wood ash as a construction material in civil engineering: A review. *Biomass Conversion and Biorefinery*, •••, 1–21. PMID: 35342681

Yadav, V. K., Gacem, A., Choudhary, N., Rai, A., Kumar, P., Yadav, K. K., Abbas, M., Khedher, N. B., Awwad, N. S., Barik, D., & Islam, S. (2022). Status of coal-based thermal power plants, coal fly ash production, utilization in India and their emerging applications. *Minerals (Basel)*, *12*(12), 1503. DOI: 10.3390/min12121503

Yilmaz, D., & Dal, L. (2015). Hydraulic properties estimation of an experimental urban soil column constructed with waste brick and compost. *International Journal of Pure and Applied Sciences*, *1*, 9–15.

Zhao, Y., Yu, M., Xiang, Y., Kong, F., & Li, L. (2020). A sustainability comparison between green concretes and traditional concrete using an emergy ternary diagram. *Journal of Cleaner Production*, *256*, 120421. DOI: 10.1016/j.jclepro.2020.120421

KEY TERMS AND DEFINITIONS

Autoclaved Aerated Concrete (AAC): Lightweight, precast concrete with high insulation properties, often used as an alternative to traditional bricks.

Carbonation: A process where concrete or bricks absorb carbon dioxide (CO_2) from the atmosphere, enhancing strength while reducing greenhouse gases.

Circular Economy: An economic system aimed at minimizing waste and maximizing resource efficiency through reuse, recycling, and sustainable production.

Compressed Stabilised Earth Blocks (CSEB): Eco-friendly bricks made from soil mixed with stabilisers like lime or cement, compacted, and cured without firing.

Fly Ash Bricks: Sustainable building units manufactured from fly ash, sand, lime, gypsum, and water; known for durability, high strength, and reduced environmental impact.

Fly Ash: A fine powder, by-product of coal combustion in thermal power plants, widely used as a supplementary material in construction.

Sustainability: Practice of using resources efficiently to reduce environmental impact, ensure long-term viability and support eco-friendly construction.

Thermal Insulation: The ability of a material to resist heat transfer, reducing heating and cooling energy needs in buildings.

Toxicity Characteristic Leaching Procedure (TCLP): A standard test developed by the U.S. Environmental Protection Agency (EPA) to evaluate the leaching potential of hazardous substances.

APPENDIX

Abbreviations

AAC – Autoclaved Aerated Concrete
CAGR – Compound Annual Growth Rate
CSEB – Compressed Stabilised Earth Blocks
CO_2 – Carbon Dioxide
EPA – Environmental Protection Agency
Mt – Million Tonnes
RSM – Response Surface Methodology
TCLP – Toxicity Characteristic Leaching Procedure

Chapter 3
Electromagnetic Desalination for Freshwater Production and Brine Valorization for a Sustainable Ecological Approach to Achieve ZLD: physical desalination Electromagnetic

Abdelbast Karbal
http://orcid.org/0009-0007-9015-2153
Renewable Energy and System Dynamics Laboratory, Faculty of Sciences Ain Chock- Hassan II University of Casablanca, Morocco

Driss Bougmoum
Condensed Matter Physics Laboratory, Hassan II University - FSBM, Casablanca, Morocco

Mustapha Faraji
http://orcid.org/0000-0002-2168-1963
Faculty of Sciences Ain Chock, Hassan II University of Casablanca, Morocco

Redouane Mghaiouini
IbnTofail University, Kénitra, Morocco

My. Youssef Elhafidi
Condensed Matter Physics Laboratory, Hassan II University - FSBM, Casablanca, Morocco

Abdeslam Elbouari
Faculty of Sciences Ain Chock, Hassan II University of Casablanca, Morocco

DOI: 10.4018/979-8-3693-6627-1.ch003

2027, IGI Global Scientific Publishing. Copying or distributing in print or electronic forms without written permission of IGI Global Scientific Publishing is prohibited. Use of this chapter to
rative artificial intelligence (AI) technologies is expressly prohibited. The publisher reserves all rights to license its use for generative AI training and machine learning model development.

ABSTRACT

The objective of this research is to present a new approach to physical desalination using Lorentz forces, electrostatic forces and flow velocity. This process involves a combination of magnetic and electric fields to separate the charged ions from the soluble salts in seawater. The added value of this desalination process lies in the downstream coupling of Magnetohydrodynamics (MHD) and Capacitive Deionization (CDI) under an external magnetic field, followed by centrifugation and crystallization for efficient brine valorization and Zero Liquid Discharge (ZLD). Our contribution is aimed at developing an environmentally sustainable process that not only produces fresh water, but also solves the problem of brine discharge and its environmental impact on the ecosystem. The recovery of residual brine as a by-product of the seawater desalination process allows valuable materials to be extracted, by recovering ions of heavy metals such as lithium, bromine, phosphorus, arsenic, zinc, copper, nickel, silver, iron and manganese through a mechanism of concentration and precipitation in the solid state. This green technology is chemical-free and sustains the environment.

1. INTRODUCTION

According to a report by the International Desalination Association (IDA), there are 19,744 desalination plants in the world, with a daily desalination capacity of around 99.7 million m^3. These plants are located in 150 various countries and meet the needs of more than 300 million people. Conventional desalination processes exist; in particular membrane and thermal treatment, but most of these techniques have drawbacks such as high energy consumption, brine discharge and high maintenance costs. A negative impact on the environment, with global brine production of 141.5 million m^3/day, mainly discharged into ocean environments (De & Son, 2023).

Seawater desalination techniques are generally divided into two types:

a) Thermal techniques, which involve a phase change;
b) Membrane techniques which do not involve a phase change;
c) Hybrid techniques that combine two methods a) and b) in a single system.

Technologies for minimizing the negative impact of discharges on the ecosystem have been developed, but their efficiency rate is not matched by high energy consumption (Gude, 2016; Park et al., 2018).

Separation by electromagnetic process has been studied and applied in various scientific and industrial fields, including: physical medicine, biomedicine, microbiology, biotechnology, agriculture for irrigation, the environment for wastewater treatment, construction for reinforced concrete, and the mining industry for separating magnetic waste from non-magnetic waste (Burn et al., 2015; Mghaiouini et al., 2022; Zadeh-Haghighi & Simon, 2022).

Our contribution to sustainable and more ecological desalination consists not only of producing fresh water, but also of resolving the problem of brine discharge and its environmental impact on natural environments at micro and macro scales.

This physical process is coupled with a centrifugation and crystallization system for extracting brine from wastewater as a by-product of the seawater desalination process;

The estimated results are:

Studies of the interactions of the influential parameters: the magnetic field gradient, the flow velocity field gradient, the current density, the applied voltage and the concentration of the solution to be treated.

Study of the various couplings between these parameters: electromagnetic, hydrodynamic and ionic.

Study of the effect of the nature of the electrodes.

Exploration of an alternative process for treating and eliminating brine by recovering the precious metals and heavy metals present in the brine (Panagopoulos, 2022).

The modeling, CFD simulation and optimization of the parameters of an experimental treatment cell thanks to the coupling between the dynamics of salt water and the physical phenomena modeled with the COMSOL Multiphysics software (Qin & Bau, 2009).

This new approach is based on Physical desalination and consists of two major principles combinations:

- MHD: Lorentz force effect consists of electromagnetic fields with downstream coupling to a centrifugation and crystallization mechanism.
- CDI: Capacitive Deionization Assisted by an External Magnetic Field

The recovery of waste brine as a by-product of the seawater desalination process enables the extraction of valuable materials, recovering heavy metal ions such as lithium, bromine, phosphorus, arsenic, zinc, copper, lead, nickel, silver, iron and manganese through a mechanism of concentration and precipitation in a solid state. This green technology is chemical-free and environmentally sustainable.

2. PHYSICAL MODEL

2.1. MHD Processes

This Mechanism is based on the theory of the dynamo effect which concerns the generation of a magnetic field from the flow of an electrically conductive fluid of an MHD instability, the Magnetic Field iterative a spontaneous self-exciting dynamo of the magnetic field can be amplified in the manner of an instability;

The growth will occur until the field is strong enough to affect the fluid flow through turbulence and Lorentz force effects until saturation Continuous generation associated with electromotive fields E_{M0} induced by the field B_{0EXT}

By iterating the induction process to determine the dynamo condition of this device:

$$B_0 \rightarrow e_0 \rightarrow i_0 \rightarrow B_1 \rightarrow e_1 \rightarrow i_1 \ldots$$

The dynamo effect is maintained from a velocity field, requires a starting magnetic field, indicating the possibility of field and current growth, the growth will occur until the field is strong enough to affect the fluid flow through the Lorentz force,

The growth must then saturate, and we end up with a balanced magneto hydrodynamic (MHD) system in which the field and flow evolve dynamically on an equal footing (Bourgoin, 2003; Nore et al., 2014).

2.2 CDI-EMF Process

a) CDI Electrochemical Process, based on the principle that most dissolved salts are positively or negatively charged and will migrate towards electrodes with an opposite charge, the electrodes exert a force on the dissolved charged particles (ions), attracting or repelling them according to their electrical charge. This process has been widely used for the desalination of seawater or brackish water, the treatment of industrial effluents, the recovery of useful materials from effluents and the production of salt (Dong et al., 2015; Cai et al., 2009; Hoffart et al., 2008).

b) Electromagnetic EMF process: the use of an oscillating electromagnetic field, the polarity of which opposes the direction of fluid flow, acts as a deflector of the charged ions. At the same time, an electric field perpendicular to the direction of flow acts as an accelerator of these charged ions. The geometric and dimensional configuration of the electromagnetic separator is adjusted according to the migration time, given the constraint linked to the passage time of the moving particles (Pammé, 2006; Lin et al., 2020; Wang).

2.3 Design of an ED-CFG Crystallization and Centrifugal Extraction System

This process involves the fusion of two physic-chemical principles: ED electrodepositing combined with a CFG centrifugation mechanism, with this method desalination represents an innovative alternative to traditional electrodepositing, with centrifugation creating increased apparent gravity.

this process is based on the mechanical separation of the constituents of a mixture according to their difference in density, subjected to a centrifugal force F_c.

Centrifugation induces an acceleration that directs the particles towards the bottom of the centrifuge (centrifugal force =Fc), with a sedimentation coefficient determining the sedimentation speed of the particles (Sherman, 2017).

The system has an inlet for the water treated downstream by the EMF process, as well as three separate outlets. The first allows crystallized salts to be extracted, the second, brine, can be re-injected at the process inlet, and the third provides drainage for fresh water (Zimmermann et al., 2015).

3. MATHEMATICAL MODEL

The mathematical approach is based on the modeling of the flow of an electrically conductive fluid generated by the motion of several charged particles 'ions' interacting with the combination of an electric magnetic field $(\vec{E}, \vec{B})$

Contribution of Fluid mechanics is defined by the Navier-Stokes equations:

$$\rho \frac{\partial}{\partial t} v + \rho(v \,.\, \nabla) v = -\nabla p + \mu \nabla^2 v + f \tag{1}$$

Contribution of electromagnetism: defined by Maxwell's equations, describing the phenomena of electromagnetism/

Magnetic flux conservation

$$div\left(\vec{B}\right) = 0 \tag{2}$$

Maxwell Gauss:

$$div\vec{E} = \rho_0/\varepsilon_0 \tag{3}$$

Faraday law:

$$\overrightarrow{rot}\vec{E} = -\partial\vec{B}/\partial t \tag{4}$$

Ampère law:

$$\overrightarrow{rot}\vec{B} = \mu_0\vec{j} + \mu_0\varepsilon_0\partial\vec{E}/\partial t \tag{5}$$

Conservation of electric charge:

$$div\vec{j} = -\frac{\partial\rho}{\partial t} \tag{6}$$

Where

$$\frac{\partial\rho_0}{\partial t} = -\sigma\varepsilon_0\rho_e où \frac{d}{dt} = \partial t + v \cdot \nabla \tag{7}$$

$$\vec{E} = -\nabla(\varphi) - \frac{\partial\vec{A}}{\partial t} \tag{8}$$

Fluid-electromagnetic coupling laws:

The dynamic evolution of the fluid through the interaction of the magnetic field and the velocity field is governed by a combination law based on two equations (one for each field), each having a coupling term with the other field. The combined electrical and magnetic forcesF_{EM}, is written as follows:

electrostatic force:

$$\vec{F}_E = q\vec{E}$$

Lorentz force:

$$\vec{F}_M = q\left(v_z \times \vec{B}\right)$$

$$\vec{F}_{EM} = \vec{F}_E + \vec{F}_M = q\left[\vec{E} + \left(v_z \times \vec{B}\right)\right] \tag{9}$$

The effect of the driving forces resulting from the induced electric and magnetic fields are defined by the above equations coupled through the expression of the Lorentz force and Ohm's law generalized

$$\vec{j} = \sigma\vec{E}^* = \sigma\left(\vec{E} + \vec{v}\mathrm{x}\vec{B}\right) = \sigma\left(-\vec{\nabla}V + \vec{v}\mathrm{x}\vec{B}\right) \tag{10}$$

$$\vec{F}_L = \vec{j} \times \vec{B} \tag{11}$$

From the law of conservation of electric charge, neglecting the displacement currents and therefore the variation in charge, and taking the divergence of Ohm's law into account, we deduce:

$$\varphi = -\nabla.\left(-\frac{\partial A}{\partial t} + \vec{v}\mathrm{x}\vec{B}\right) \tag{12}$$

MHD mechanism:

Figure 1. The feedback of the induced field Bi on the applied field

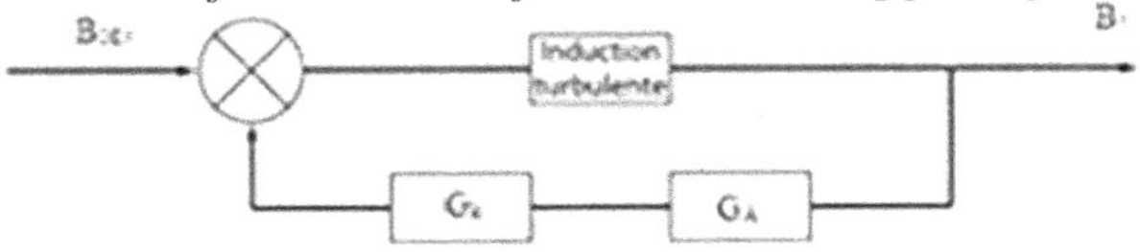

The various transfer functions are determined by the following relationships Induction: on average, the induced field BI is proportional to the applied field B_{0EX} Amplification: the amplifier output voltage U_{0EX} is proportional to the induced field (Verhille et al., 2011)

$$U_{0EX} = G_A . B_I \tag{13}$$

Coil response the applied field is done by the voltage U_{0EX}

$$B_{0EX} = G_K . U_{0EX} \tag{14}$$

$$B_{0EX} = G_K . G_A . B_I \tag{15}$$

U_{0EX}: control voltage for the current flowing through the coils

$$F_c = m\gamma F_g = mg \tag{16}$$

$$\gamma = r\omega^2\omega = \frac{v}{r} = \frac{2\pi N}{60} \tag{17}$$

$$\frac{Fc}{F_g} = \frac{\mathrm{mr}\ \omega\ 2}{mg} = \frac{r}{g}(2\pi N/60)^2 = 0.001118 r N^2 \tag{18}$$

So the centrifugal force is greater than the force of gravity of $\frac{r\omega^2}{g}$ times

4. MODELING AND SIMULATION WITH COMSOL MULTIPHYSICS

4.1. The Mutual Effects of Different Couplings on the Performance of the MHD Cell (Mosin & Ignatov, 2015; Smolyanov, 2021).

Table 1.

Coupling	Electromagnetic $(\vec{J},\vec{B},V)$	Hydrodynamic $(\vec{u})$	Ionic (C_i)
Electromagnetic $(\vec{J},\vec{B},V)$	—	Effect of $\vec{J}$ *and* $\vec{B}$ *on* $\vec{u}$	Effect of V *on* C_i
Hydrodynamic $(\vec{u})$	Effect of $\vec{u}$ *on* $\vec{J}$	—	Effect of $\vec{u}$ *on* C_i
Ionic(C_i)	Effect of C_i on the electromagnetic properties	Effect of C_i on the hydrodynamic properties	—

MHD modeling is based on the coupling of a system of two equations from electromagnetism and fluid mechanics, equation (1) the convection volume force $(\vec{F})$ becomes the Laplace force.

Resulting from the interaction of the current density $(\vec{J})$ injected at the electrodes with the magnetic field lines $(\vec{B})$. According to the Navier-Stokes equation, this force is responsible for pumping the salt water even in the absence of an external pressure difference between the cell inlet and outlet. This coupling is based on:

- Electromagnetic - Hydrodynamic coupling, which is presented by a system of equations whose resolution, is not independent. They are linked by the magneto hydrodynamic forces acting on the fluids and the current densities linked to the velocity field by Ohm's law.

a) Electromagnetic equations, to determine the induced current density
b) Hydrodynamic equations, the flow velocity is determined.

- Electromagnetic coupling - Ion transport:

The process of transport and density of the flow of matter was carried out under the action of several driving forces.

The fact that the total molar ionic flux (N_i) of a species (i) is the sum of several fluxes:

- Diffusion: due to the ionic concentration gradient, the diffusion flux of a species i
- Migration: due to the electric potential gradient, the migration flux of a species i
- Forced and/or natural convection: the movement of the solution is accompanied by a convective transfer phenomenon for each of the species. The flow generated by this convection
- Coupling Hydrodynamics - Ion transport: the equation of spatio-temporal variation of the transport of the species present in the fluid, the ionic concentration (c_i) is linked to the velocity field ($\vec{u}_i$) via the convection flux ($\vec{N}_i^{conv}$) and that the volume density responds to the law of variation of the concentration of the species present in the solution as follows:

$$\rho = \sum_i M_i \times c_i.$$

4.2. MHD Simulation of the Effect of the Earth's Magnetic Field on Flow Velocity Using Comsol Multiphysics Software

Figure 2.

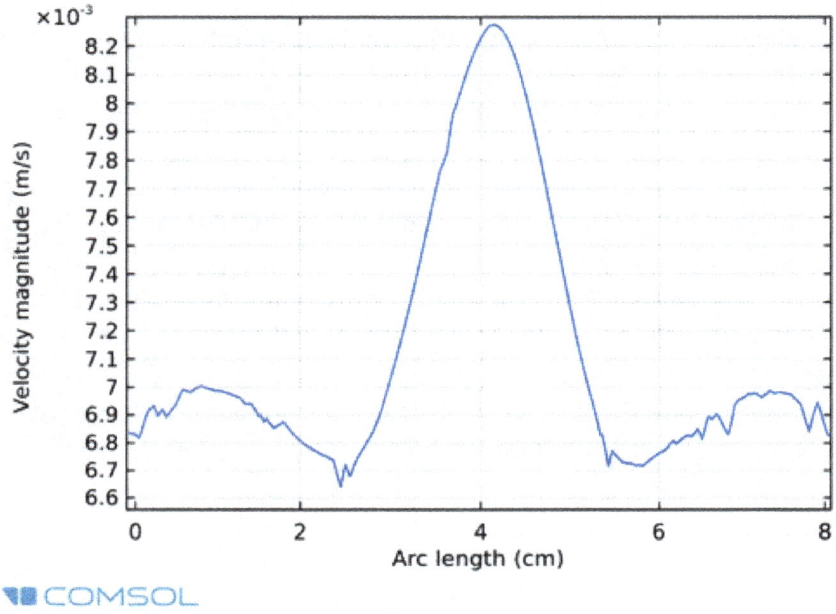

Figure 2 illustrates the effect of the Earth's magnetic field on flow velocity.

Analysis of this figure shows that in the absence of an external magnetic field, the fluid flow velocity is at its maximum in the middle of the treatment channel.

4.3. MHD Simulation of the Effect of the Magnetic Field on Flow Velocity Using Comsol Multiphysics Software (Pulugundla, 2013).

Figure 3.

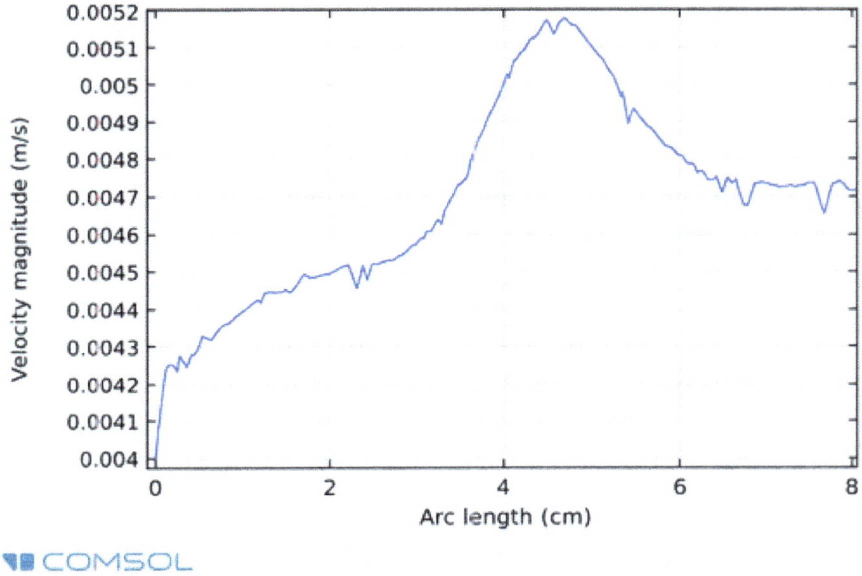

Figure 4.

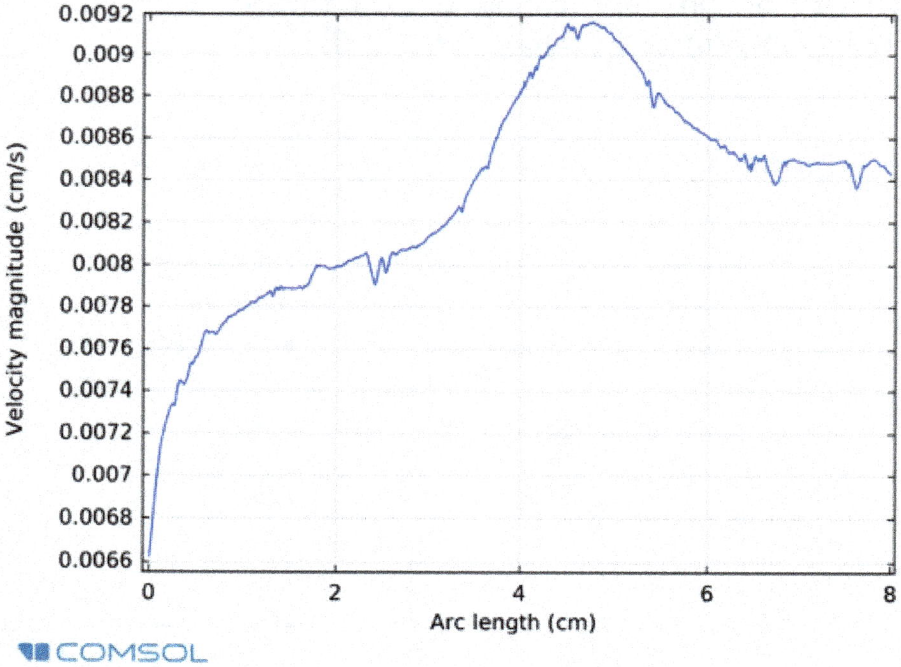
Velocity magnitude (cm/s)
0.0092
0.009
0.0088
0.0086
0.0084
0.0082
0.008
0.0078
0.0076
0.0074
0.0072
0.007
0.0068
0.0066
0
2
4
6
8
Arc length (cm)
COMSOL

Figure 5.

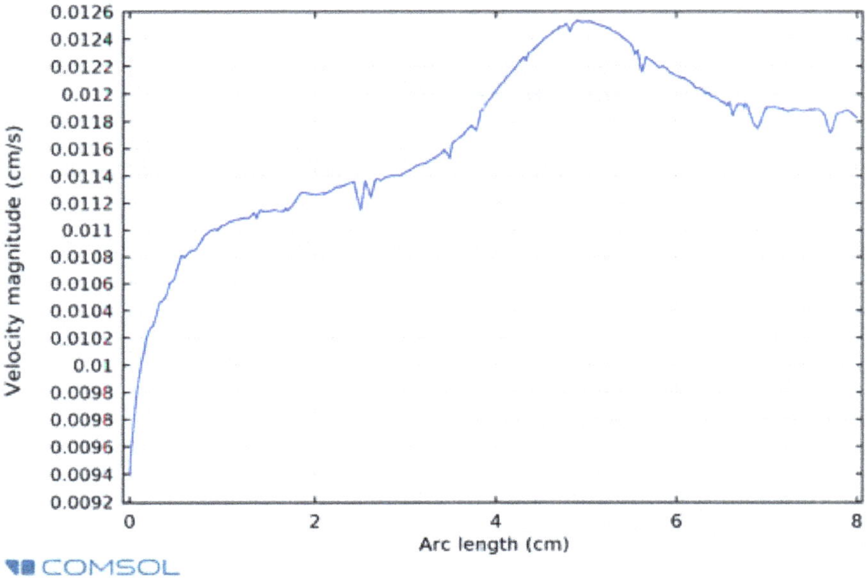

Interpretation:

Figures 3, 4 and 5 shows a comparative simulation with a constant voltage of 1V, varying the value of the magnetic field by 0.5T, 1T and 1.5T. The effect of the value of the magnetic field on the speed of the fluid is remarkable, the greater the magnetic field, the greater the fluid velocity.

The effect of the magnetic field is greatest in the middle of the channel.

5. CRYSTALLIZATION

To crystallise salt from a brine solution, one must overcome the enthalpy of hydration, which is the lower chemical potential energy present in the solution when the ions are dissolved in water and secondly, energy must be added to overcome the lattice enthalpy (or lattice energy) required for the ions to aggregate sufficiently to crystallise.In current technology, the nucleation energy needed to overcome both barriers is provided by electromagnetic fields via

MHD. Current technology is based on the theory that an electrocrystallisation process can extract salt crystals from the water solution to be treated. The presence of an electric fied in a solution has been shown to promote crystal nucleation by affecting molecular orientation and density fluctuations, and nucleation is also thought to be affected by the reduction in free energy of small crystallites in the presence of a field, due to their polar nature The crystallization phase in the solid state follows the following process:

Step 0: Stabilizing the Aqueous Solution

At this stage the aqueous solution is pre-balanced to ensure a stable chemical composition and uniformity of parameters such as ionic concentration, pH and temperature. This creates the initial conditions required for the rest of the process.

$$F(n) = -n\Delta\mu + \gamma an2/3 \tag{19}$$

Step 1: Driving Forces - Disturbance of the Solution's Equilibrium

Once Magneto hydrodynamics (MHD) is induced, significant changes in the driving forces of the solution occur.

MHD, generated by the application of magnetic and electric fields, breaks the equilibrium of the solution, initiated by external forces..

$$\Delta\mu\varepsilon = \Delta\mu + c\varepsilon E2 \tag{20}$$

$$C_\epsilon = \frac{3\epsilon_0\epsilon_m(\epsilon_c - \epsilon_m)V_0}{2(\epsilon_c - \epsilon_m)} \tag{21}$$

Step 2: Core Growth With EMF for Priming

Electromotive Force (EMF) is used to stimulate the growth of crystalline nuclei, thereby initiating the crystallization process. This step is crucial to guarantee homogeneous and controlled crystallization.

$$\Delta td(n)/dt \ll csatV(S-1) \tag{22}$$

Step 3: Crystallization - Fixation Frequency for Nucleation, Followed by Centrifugation [7]

Crystallization takes place at an appropriate fixing frequency for crystal nucleation. This step is then followed by centrifugation, a mechanical process that separates the components according to their density, helping the crystals to precipitate.

$$D = 4\pi r + +Dv\ (18),\ D = 4\pi(r + +)2 \tag{23}$$

Step 4: Assessment of Nucleation Rate

The final step involves a precise assessment of the nucleation rate, measuring the speed at which new crystals form. This assessment is essential for understanding the kinetics of crystallization and optimizing the overall process

6. MECHANICAL CENTRIFUGAL EXTRACTION

The centrifugation process, by exploiting centrifugal acceleration on the difference in density of the phases present, involves the field of centrifugal acceleration and the force associated with this process is applicable in the event of saturation. During centrifugation, the components with the highest density settle at the periphery of the rotating bowl. A high settling velocity is critical to the speed and efficiency of the process. Separation efficiency depends on the difference in density of the fluids to being separated, particle size, viscosity and temperature. Most dissolved salts are positively or negatively charged and will migrate towards electrodes of opposite charge. The electrodes exert a force on the dissolved charged particles (ions), attracting them.

7. METHODS PROCESS

7.1. Block Diagram

Figure 6. Treatment process flowchart

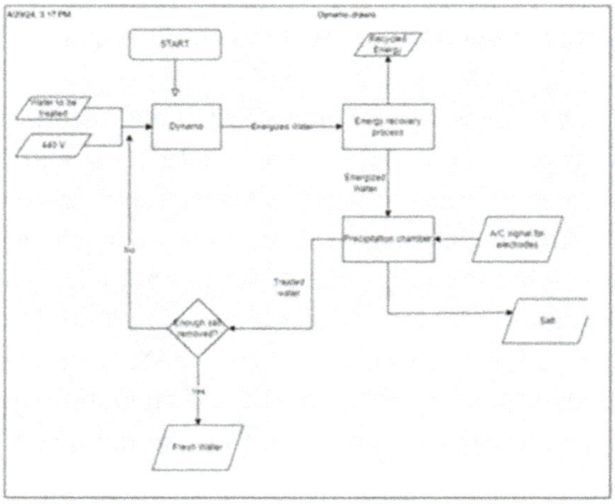

Figure 7. Treatment process summary

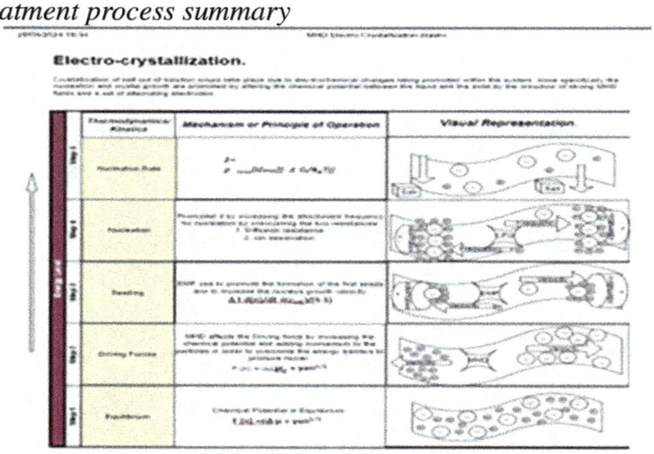

Figure 8. Bowl centerfuge

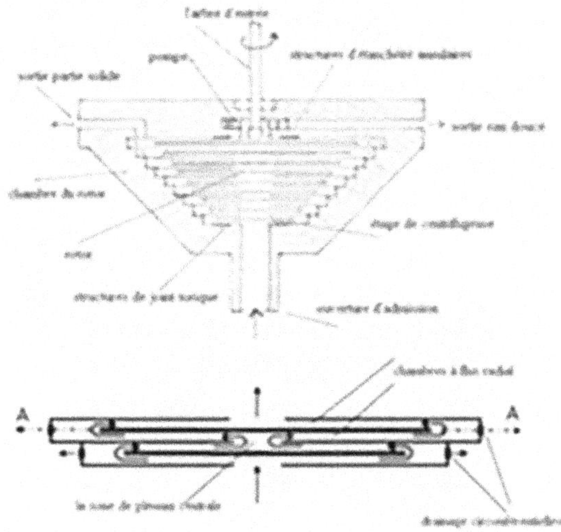

7.2. Description

7.2.1 Coil Dimensioning

The coil used in this experiment to create the magnetic field was designed and manufactured according to Biot and Savart's law. It is made up of double turns N=5000n of dimensions L=600mm and Ø=70mm supplied with a direct current of 30VDC.

The coil generates a magnetic induction of between 100mT, 500mT,1000Mt and 1500mT

The magnetic circuit consists of a ferromagnetic core laminated with iron powder and cooled by insulating mineral oil.

Figure 9. coil

7.2.2. MHD

To validate our experiment (Alnaimat et al., 2016), we chose a rectangular channel 20mm wide and 200mm long powered at 1VDC, with a permanent magnet in the shape of a parallelepiped (200x50x50)

Figure 10. MHD system

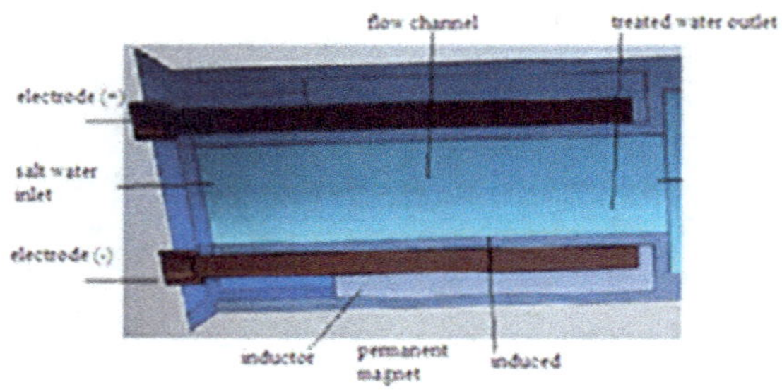

7.2.3. Capacitive Deionization (CDI) Cell

Figure 11. Magnetically Assisted Capacitive Deionization (CDI) Cell

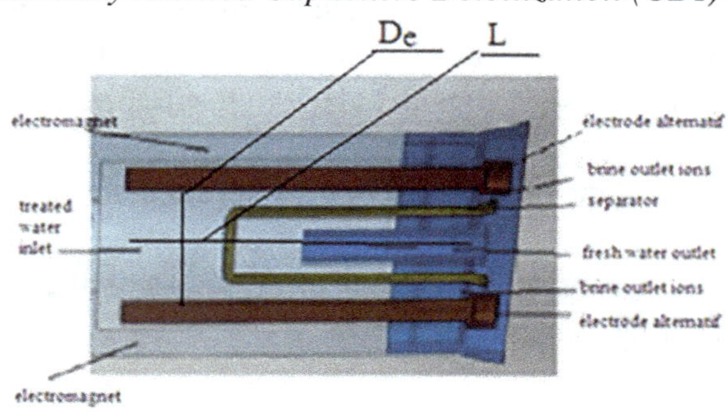

The electrochemical (Imris & Schenk, 2007), bifurcation cell has a cylindrical geometry, is made of an electrically non-conductive material and is designed to conduct seawater or brackish water for continuous treatment;

This cell has an inlet for salt water and an outlet in the form of a concentric cylinder whose axis is for fresh water and whose outer skin is for concentrated brine;

Equipped with two curved half-plate electrodes, the anode and the cathode, which are connected to an alternating current generator

7.2.4. Electrodes: Carbon, Graphite. Platinum, Titanium

7.2.5.EMF Treatment Equipment

Figure 12. Experimental Prototype of the Magnetically Assisted MHD–CDI Desalination Unit Developed for Sustainable Freshwater Production and Zero Liquid Discharge (ZLD).

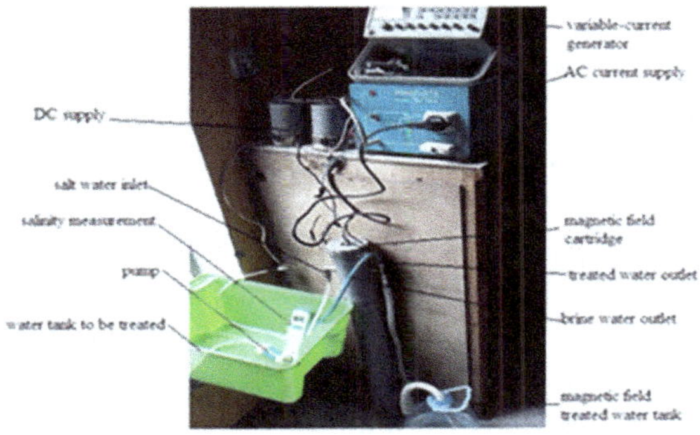

7.3. Results

7.3.1 Samples

Table 1 different sample of the solutions to be treated:

Seawater: sampling tank containing 5 liters of seawater with a salinity of E_{M0} is S_0=35.5 g/L at an ambient temperature of 20°C, with a flow rate of 24 l/min.

Aqueous solution NaCl at room temperature of 20 ° C preparation of 5 liters of Salinity, with a flow rate of 24 l/min

P_1: S_1 = 17, 8 g/L

P_2: S_2= 9, 9 g/L

P_3: S_3= 6,4g/ L

The salt water sample passes through the system for 5 and 10 minutes. Water was collected after each period and its salinity value were noted by the HANNA measuring instrument, which indicates an increase in temperature and electrical conductivity value.

Table 2. Treatment result and Rate (%)

test tube	S_i g/L	S_f g/L	$-\Delta S$g/L	Rate%
EM_0	35,5	32,2	3,3	9,29
P_1	17, 8	16,1	1,7	9, 55
P_2	9, 9	7, 9	2,0	20,20
P_3	6,4	3,6	2,8	43,75

Table 2 Analysis of this table show: Analysis of these results shows that the lower the NaCl concentration, the higher the desalination rate.

7.3.2 Heavy Metal Removal

Table 3. Results in % of heavy metal elimination

Intensity of Electromagnetic Field(mT)	Duration of Electrochemical Treatment(min)	Electrode Composition	Heavy Metal Removal Rate (%)
100	30	Platinum	80
100	60	Graphite	75
100	120	Carbon	85
100	180	Titanium	78
500	30	Graphite	88
500	60	Platinum	92
500	120	Titanium	90
500	180	Carbon	89
1000	30	Carbon	93
1000	60	Titanium	95
1000	120	Platinum	94
1000	180	Graphite	91
1500	30	Titanium	96
1500	60	Carbon	97
1500	120	Graphite	98
1500	180	Platinum	99

The effects of Intensity of Electromagnetic field, duration of electrochemical treatment, and electrode composition on process responses are analyzed. This approach allowed us to investigate the interactions and independent influences of these factors in a comprehensive manner while minimizing the number of trials and time required, similar to a completed factorial design. The results of this analysis are presented in Table 2.

The results obtained using the Taguchi methods are summarized in Table2. The heavy metal removal rate was analyzed using four levels. The optimal combination that yielded the highest heavy metal removal rate was achieved with intensity of electromagnetic field (level 4), duration of electrochemical treatment (level 3), and Electrode composition

(level 1) (El-Shamy et al., 2017; Roy, 2001).

8. CONCLUSION AND PERSPECTIVES

This study demonstrates that coupling Magnetohydrodynamic (MHD) flow control with magnetically assisted Capacitive Deionization (CDI) provides an effective and environmentally sustainable approach to seawater desalination. The combined action of the applied magnetic field and electrical potential increases fluid velocity within the treatment channel, with the peak velocity occurring at the center, while multi-stage magnetic configurations maintain a uniform flow profile.

This hydrodynamic control promotes efficient ion transport toward the CDI electrodes, enhancing the removal of Na+ and Cl-1 ions.

A key advantage of this architecture is the downstream integration of centrifugation and crystallization. Rather than treating concentrated brine as a waste stream, the process enables the recovery of valuable minerals, including lithium, thereby aligning the system with Zero Liquid Discharge (ZLD) principles.

Furthermore, the findings highlight that electrode material selection critically influences desalination performance. The MHD-CDI system also proves highly effective as a pretreatment stage for reverse osmosis, mitigating scaling and membrane fouling to lower overall maintenance requirements.

Although this technology shows promising potential, transitioning from laboratory scale validation to practical industrial applications requires addressing key operational challenges. Future research will focus on quantifying and improving the net energy efficiency of the coupled MHD-CDI process, optimizing electrode materials under dynamic operating conditions, and refining the kinetics of selective ion recovery during downstream crystallization.

REFERENCES

Alnaimat, F., Alhseinat, E., Banat, F., & Mittal, V. (2016). Electromagnetic–mechanical desalination: Mathematical modeling. *Desalination, 380*, 75–84. DOI: 10.1016/j.desal.2015.11.012

Bourgoin, M. (2003). *Études en magnétohydrodynamique, application à l'effet dynamo* (Doctoral dissertation, Ecole normale supérieure de lyon-ENS LYON).

Burn, S., Hoang, M., Zarzo, D., Olewniak, F., Campos, E., Bolto, B., & Barron, O. (2015). Desalination techniques—A review of the opportunities for desalination in agriculture. *Desalination, 364*, 2–16. DOI: 10.1016/j.desal.2015.01.041

Cai, R., Yang, H., He, J., & Zhu, W. (2009). The effects of magnetic fields on water molecular hydrogen bonds. *Journal of Molecular Structure, 938*(1-3), 15–19. DOI: 10.1016/j.molstruc.2009.08.037

De, S., & Son, M. (2023). REVUE BIBLIOGRAPHIQUE ET ÉTUDE DE CAS SUR LE DESSALEMENT D'EAU DE MER ET SON IMPACT SUR LE MILIEU MARIN.

Dong, K., Zhang, S., & Wang, Q. (2015). A new class of ion-ion interaction: Z-bond. *Science China. Chemistry, 58*(3), 495–500. DOI: 10.1007/s11426-014-5147-2

El-Shamy, A. M., Farag, H. K., & Saad, W. (2017). Comparative study of removal of heavy metals from industrial wastewater using clay and activated carbon in batch and continuous flow systems. *Egyptian Journal of Chemistry, 60*(6), 1165–1175.

Gude, V. G. (2016). Desalination and sustainability–an appraisal and current perspective. *Water Research, 89*, 87–106. DOI: 10.1016/j.watres.2015.11.012 PMID: 26641014

Hoffart, D. J., Tiburcio, J., de la Torre, A., Knight, L. K., & Loeb, S. J. (2008). Cooperative ion-ion interactions in the formation of interpenetrated molecules. *Angewandte Chemie International Edition, 47*(1), 97–101. DOI: 10.1002/anie.200703019 PMID: 18008294

Imris, P., & Schenk, D. (2007). *Method and device for the desalination of in particular seawater with the aid of alternating fields of identical frequencies* (European Patent No. EP1828060A1). European Patent Office. https://patents.google.com/patent/EP1828060A1/en

Lin, L., Jiang, W., Xu, X., & Xu, P. (2020). A critical review of the application of electromagnetic fields for scaling control in water systems: Mechanisms, characterization, and operation. *NPJ Clean Water*, *3*(1), 25. DOI: 10.1038/s41545-020-0071-9

Mghaiouini, R., Saifaoui, D., & Salah, M. (2022). New ecological and economic process to save energy by electromagnetic fields. *Materials Today: Proceedings*, *66*, 11–16. DOI: 10.1016/j.matpr.2022.02.568

Mosin, O., & Ignatov, I. (2015). Magnetohydrodynamic cell for magnetic water treatment. *Nanotechnology Research and Practice*, *6*(2), 81–92. DOI: 10.13187/ejnr.2015.6.81

Nore, C., Guermond, J. L., Léorat, J., & Luddens, F. (2014, July). Effet dynamo dans des géométries cylindriques. In *Symposium de Génie Électrique 2014*.

Pamme, N. (2006). Magnetism and microfluidics. *Lab on a Chip*, *6*(1), 24–38. DOI: 10.1039/B513005K PMID: 16372066

Panagopoulos, A. (2022). Brine management (saline water & wastewater effluents): Sustainable utilization and resource recovery strategy through Minimal and Zero Liquid Discharge (MLD & ZLD) desalination systems. *Chemical Engineering and Processing*, *176*, 108944. DOI: 10.1016/j.cep.2022.108944

Park, K., Heo, H., & Yang, D. R. (2018). Feasibility study of a forward osmosis/crystallization/reverse osmosis hybrid process with high-temperature operation: Modeling, experiments, and energy consumption. *Journal of Membrane Science*, *555*, 206–219. DOI: 10.1016/j.memsci.2018.03.031

Pulugundla, G. (2013). *Numerical modelling of liquid metal flows interacting with strongly inhomogeneous magnetic fields* (Doctoral dissertation, TU Ilmenau).

Qin, M., & Bau, H. H. (2009). Magneto-hydrodynamic flow in electrolyte solutions. In *Comsol conference*.

Roy, R. K. (2001). *Design of experiments using the Taguchi approach: 16 steps to product and process improvement*. John Wiley & Sons.

Sherman, K. L. (2017). *Continuous flow centrifuge with application of water desalination and purification* (W.O. Patent No. WO2017030598A1). World Intellectual Property Organization. https://patents.google.com/patent/WO2017030598A1/en

. Smolyanov, I. (2021). Numerical Approaches to Analyzing of MHD Processes Occuring to Induction Pump.

Verhille, G., Plihon, N., Khalilov, R., Pinton, J. F., & Frick, P. (2011, August). Modification of a von Kármán turbulent flow by a magnetic field. In *CFM 2011 - 20th French Congress of Mechanics*. AFM, Maison de la Mécanique, 39/41 rue Louis Blanc - 92400 Courbevoie.

Wang, Y., Wei, H., & Li, Z. (2018). Effect of magnetic field on the physical properties of water. *Results in Physics*, *8*, 262–267. DOI: 10.1016/j.rinp.2017.12.022

Zadeh-Haghighi, H., & Simon, C. (2022). Magnetic field effects in biology from the perspective of the radical pair mechanism. *Journal of the Royal Society, Interface*, *19*(193), 20220325. DOI: 10.1098/rsif.2022.0325 PMID: 35919980

Zimmermann, N. E., Vorselaars, B., Quigley, D., & Peters, B. (2015). Nucleation of NaCl from aqueous solution: Critical sizes, ion-attachment kinetics, and rates. *Journal of the American Chemical Society*, *137*(41), 13352–13361. DOI: 10.1021/jacs.5b08098 PMID: 26371630

Chapter 4
Harnessing the Power of Advanced Optical Materials for Real-World Applications

J. Mangaiyarkkarasi
http://orcid.org/0000-0003-1431-9584
Nadar Mahajana Sangam S. Vellaichamy Nadar College, India

Shanthalakshmi Revathy J.
http://orcid.org/0000-0003-1724-7117
University College of Engineering, Ramanathapuram, India

ABSTRACT

Advanced optical materials have driven innovations in telecommunications, biomedicine, and energy by manipulating light at the nanoscale. These materials enhance absorption, emission, and nonlinear optical responses. Quantum dots and photonic crystals boost data transmission and device efficiency, while quantum dots and plasmonic nanoparticles advance imaging and diagnostics in biomedicine. In energy, quantum dot solar cells and plasmonic nanostructures improve solar energy harvesting. Additionally, these materials are utilized in environmental monitoring, high-resolution displays, and security systems, aiding in pollution detection, energy efficiency, and national security. Despite progress, challenges in stability, scalability, and cost-effectiveness persist. Ongoing interdisciplinary research and collaboration are essential for overcoming these challenges and expanding applications. Advanced optical materials promise to address societal issues and drive technological progress.

DOI: 10.4018/979-8-3693-6627-1.ch004

© 2027, IGI Global Scientific Publishing. Copying or distributing in print or electronic forms without written permission of IGI Global Scientific Publishing is prohibited. Use of this chapter to [...]rative artificial intelligence (AI) technologies is expressly prohibited. The publisher reserves all rights to license its use for generative AI training and machine learning model development.

1. INTRODUCTION

Advanced optical materials have emerged as a cornerstone of modern technological innovation, driving significant advancements across many fields. These materials, engineered to manipulate light at the nanoscale, exhibit a range of unique properties such as enhanced light absorption, emission, transmission, and exceptional nonlinear optical responses. Their ability to control and interact with light unprecedentedly has opened up new avenues for research and application, making them indispensable in addressing contemporary scientific and engineering challenges. The telecommunications industry, for instance, has benefited immensely from these innovations. Advanced optical materials like semiconductor quantum dots and photonic crystals have revolutionized data transmission, enabling higher data rates and longer transmission distances (Kumbhakar et al., 2021). Quantum dot-based lasers and other photonic devices leverage the precise control over light propagation provided by these materials, resulting in more efficient and compact optical communication systems. In the realm of biomedicine, the impact of advanced optical materials is equally transformative. Quantum dots serve as highly effective fluorescent probes for bioimaging and cellular labeling, thanks to their tunable emission spectra and excellent photostability. Plasmonic nanoparticles, through their localized surface plasmon resonance, significantly boost the sensitivity and specificity of biomolecular detection. This advancement paves the way for breakthroughs in diagnostics, drug delivery, and therapeutic interventions (Duan et al., 2023). The field of energy also sees substantial contributions from these materials, particularly in solar energy harvesting and conversion. Innovations such as quantum dot solar cells and plasmonic-enhanced thin-film solar cells exploit the unique properties of semiconductor nanocrystals and nanostructures to achieve higher efficiencies, supporting the progress of supportable and renewable energy technologies. Beyond telecommunications, biomedicine, and energy, advanced optical materials find applications in environmental monitoring, display technologies, security systems, and quantum technologies. Their diverse functionalities make them invaluable tools for addressing real-world challenges and fostering technological progress(Dai et al., 2024). This chapter delves into the remarkable properties and multifaceted applications of advanced optical materials, exploring how they are harnessed to drive innovation across various sectors. By understanding these cutting-edge materials, we can better appreciate their potential to revolutionize our world.

1.1 Principles Underlying the Usage of Optical Materials

Optical materials are fundamental to modern photonic and optoelectronic technologies, as their interaction with electromagnetic radiation governs light generation, propagation, modulation, and detection. The effective use of optical materials in real-world applications is based on several core physical and chemical principles that dictate their performance, reliability, and integration into devices (Chauhan, 2024).

A primary principle underlying optical materials is light–matter interaction, which describes how photons interact with electrons, phonons, and excitons within a material. Depending on the electronic structure and band gap, materials may absorb, transmit, reflect, or emit light over specific wavelength ranges. For example, wide-band-gap semiconductors such as ZnO and GaN efficiently interact with ultraviolet light, making them suitable for UV detectors and LEDs, while narrow-band-gap materials are preferred for infrared sensing and communication technologies.

Electronic band structure engineering is another critical principle guiding the usage of optical materials. By tailoring band gaps through doping, alloying, or quantum confinement, researchers can tune optical absorption and emission characteristics. This principle is extensively exploited in optoelectronic devices, where controlled energy levels enable efficient charge carrier generation, recombination, and transport. Defect engineering further refines optical response by introducing or suppressing localized states that influence luminescence efficiency and carrier lifetime.

The principle of refractive index control plays a central role in optical waveguiding and light confinement. Variations in refractive index enable the manipulation of light propagation through reflection, refraction, and total internal reflection. This principle underpins the operation of lenses, optical fibers, photonic crystals, and integrated photonic circuits. Precise control over refractive index contrast is essential for minimizing optical losses and achieving compact device architectures (Park & Jun, 2023).

Nonlinear optical response is a key principle for advanced photonic applications, particularly in high-intensity light environments. In nonlinear optical materials, the polarization response depends nonlinearly on the applied electromagnetic field, enabling frequency conversion processes such as second-harmonic generation and optical parametric amplification. These effects are vital for laser technology, ultrafast optics, and optical signal processing, where wavelength flexibility and high-speed modulation are required.

At the nanoscale, plasmonic and excitonic effects become dominant. Metallic nanostructures support localized surface plasmon resonances, leading to strong electromagnetic field enhancement and subwavelength light confinement. This principle is exploited in sensing, spectroscopy, and photothermal applications. Similarly, excitonic interactions in low-dimensional and organic optical materials enhance light absorption and emission, enabling high-efficiency optoelectronic devices (An et al., 2025)

The structure–property relationship is a fundamental principle underlying the usage of optical materials. Crystallinity, grain size, surface morphology, and dimensionality strongly influence optical transparency, scattering, and emission behavior. Advanced synthesis and fabrication techniques are employed to achieve controlled microstructures that optimize optical performance while ensuring mechanical and thermal stability.

Finally, material stability and compatibility are essential principles for practical deployment. Optical materials must maintain performance under prolonged exposure to light, heat, and environmental conditions. Chemical stability, resistance to photodegradation, and compatibility with existing fabrication processes determine their suitability for industrial applications.

The usage of optical materials is governed by interconnected principles involving light–matter interaction, electronic structure, refractive index control, nonlinear response, nanoscale effects, structural optimization, and long-term stability. A thorough understanding of these principles enables the rational design and application of optical materials across photonics, energy, sensing, and communication technologies.

2. TELECOMMUNICATIONS: ENHANCING DATA TRANSMISSION WITH QUANTUM DOTS AND PHOTONIC CRYSTALS

The telecommunications landscape is perpetually evolving, driven by the relentless pursuit of faster, more efficient, and more secure data transmission methods. Among the cutting-edge advancements pushing the boundaries of what's possible in this field are quantum dots and photonic crystals (Mehtab et al., 2022). These technologies promise to revolutionize data transmission, offering unprecedented improvements in speed, capacity, and reliability.

2.1 Quantum Dots: The Quantum Leap in Data Transmission

Quantum dots (QDs) are semiconductor nanoparticles that exhibit exclusive optical and electronic properties because of their quantum mechanical characteristics. Figure 1 shows the structure of the quantum dot.

Figure 1. Structure of quantum dot

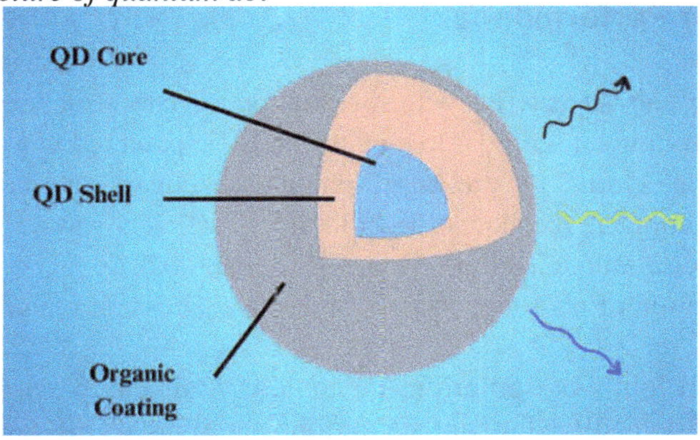

These properties arise from the quantum confinement effect, which occurs when electrons within the quantum dots are confined to a very small volume, typically in the range of a few nanometers. This confinement leads to discrete energy levels, akin to those found in atoms, which can be precisely controlled by adjusting the size and composition of the quantum dots (Liu et al., 2022). In telecommunications, quantum dots have the potential to enhance data transmission in several ways:

High-Speed Data Transfer: Quantum dots can be used to develop ultra-fast lasers and single-photon sources, which are essential for high-speed optical communication systems. Their ability to emit photons at precise wavelengths with high efficiency makes them ideal for creating highly coherent light sources that can be modulated at extremely high frequencies, enabling faster data transmission rates.

Increased Bandwidth: Quantum dots can be engineered to produce light at different wavelengths simultaneously. This capability is especially valuable in wavelength-division multiplexing (WDM) systems, where multiple data channels are transmitted through the same optical fiber using different wavelengths. By incorporating quantum dots, the number of available channels can be significantly increased, thereby enhancing the overall bandwidth of the communication system(Huang et al., 2020).

Improved Signal-to-Noise Ratio: The coherent light emitted by quantum dots can improve the signal-to-noise ratio in optical communication systems. This leads to clearer signals over long distances, reducing the need for signal amplification and repeaters, and thereby lowering the overall cost and complexity of the network.

2.2 Photonic Crystals: Structuring Light for Superior Performance

Photonic crystals are materials with a periodic optical nanostructure that influences the movement of photons similar to how the periodic potential in a semiconductor crystal affects electrons. By creating a photonic bandgap, photonic crystals can control and manipulate the flow of light in highly efficient ways, which is crucial for advanced telecommunications(Kumar & Kumar, 2023).

The application of photonic crystals in data transmission includes several key advantages:

Enhanced Light Manipulation: Photonic crystals can precisely control the propagation of light, allowing for the development of highly efficient waveguides, filters, and resonators. This precise control facilitates the development of compact and integrated photonic circuits that can process data at the speed of light, leading to faster and more efficient communication systems. Figure 2 shows the flow of light through photonic crystal

Figure 2. Flow of light through photonic crystal

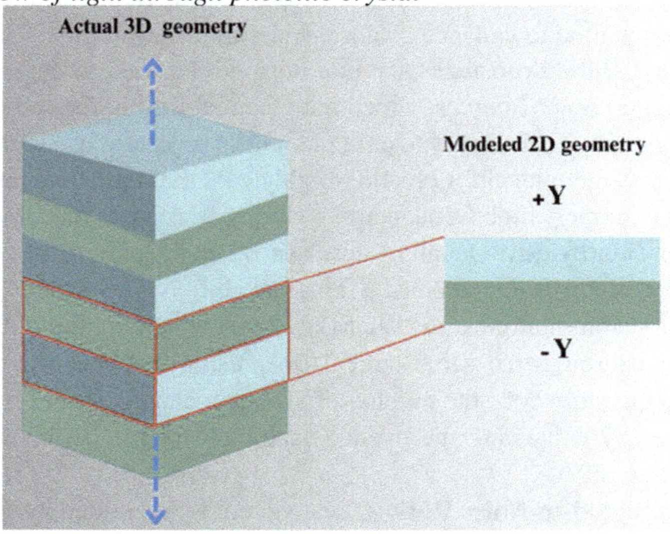

Reduction of Losses: The exceptional properties of photonic crystals permit the minimization of optical losses. By guiding light more effectively and reducing scattering and absorption, photonic crystals ensure that more of the transmitted data reaches its destination without degradation. This is particularly beneficial for long-distance data transmission.

Integration with Silicon Technology: Photonic crystals can be integrated with existing silicon-based technologies, enabling the progress of hybrid systems that combine the advantages of both photonics and electronics. This integration is pivotal for creating high-speed, low-power communication devices that can be easily incorporated into current telecommunication infrastructures(L. Li et al., 2022).

2.3 Synergistic Potential: Quantum Dots and Photonic Crystals

The combination of quantum dots and photonic crystals holds even greater promise for the future of telecommunications. Quantum dots can be embedded within photonic crystal structures to create sources of single photons or entangled photon pairs, which are essential for quantum communication protocols that promise unparalleled security. Additionally, the integration of quantum dots with photonic crystal waveguides and cavities can lead to highly efficient and compact devices for data processing and transmission. The integration of quantum dots and photonic crystals represents a significant leap forward in the quest for advanced telecommunication technologies. By harnessing the unique properties of these materials, researchers and engineers can develop systems that offer faster data transmission rates, increased bandwidth, improved signal integrity, and enhanced security. As these technologies advance, they will undoubtedly be pivotal in shaping the future of global communications, leading to more connected and efficient digital societies (X. Wang et al., 2024).

3. BIOMEDICINE: REVOLUTIONIZING IMAGING AND DIAGNOSTICS WITH QUANTUM DOTS AND PLASMONIC NANOPARTICLES

In the rapidly progressing field of biomedicine, the integration of nanotechnology has unlocked new pathways for innovation, especially in imaging and diagnostics. Quantum dots and plasmonic nanoparticles are at the forefront of these advancements, offering unprecedented capabilities in terms of sensitivity, resolution, and functionality (Yao et al., 2018). These nanomaterials are poised to revolutionize how we visualize and diagnose diseases, leading to earlier detection and more active treatments.

3.1 Quantum Dots: Illuminating New Possibilities in Imaging

Quantum dots (QDs) are semiconductor nanocrystals with unique optical and electronic properties, including size-tunable light emission and high brightness. Their potential applications in biomedical imaging are vast and transformative:

Enhanced Fluorescence Imaging: Quantum dots are known for their intense and stable fluorescence. Unlike traditional organic dyes, which tend to photobleach over time, quantum dots provide sustained illumination. This makes them ideal for long-term imaging studies, allowing researchers to track biological processes over extended periods without signal degradation(Urmi et al., 2024).

Multiplexed Imaging: The emission wavelengths of quantum dots can be finely adjusted by altering their size and composition. This capability allows for the immediate detection of multiple biomarkers in a sole sample by using quantum dots of different colors. Such multiplexed imaging is invaluable for comprehensive diagnostic assays, enabling the identification of several disease markers concurrently.

Deep Tissue Imaging: Quantum dots can be planned to emit light in the near-infrared (NIR) spectrum, which penetrates tissues more effectively than visible light. This property enhances the ability to image deep within the body, providing clearer and more detailed images of internal structures and tumors. NIR-emitting quantum dots are particularly useful for in vivo imaging applications, where high-resolution visualization of deep tissues is essential.

3.2 Plasmonic Nanoparticles: Harnessing Light for Superior Diagnostics

Plasmonic nanoparticles, typically made of noble metals like gold and silver, exhibit unique optical properties due to restricted surface plasmon resonances (LSPRs). These resonances occur when conduction electrons on the nanoparticle's surface oscillate in response to light, leading to enhanced electromagnetic fields at specific wavelengths (Mahmoudpour et al., 2021). This phenomenon is leveraged in various diagnostic applications:

Surface-Enhanced Raman Scattering (SERS): Plasmonic nanoparticles greatly amplify the Raman scattering signal of molecules adsorbed on their surfaces. This enhancement enables highly sensitive detection of biomolecules, even at extremely low concentrations. SERS-based diagnostics can identify molecular signatures of diseases with great precision, making it a powerful tool for early diagnosis.

Localized Plasmonic Sensing: The sensitivity of plasmonic nanoparticles to changes in their local environment makes them excellent sensors for detecting biomolecular interactions. When a target molecule binds to the nanoparticle's surface, it induces a measurable shift in the plasmon resonance frequency. This shift can be detected and quantified, providing real-time monitoring of biochemical processes and interactions.

Photothermal Therapy and Imaging: Plasmonic nanoparticles can convert absorbed light into heat, a feature utilized in photothermal therapy. By targeting cancer cells with these nanoparticles and irradiating them with near-infrared light, localized heating can selectively destroy tumor cells while sparing the surrounding healthy tissue. Additionally, the heat generated can be used for imaging purposes, providing a dual-functionality in both treatment and diagnosis (Xie et al., 2021).

3.3 Synergistic Potential: Combining Quantum Dots and Plasmonic Nanoparticles

The integration of quantum dots and plasmonic nanoparticles holds immense potential for enhancing biomedical imaging and diagnostics. For example, combining the bright fluorescence of quantum dots with the signal amplification properties of plasmonic nanoparticles can create highly sensitive and multiplexed diagnostic platforms. Such hybrid systems can achieve unparalleled levels of detection sensitivity and specificity, crucial for identifying diseases at their earliest stages. Quantum dots and plasmonic nanoparticles represent the cutting edge of nanotechnology applications in biomedicine(Yang et al., 2022). Their unique optical properties enable significant advancements in imaging and diagnostics, from enhanced fluorescence imaging and deep tissue visualization to highly sensitive biosensing and targeted photothermal therapy. As research and development in this field continue to progress, these nanomaterials will likely play a pivotal role in revolutionizing medical diagnostics and treatments, leading to better patient outcomes and more personalized healthcare solutions.

4. ENERGY: ADVANCEMENTS IN SOLAR ENERGY HARVESTING AND CONVERSION

Solar energy is offering an inexhaustible supply of clean power. Recent advancements in solar energy harvesting and conversion have significantly enhanced the efficacy and viability of solar technologies, making them more competitive with traditional fossil fuels (Hao et al., 2022). These advancements span materials science, photovoltaic technologies, and innovative solar harvesting methods.

4.1 Breakthroughs in Photovoltaic Materials

The main function of solar energy conversion lies in photovoltaic (PV) materials that convert sunlight directly into electricity. Traditional silicon-based solar cells have dominated the market due to their reliability and relatively high efficiency. However, they face limitations in terms of efficiency and production costs. Recent research has focused on developing new materials that can overcome these barriers.

Perovskite Solar Cells: One of the most exhilarating progresses in solar technology is the emergence of perovskite solar cells. These materials have shown remarkable efficiency improvements, reaching over 25% in laboratory settings. Perovskites are cheaper and easier to produce than silicon, and they can be manufactured using solution-based processes, which are less energy-intensive(Pan et al., 2021). Additionally, perovskite solar cells can be made semi-transparent, opening up new applications in building-integrated photovoltaics (BIPV) where windows and facades can generate electricity. Figure 3 depicts perovskite solar cell

Figure 3. Perovskite solar cell

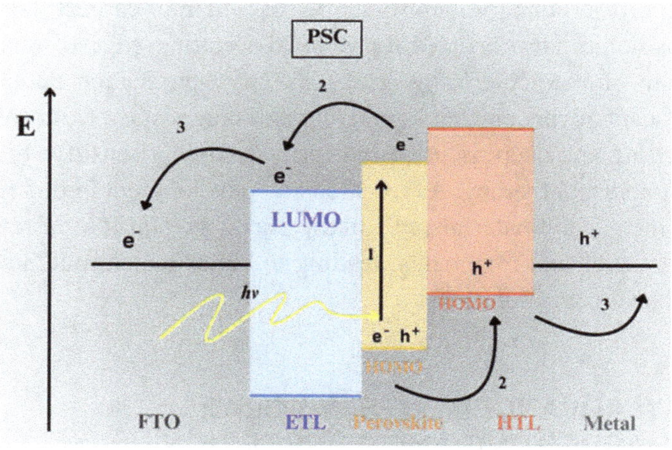

Organic Photovoltaics (OPVs): Organic solar cells use organic molecules or polymers to absorb light and convert it into electricity. While their efficiency is currently lower than silicon or perovskite cells, they offer advantages such as flexibility, lightweight, and the potential for low-cost production through roll-to-roll printing techniques. Advances in the stability and efficiency of OPVs continue to make them a capable candidate for numerous applications, including portable and wearable solar devices.

Tandem Solar Cells: These cells combine multiple layers of different photovoltaic materials to get a broader spectrum of sunlight. For example, tandem cells might stack a perovskite layer on top of a silicon layer, with each material optimized to absorb different parts of the solar spectrum(Azam et al., 2021). This layered approach can surpass the efficiency limits of single-junction solar cells, with some tandem cells achieving efficiencies above 30%.

4.2 Innovations in Solar Harvesting Techniques

Beyond material advancements, new methods of solar energy harvesting have been developed to improve efficiency and practicality.

Concentrated Solar Power (CSP): CSP systems employ mirrors or lenses to focus sunlight onto a small area, generating high temperatures. These temperatures can then be harnessed to produce electricity via a heat engine or turbine. Innovations in CSP include the development of advanced thermal storage systems that allow for energy generation even when there is no sunlight, providing a stable and consistent power supply. Molten salt storage, for example, can retain heat for several hours, enabling CSP plants to produce electricity during peak demand periods.

Floating Solar Farms: Placing solar panels on bodies of water, such as reservoirs and lakes, addresses the land use issues associated with large-scale solar installations. Floating solar farms benefit from the cooling effect of water, which can increase panel efficiency. Additionally, they reduce water evaporation and algae growth, providing environmental benefits. China and Japan have already executed large-scale floating solar projects, demonstrating their feasibility and effectiveness(Sun et al., 2021).

Solar Windows and Building-Integrated Photovoltaics (BIPV): Integrating solar cells into building materials, like windows, roofs, and facades, allows for the generation of electricity without requiring additional space. Solar windows, which incorporate transparent or semi-transparent PV materials, can generate power while still allowing light to pass through. BIPV solutions not only produce energy but also enhance the aesthetic appeal and energy efficiency of buildings, making them an attractive option for urban environments.

4.3 Advances in Energy Storage and Conversion Efficiency

Improving energy storage and conversion efficiency is critical for maximizing the utility of solar power. Recent advancements include:

Advanced Battery Technologies: The advancement of high-capacity and long-lasting batteries, including lithium-sulfur and solid-state batteries, improves the capacity to store solar energy for use during periods of low sunlight. These technologies provide higher energy densities and enhanced safety compared to traditional lithium-ion batteries.

Power Electronics: Innovations in power electronics, such as more efficient inverters and smart grid technologies, facilitate the efficient conversion and distribution of solar energy. These advancements help integrate solar power into the grid more effectively, reducing losses and improving reliability(Wei et al., 2021). The advancements in solar energy harvesting and conversion are driving a significant transformation in the renewable energy landscape. From cutting-edge photovoltaic materials like perovskites and organic solar cells to innovative harvesting techniques like floating solar farms and BIPV, these developments are making solar power more effective, cost-effective, and useful(Libra et al., 2021). Coupled with improvements in energy storage and conversion technologies, these innovations are poised to make solar energy a cornerstone of the global transition to a bearable and clean energy future.

5. ENVIRONMENTAL MONITORING AND DISPLAY TECHNOLOGIES: APPLICATIONS AND INNOVATIONS OF ADVANCED OPTICAL MATERIALS

Environmental monitoring and display technologies play a critical role in assessing and addressing the myriad challenges posed by pollution, climate change, and ecosystem degradation. Advanced optical materials, such as plasmonic nanoparticles and quantum dots, are at the forefront of innovations in this field, offering enhanced sensitivity, specificity, and multifunctionality(Yan et al., 2021). This essay explores the applications and innovations of these materials in environmental monitoring and display technologies. Table 1 describes the comparative Advantages of Advanced Optical Materials for Real-world Applications.

Table 1. Comparative Advantages of Advanced Optical Materials for Real-world Applications

Optical Material Class	Key Optical Advantages	Distinct Functional Benefits	Representative Real-world Applications
Semiconducting Optical Materials (ZnO, TiO_2, GaN)	Tunable band gap, high exciton binding energy, strong UV–visible absorption	Efficient charge carrier generation, high optical sensitivity	Photodetectors, LEDs, solar cells, photocatalysis, environmental monitoring
Nonlinear Optical (NLO) Materials	High second- and third-order nonlinear susceptibility	Frequency conversion, optical switching, ultrafast response	Laser systems, optical signal processing, telecommunication devices
Plasmonic Materials (Au, Ag, Cu nanostructures)	Strong localized surface plasmon resonance (LSPR)	Enhanced light–matter interaction, local field amplification	Biosensing, surface-enhanced Raman spectroscopy (SERS), photothermal therapy
Photonic Crystals	Photonic band gap control, strong light confinement	Tailored light propagation, reduced optical losses	Optical filters, waveguides, low-threshold lasers
Metamaterials & Metasurfaces	Negative refractive index, subwavelength light manipulation	Unconventional electromagnetic responses	Superlenses, invisibility cloaks, beam steering, compact optical devices
Organic Optical Materials	High optical gain, flexibility, solution processability	Low-cost fabrication, mechanical flexibility	OLEDs, organic lasers, flexible displays
Hybrid Optical Nanocomposites	Synergistic optical enhancement from multi-component systems	Improved stability, multifunctionality	Smart coatings, sensors, energy harvesting devices
Rare-earth-doped Optical Materials	Sharp emission lines, long luminescence lifetimes	High color purity, minimal photobleaching	Solid-state lasers, optical amplifiers, bioimaging
2D Optical Materials (Graphene, MoS_2, WS_2)	Strong light absorption at atomic thickness	Ultrafast carrier dynamics, broadband response	Ultrafast photonics, optical modulators, photodetectors

5.1 Applications of Advanced Optical Materials in Environmental Monitoring

Air Quality Monitoring

Plasmonic Nanoparticles: Leveraging the surface-enhanced Raman scattering (SERS) characteristics of plasmonic nanoparticles enables the detection of trace amounts of airborne pollutants, such as volatile organic compounds, nitrogen oxides, and sulfur oxides. This high sensitivity allows for real-time monitoring and early warning systems in urban and industrial areas.

Quantum Dots: Quantum dots' fluorescence properties are utilized in sensors to detect specific gas molecules. When these dots interact with target pollutants, they exhibit changes in their emission spectra, providing precise measurements of air quality.

Water Quality Monitoring

Plasmonic Nanoparticles: These nanoparticles are used in sensors to detect contaminants like heavy metals, pesticides, and pathogens in water bodies. Their high surface area and reactive surface improve the sensitivity and selectivity of detection, permitting continuous and real-time water quality assessment(Chapman et al., 2020).

Quantum Dots: Quantum dots can be functionalized to selectively bind with specific waterborne contaminants, such as lead or mercury ions. The resulting changes in their optical properties facilitate the detection and quantification of these hazardous substances at very low concentrations.

Soil and Agricultural Monitoring

Plasmonic Nanoparticles: These are employed in sensors to monitor soil health and detect contaminants like pesticides and heavy metals. The ability to detect low-level contaminants helps in managing soil quality and ensuring safe agricultural practices.

Quantum Dots: In agricultural monitoring, quantum dots are used to track nutrient levels and detect harmful chemicals in soil, aiding in precision farming and sustainable agriculture practices.

5.2 Innovations in Display Technologies for Environmental Data

Interactive Dashboards and Maps

Data Visualization: Advanced optical materials enable the creation of highly sensitive sensors that feed real-time data into interactive dashboards and maps. An interactive dashboard model is These tools visualize complex environmental data, such as pollution levels and climate variables, in an accessible and user-friendly manner, facilitating better decision-making and public awareness(Shabir et al., 2023).

Augmented Reality (AR) and Virtual Reality (VR)

Environmental Awareness: AR and VR technologies, enhanced by the data from sensors using advanced optical materials, offer immersive experiences that can educate and inform the public about environmental issues. For example, AR applications can overlay real-time pollution data on a user's surroundings, providing a tangible understanding of air quality in specific locations.

Training and Simulation: VR environments can simulate the impacts of environmental changes, such as rising sea levels or deforestation, providing valuable tools for training and planning in environmental management and disaster response.

Wearable Environmental Sensors

Personal Exposure Monitoring: Wearable devices equipped with sensors using plasmonic nanoparticles and quantum dots allow individuals to see their exposure to pollutants and other conservational factors in real time. These wearables can provide alerts and recommendations based on personal exposure data, promoting healthier lifestyles and raising environmental awareness.

Public Display Systems

Community Engagement: Public display systems, such as digital billboards and kiosks, can present real-time environmental data collected by advanced optical sensors. These systems can show air and water quality indices, weather conditions, and other relevant environmental information, encouraging community involvement and informed decision-making. Table 2 explains various limitations of Advanced Optical Materials in Real-world Applications.

Table 2. Limitations of Advanced Optical Materials in Real-world Applications

Optical Material Class	Key Limitations	Practical Challenges
Semiconducting Optical Materials (ZnO, TiO_2, GaN)	Wide band gaps limit visible-light utilization; defect-related recombination	Doping control, stability under prolonged illumination
Nonlinear Optical (NLO) Materials	Low damage threshold; phase-matching constraints	Complex crystal growth, limited wavelength tunability
Plasmonic Materials (Au, Ag, Cu)	High optical losses due to Ohmic heating	Thermal instability, oxidation (especially Cu, Ag)
Photonic Crystals	Narrow operational bandwidth	Fabrication complexity and high sensitivity to structural defects

continued on following page

Table 2. Continued

Optical Material Class	Key Limitations	Practical Challenges
Metamaterials & Metasurfaces	High intrinsic losses; limited scalability	Expensive nanofabrication and poor large-area uniformity
Organic Optical Materials	Low thermal and environmental stability	Short operational lifetime, photodegradation
Hybrid Optical Nanocomposites	Interfacial incompatibility	Phase segregation and reproducibility issues
Rare-earth-doped Optical Materials	Weak absorption cross-sections	High pumping power required; limited host compatibility
2D Optical Materials (Graphene, MoS_2)	Large-area synthesis challenges	Transfer defects, substrate dependence

Table 3 lists different types of technical, Economic, and Environmental Barriers to the Widespread Adoption of Advanced Optical Materials

Table 3. Technical, Economic, and Environmental Barriers to the Widespread Adoption of Advanced Optical Materials

Optical Material Class	Technical Barriers	Economic Barriers	Environmental & Sustainability Barriers
Semiconducting Optical Materials (ZnO, TiO_2, GaN)	Defect-induced recombination, limited visible-light absorption, dopant inhomogeneity	High-purity precursors and controlled synthesis increase production cost	Energy-intensive synthesis routes; limited recycling pathways
Nonlinear Optical (NLO) Materials	Phase-matching constraints, low laser damage threshold	Expensive crystal growth and polishing	Use of toxic or scarce elements in some NLO crystals
Plasmonic Materials (Au, Ag, Cu)	Ohmic losses, thermal instability, oxidation (Ag, Cu)	High cost of noble metals; limited cost-effective alternatives	Mining-related environmental impacts; nanoparticle toxicity concerns
Photonic Crystals	Narrow operational bandwidth, defect sensitivity	Complex nanofabrication processes limit scalability	High material waste during fabrication
Metamaterials & Metasurfaces	High intrinsic losses, narrow frequency range	Costly lithographic techniques; poor large-area manufacturability	High energy consumption during nanofabrication
Organic Optical Materials	Photodegradation, low thermal and moisture stability	Short device lifetime increases replacement cost	Disposal challenges; potential release of organic solvents
Hybrid Optical Nanocomposites	Interfacial instability, reproducibility issues	Multistep synthesis increases manufacturing complexity	Limited life-cycle assessment data; difficult recyclability

continued on following page

Table 3. Continued

Optical Material Class	Technical Barriers	Economic Barriers	Environmental & Sustainability Barriers
Rare-earth-doped Optical Materials	Low absorption efficiency, concentration quenching	Dependence on critical rare-earth supply chains	Environmentally damaging mining and geopolitical supply risks
2D Optical Materials (Graphene, MoS_2)	Defect control and large-area uniformity remain challenging	High cost of scalable CVD growth and transfer processes	Chemical-intensive synthesis; waste generation during transfer

5.3 Stability and Durability

Chemical and Physical Stability: Ensuring the long-term stability of advanced optical materials in various environmental conditions is crucial. Innovations in protective coatings and material composition are necessary to enhance the durability and reliability of sensors(Akarsu & Serdar Genç, 2022).

Scalability and Cost-Effectiveness

Manufacturing Scale-Up: Developing cost-effective manufacturing processes that maintain the high performance of these materials at scale is essential for widespread adoption. Research into scalable synthesis methods and the use of more affordable raw materials can drive down costs.

Interdisciplinary Collaboration

Cross-Field Synergy: The successful development and deployment of these technologies require collaboration between material scientists, environmental scientists, engineers, and policymakers. Interdisciplinary initiatives can foster innovation and ensure that technologies meet practical needs and regulatory standards(Gholami et al., 2022). Advanced optical materials like plasmonic nanoparticles and quantum dots are revolutionizing environmental monitoring and display technologies. Their superior sensitivity and multifunctionality enable precise and real-time detection of pollutants, contributing significantly to air, water, and soil quality monitoring. Innovations in display technologies, from interactive dashboards to wearable sensors, enhance the accessibility and impact of environmental data, fostering greater public awareness and engagement. Addressing challenges related to stability, scalability, and interdisciplinary collaboration will be key to fully realizing the potential of these advanced materials in promoting a sustainable and healthy environment.

6. SECURITY AND DEFENSE: ENHANCING THREAT DETECTION WITH PLASMONIC NANOPARTICLES AND QUANTUM DOTS

Top of FormBottom of FormThe fields of security and defense are continuously evolving, driven by the need to protect against increasingly sophisticated threats. Among the cutting-edge technologies advancing threat detection capabilities are plasmonic nanoparticles and quantum dots. These nanomaterials offer unprecedented sensitivity and specificity in spotting a widespread range of threats, from chemical and biological agents to explosives and illicit drugs. This essay explores the applications and innovations of plasmonic nanoparticles and quantum dots in enhancing threat detection(Chadge et al., 2022).

6.1 Plasmonic Nanoparticles: Harnessing Light for Superior Sensing

Plasmonic nanoparticles, typically made from noble metals like gold and silver, exhibit unique optical properties due to localized surface plasmon resonances. These resonances occur when conduction electrons on the nanoparticle's surface oscillate in response to light, resulting in enhanced electromagnetic fields at specific wavelengths. This phenomenon is exploited in various security applications:

Surface-Enhanced Raman Scattering (SERS): SERS is a dominant technique that uses plasmonic nanoparticles to amplify the Raman scattering signals of molecules adsorbed on their surfaces. This amplification enables the detection of trace amounts of substances with high specificity and sensitivity. In security and defense, SERS can identify chemical warfare agents, explosives, and illicit drugs at very low concentrations, making it a valuable tool for field-based threat detection.

Biosensing and Medical Diagnostics: Plasmonic nanoparticles are also used in biosensing applications to detect biological threats, such as pathogens and toxins. By functionalizing the nanoparticles with specific antibodies or DNA sequences, they can selectively bind to target biomolecules. The binding event causes a measurable shift in the plasmon resonance frequency, indicating the presence of the threat. This approach is crucial for the rapid and accurate detection of biological agents in security screenings and medical diagnostics(Pham et al., 2021).

Colorimetric Sensors: Plasmonic nanoparticles can change color in response to the binding of specific analytes, providing a simple and rapid visual indication of the presence of a threat. These colorimetric sensors are highly sensitive and can be used for on-site detection of various hazardous substances, enhancing the capability to quickly assess and respond to potential threats.

6.2 Quantum Dots: High-Resolution Fluorescent Detection

Quantum dots (QDs) are semiconductor nanoparticles characterized by their unique optical and electronic properties, such as size-tunable light emission and high brightness. Their applications in security and defense include:

Fluorescence-Based Detection: Quantum dots exhibit strong and stable fluorescence, making them ideal for detecting trace amounts of chemical and biological agents. When conjugated with specific ligands, quantum dots can bind to target molecules and emit fluorescence upon excitation. This property enables high-resolution imaging and quantification of threats, providing valuable information for both detection and forensic analysis(Bakier et al., 2020).

Multiplexed Detection: The emission wavelengths of quantum dots can be precisely controlled by adjusting their size and composition, allowing for the simultaneous detection of multiple targets. This capability is particularly useful in security applications where multiple threats may be present. For example, a single assay could detect various chemical warfare agents, biological toxins, and explosive residues, providing a comprehensive threat assessment.

Quantum Dot-Enhanced Immunoassays: Quantum dots can enhance traditional immunoassays by providing brighter and more stable signals than conventional fluorophores. This improvement increases the sensitivity and reliability of assays used to detect pathogens, toxins, and other hazardous substances. Such enhanced immunoassays are critical for rapid and accurate threat detection in both field and laboratory settings(Q. Li et al., 2020).

6.3 Integration and Future Directions

The integration of plasmonic nanoparticles and quantum dots with advanced detection systems holds immense potential for enhancing security and defense capabilities. Combining these nanomaterials with microfluidics, for example, can create portable and highly sensitive detection platforms that provide real-time analysis of environmental samples. These integrated systems can be deployed in various scenarios, from border security and military operations to public health monitoring and environmental surveillance. AI-driven systems can identify patterns and correlations that may be missed by human analysts, providing more robust threat detection and response strategies.

Plasmonic nanoparticles and quantum dots represent significant advancements in the field of threat detection, offering superior sensitivity, specificity, and multiplexing capabilities. These nanomaterials are revolutionizing how security and defense sectors detect and respond to chemical, biological, and explosive threats(Zhan et al., 2024). By harnessing the unique optical properties of plasmonic nanoparticles and quantum dots, researchers and engineers are developing innovative detection platforms that enhance situational awareness and improve response times. As these technologies continue to evolve and integrate with other advanced systems, they will play a crucial role in safeguarding societies against an array of emerging threats.

7. FUTURE DIRECTIONS AND CHALLENGES: STABILITY, SCALABILITY, AND INTERDISCIPLINARY COLLABORATION OF ADVANCED OPTICAL MATERIALS

Advanced optical materials, such as plasmonic nanoparticles and quantum dots, are at the forefront of numerous technological advancements in fields ranging from security and defense to environmental monitoring and biomedicine. However, realizing the full potential of these materials involves overcoming significant challenges related to stability, scalability, and interdisciplinary collaboration. Addressing these challenges will be critical for their successful integration into commercial applications and broader societal benefits. Table 4 exhibits the Future Directions and Challenges of Advanced Optical Materials.

Table 4. Industrial Applications and Technical Challenges of Advanced Optical Materials

Optical Material Class	Key Industrial Applications	Critical Technical Challenges	Manufacturing & Scale-up Constraints	Impact on Industrial Adoption
Semiconducting Optical Materials (ZnO, TiO_2, GaN)	UV/visible LEDs, laser diodes, photodetectors, photocatalytic reactors	Defect-mediated non-radiative recombination, carrier trapping, limited p-type doping (ZnO, GaN)	Epitaxial growth control, wafer-scale uniformity, dopant reproducibility	Performance variability and yield loss in optoelectronic devices
Nonlinear Optical (NLO) Materials	Frequency doublers, optical parametric oscillators, ultrafast laser systems	Phase-matching constraints, optical damage under high-power operation	Single-crystal growth, orientation-dependent cutting and polishing	Restricted wavelength flexibility and high component cost
Plasmonic Materials (Au, Ag, Cu)	Chemical and biosensors, SERS substrates, photothermal devices	Ohmic losses, plasmon damping, thermal degradation	Nanopatterning precision, oxidation control (Ag, Cu)	Reduced device efficiency and limited long-term reliability

continued on following page

Table 4. Continued

Optical Material Class	Key Industrial Applications	Critical Technical Challenges	Manufacturing & Scale-up Constraints	Impact on Industrial Adoption
Photonic Crystals	Optical waveguides, filters, on-chip photonics	Narrow photonic bandgap, sensitivity to nanoscale defects	High-resolution lithography and etching complexity	Scalability challenges in integrated photonic circuits
Metamaterials & Metasurfaces	Beam steering, flat optics, imaging systems	Intrinsic losses, limited broadband response	Large-area nanofabrication and alignment accuracy	High production cost and limited commercial deployment
Organic Optical Materials	OLED displays, organic lasers, flexible photonics	Exciton quenching, photo-oxidation, low thermal stability	Encapsulation requirements and solvent-based processing	Short operational lifetime and reliability concerns
Hybrid Optical Nanocomposites	Smart coatings, multifunctional sensors, energy harvesting	Interfacial charge transfer inefficiency, phase instability	Multistep synthesis and reproducibility control	Difficulty in standardization for industrial certification
Rare-earth-doped Optical Materials	Solid-state lasers, optical amplifiers, fiber optics	Low absorption cross-section, concentration quenching	Precise dopant control and host material compatibility	Increased pumping power and higher system cost
2D Optical Materials (Graphene, MoS_2, WS_2)	Optical modulators, photodetectors, transparent electrodes	Grain boundaries, substrate-induced scattering	CVD growth uniformity and defect-free transfer	Inconsistent device performance at wafer scale

7.1 Stability of Advanced Optical Materials

Chemical Stability: One of the primary concerns with advanced optical materials, especially quantum dots and plasmonic nanoparticles, is their chemical stability. Quantum dots, for instance, can be prone to oxidation and photobleaching, which degrade their optical properties over time. Developing robust surface coatings and passivation techniques can mitigate these issues, extending the lifespan and reliability of these materials in various applications(G. Wang et al., 2020).

Thermal Stability: Many advanced optical materials, particularly those used in high-intensity light applications, need to maintain their properties under elevated temperatures. Research into thermal stabilization methods, such as the development of heat-resistant coatings or the incorporation of materials with higher melting points, is essential for ensuring performance stability in harsh environments.

Environmental Stability: Ensuring that these materials remain stable in different environmental conditions, such as varying humidity, pH levels, and exposure to reactive chemicals, is crucial for their deployment in real-world applications. This involves extensive testing and the development of protective measures that shield the materials from environmental degradation.

7.2 Scalability of Production Processes

Manufacturing Techniques: Scaling up the production of advanced optical materials from the laboratory to industrial scale is a significant challenge. Current manufacturing processes, such as chemical synthesis and lithography, often face limitations in terms of cost, complexity, and yield. Innovations in manufacturing technologies, like roll-to-roll processing for organic photovoltaics or large-scale chemical vapor deposition for nanomaterials, are necessary to produce these materials efficiently and economically.

Quality Control: Maintaining consistent quality and performance of advanced optical materials during large-scale production is another critical issue. Variations in size, shape, and composition can significantly impact their optical properties. Developing rigorous quality control protocols and advanced characterization techniques will ensure that large-scale production meets the required standards for specific applications.

Cost Reduction: The high cost of producing advanced optical materials can hinder their widespread adoption. Research into more cost-effective raw materials, streamlined production processes, and recycling strategies for rare and expensive components will be key to making these technologies economically viable for large-scale use.

7.3 Interdisciplinary Collaboration

Cross-Disciplinary Research: The development and application of advanced optical materials inherently require expertise from multiple disciplines.. Promoting cross-disciplinary research initiatives and collaborative projects can accelerate innovation and address complex challenges more effectively. Institutions and funding agencies should encourage and support such collaborative efforts.

Integration with Existing Technologies: Successfully integrating advanced optical materials with existing technologies and systems often requires collaboration between academia, industry, and government agencies. For example, integrating quantum dots into current semiconductor manufacturing processes or incorporating plasmonic nanoparticles into medical diagnostic devices requires a holistic approach that leverages the strengths of various fields.

Regulatory and Ethical Considerations: The development and deployment of advanced optical materials also involve regulatory and ethical considerations, particularly regarding environmental impact and human health. Collaboration with regulatory bodies, policymakers, and bioethicists is essential to establish guidelines and standards that ensure safe and responsible use of these technologies.

7.4 Future Directions

Multifunctional Materials: The future of advanced optical materials lies in the development of multifunctional materials that combine several beneficial properties, such as optical, electrical, and thermal characteristics. This integration can lead to more efficient and versatile applications, from smart sensors to integrated photonic circuits.

Sustainable Materials: As sustainability becomes increasingly important, research into eco-friendly and biodegradable optical materials is gaining momentum. Developing materials that minimize environmental impact, both in production and disposal, will be crucial for the long-term success and acceptance of these technologies(Kollipara et al., 2020).

Artificial Intelligence and Machine Learning: Leveraging AI and machine learning to optimize the design, synthesis, and application of advanced optical materials can significantly enhance their performance and functionality. These technologies can predict material properties, optimize manufacturing processes, and identify new applications, driving innovation at an unprecedented pace. The future of advanced optical materials holds immense promise, but realizing this potential requires addressing challenges related to stability, scalability, and interdisciplinary collaboration. By focusing on improving material stability, developing scalable production processes, and fostering cross-disciplinary research and integration, we can unlock new possibilities in various fields, from security and defense to environmental monitoring and biomedicine. As we advance, the synergy between scientific innovation, industrial application, and ethical considerations will be crucial in shaping a sustainable and technologically advanced future.

CONCLUSION

Advanced optical materials, including plasmonic nanoparticles and quantum dots, are revolutionizing various sectors by enhancing capabilities in imaging, diagnostics, environmental monitoring, and security. Their unique properties enable high sensitivity, specificity, and multifunctionality, offering unprecedented advantages over traditional materials. As these materials continue to evolve, they hold the promise of transforming real-world applications, driving significant technological advancements, and addressing critical global challenges. To fully harness the power of these advanced optical materials, addressing the challenges of stability, scalability, and interdisciplinary collaboration is essential. Ensuring chemical, thermal, and environmental stability will enhance their reliability and longevity in practical applications. Developing cost-effective, high-quality manufacturing processes will facilitate their large-scale deployment, making these materials accessible for widespread use. Interdisciplinary collaboration will be pivotal, fostering innovation through the integration of expertise from diverse fields and ensuring that new technologies meet societal needs and regulatory standards. The future directions for advanced optical materials include the development of multifunctional and sustainable materials, leveraging artificial intelligence for material optimization, and fostering cross-sector partnerships to drive innovation. By focusing on these areas, we can unlock the full potential of plasmonic nanoparticles and quantum dots, leading to breakthroughs in various domains and contributing to a more sustainable and technologically advanced society. The advanced optical materials are at the forefront of a new era in technology and science. By overcoming current challenges and leveraging their unique properties, we can harness their power to create innovative solutions for real-world applications, ultimately improving the quality of life and addressing some of the most pressing issues of our time.

REFERENCES

Akarsu, B., & Serdar Genç, M. (2022). Optimization of electricity and hydrogen production with hybrid renewable energy systems. *Fuel, 324*, 124465. DOI: 10.1016/j.fuel.2022.124465

An, R., Chu, D., Xie, C., Pan, S., & Yang, Z. (2025). Navigating the Landscape of Next-Generation Nonlinear Optical Materials Discovery: Opportunities and Challenges Driven by Artificial Intelligence. *Accounts of Materials Research*. Advance online publication. DOI: 10.1021/accountsmr.5c00245

Azam, A., Ahmed, A., Wang, H., Wang, Y., & Zhang, Z. (2021). Knowledge structure and research progress in wind power generation (WPG) from 2005 to 2020 using CiteSpace based scientometric analysis. *Journal of Cleaner Production, 295*, 126496. DOI: 10.1016/j.jclepro.2021.126496

Bakier, Y. M., Ghali, M., & Zahra, W. K. (2020). Highly sensitive fluorescent detection of pyridine using small size carbon quantum dots derived from folic acid. *Journal of Physics. D, Applied Physics, 53*(40), 405103. DOI: 10.1088/1361-6463/ab985e

Chadge, R. B., Sunheriya, N., Giri, J. P., & Shrivastava, N. (2022). Experimental investigation of different electrical configurations and topologies for Photovoltaic system. *Materials Today: Proceedings, 57*, 316–320. DOI: 10.1016/j.matpr.2022.03.097

Chapman, J., Truong, V. K., Elbourne, A., Gangadoo, S., Cheeseman, S., Rajapaksha, P., Latham, K., Crawford, R. J., & Cozzolino, D. (2020). Combining Chemometrics and Sensors: Toward New Applications in Monitoring and Environmental Analysis. *Chemical Reviews, 120*(13), 6048–6069. DOI: 10.1021/acs.chemrev.9b00616 PMID: 32364371

Chauhan, A. (2024). *Exploring the Principles of Optics Through Modern Technologies*. DOI: 10.22541/au.172660512.22821656/v1

Dai, D., Zhang, Y., Yang, S., Kong, W., Yang, J., & Zhang, J. (2024). Recent Advances in Functional Materials for Optical Data Storage. *Molecules (Basel, Switzerland), 29*(1), 254. DOI: 10.3390/molecules29010254 PMID: 38202837

Duan, X.-F., Zhou, L.-P., Li, H.-R., Hu, S.-J., Zheng, W., Xu, X., Zhang, R., Chen, X., Guo, X.-Q., & Sun, Q.-F. (2023). Excited-Multimer Mediated Supramolecular Upconversion on Multicomponent Lanthanide-Organic Assemblies. *Journal of the American Chemical Society, 145*(42), 23121–23130. DOI: 10.1021/jacs.3c06775 PMID: 37844009

Gholami, A., Ameri, M., Zandi, M., & Gavagsaz Ghoachani, R. (2022). Electrical, thermal and optical modeling of photovoltaic systems: Step-by-step guide and comparative review study. *Sustainable Energy Technologies and Assessments*, *49*, 101711. DOI: 10.1016/j.seta.2021.101711

Hao, D., Qi, L., Tairab, A. M., Ahmed, A., Azam, A., Luo, D., Pan, Y., Zhang, Z., & Yan, J. (2022). Solar energy harvesting technologies for PV self-powered applications: A comprehensive review. *Renewable Energy*, *188*, 678–697. DOI: 10.1016/j.renene.2022.02.066

Huang, Y.-M., Singh, K. J., Liu, A.-C., Lin, C.-C., Chen, Z., Wang, K., Lin, Y., Liu, Z., Wu, T., & Kuo, H.-C. (2020). Advances in Quantum-Dot-Based Displays. *Nanomaterials (Basel, Switzerland)*, *10*(7), 1327. DOI: 10.3390/nano10071327 PMID: 32640754

Kollipara, P. S., Li, J., & Zheng, Y. (2020). Optical Patterning of Two-Dimensional Materials. *Research*, *2020*, 2020/6581250. Advance online publication. DOI: 10.34133/2020/6581250 PMID: 32043085

Kumar, A., & Kumar, S. (2023). Light-controlled receptors for environmentally and biologically relevant anions. *Chemical Engineering Journal*, *474*, 145493. DOI: 10.1016/j.cej.2023.145493

Kumbhakar, P., Chowde Gowda, C., & Tiwary, C. S. (2021). Advance Optical Properties and Emerging Applications of 2D Materials. *Frontiers in Materials*, *8*, 721514. Advance online publication. DOI: 10.3389/fmats.2021.721514

Li, L., Kong, W., & Chen, F. (2022). Femtosecond laser-inscribed optical waveguides in dielectric crystals: A concise review and recent advances. *Advanced Photonics*, *4*(02). Advance online publication. DOI: 10.1117/1.AP.4.2.024002

Li, Q., Guo, Z., Zhao, X., Zhang, T., Chen, J., & Wei, Y. (2020). One-pot synthesis of 2,2′-dipicolylamine derived highly photoluminescent nitrogen-doped carbon quantum dots for Fe 3+ detection and fingermark detection. *Nanotechnology*, *31*(33), 335501. DOI: 10.1088/1361-6528/ab8f4b PMID: 32357348

Libra, M., Petrik, T., Poulek, V., Tyukhov, I. I., & Kourim, P. (2021). Changes in the Efficiency of Photovoltaic Energy Conversion in Temperature Range With Extreme Limits. *IEEE Journal of Photovoltaics*, *11*(6), 1479–1484. DOI: 10.1109/JPHOTOV.2021.3108484

Liu, S., Yan, L., Huang, J., Zhang, Q., & Zhou, B. (2022). Controlling upconversion in emerging multilayer core–shell nanostructures: From fundamentals to frontier applications. *Chemical Society Reviews*, *51*(5), 1729–1765. DOI: 10.1039/D1CS00753J PMID: 35188156

Mahmoudpour, M., Ding, S., Lyu, Z., Ebrahimi, G., Du, D., Ezzati Nazhad Dolatabadi, J., Torbati, M., & Lin, Y. (2021). Aptamer functionalized nanomaterials for biomedical applications: Recent advances and new horizons. *Nano Today*, *39*, 101177. DOI: 10.1016/j.nantod.2021.101177

Mehtab, A., Ahmed, J., Alshehri, S. M., Mao, Y., & Ahmad, T. (2022). Rare earth doped metal oxide nanoparticles for photocatalysis: A perspective. *Nanotechnology*, *33*(14), 142001. DOI: 10.1088/1361-6528/ac43e7 PMID: 34915455

Pan, H., Qi, L., Zhang, Z., & Yan, J. (2021). Kinetic energy harvesting technologies for applications in land transportation: A comprehensive review. *Applied Energy*, *286*, 116518. DOI: 10.1016/j.apenergy.2021.116518

Park, S.-M., & Jun, B.-H. (2023). Synthesis and Applications of Optical Materials. *Nanomaterials (Basel, Switzerland)*, *13*(2), 297. DOI: 10.3390/nano13020297 PMID: 36678049

Pham, A. T. T., Wallace, A., Zhang, X., Tohl, D., Fu, H., Chuah, C., Reynolds, K. J., Ramsey, C., & Tang, Y. (2021). Optical-Based Biosensors and Their Portable Healthcare Devices for Detecting and Monitoring Biomarkers in Body Fluids. *Diagnostics (Basel)*, *11*(7), 1285. DOI: 10.3390/diagnostics11071285 PMID: 34359368

Shabir, M., Hussain, I., Işık, Ö., Razzaq, K., & Mehroush, I. (2023). The role of innovation in environmental-related technologies and institutional quality to drive environmental sustainability. *Frontiers in Environmental Science*, *11*, 1174827. Advance online publication. DOI: 10.3389/fenvs.2023.1174827

Sun, J., Wu, M., Jiang, H., Fan, X., & Zhao, T. (2021). Advances in the design and fabrication of high-performance flow battery electrodes for renewable energy storage. *Advances in Applied Energy*, *2*, 100016. DOI: 10.1016/j.adapen.2021.100016

Urmi, R., Banerjee, P., Singh, M., Singh, R., Chhillar, S., Sharma, N., Chandra, A., Singh, N., & Qamar, I. (2024). Revolutionizing biomedicine: Aptamer-based nanomaterials and nanodevices for therapeutic applications. *Biotechnology Reports (Amsterdam, Netherlands)*, *42*, e00843. DOI: 10.1016/j.btre.2024.e00843 PMID: 38881649

Wang, G., Wang, F., Shen, F., Jiang, T., Chen, Z., & Hu, P. (2020). Experimental and optical performances of a solar CPV device using a linear Fresnel reflector concentrator. *Renewable Energy*, *146*, 2351–2361. DOI: 10.1016/j.renene.2019.08.090

Wang, X., Luo, Y., Liao, J., Joselevich, E., & Xu, J. (2024). Selective-Area Growth of Aligned Organic Semiconductor Nanowires and Their Device Integration. *Advanced Functional Materials*, *34*(7), 2308708. Advance online publication. DOI: 10.1002/adfm.202308708

Wei, M., Lee, S. H., Hong, T., Conlon, B., McKenzie, L., Hendron, B., & German, A. (2021). Approaches to cost-effective near-net zero energy new homes with time-of-use value of energy and battery storage. *Advances in Applied Energy*, *2*, 100018. DOI: 10.1016/j.adapen.2021.100018

Xie, S., Ai, L., Cui, C., Fu, T., Cheng, X., Qu, F., & Tan, W. (2021). Functional Aptamer-Embedded Nanomaterials for Diagnostics and Therapeutics. *ACS Applied Materials & Interfaces*, *13*(8), 9542–9560. DOI: 10.1021/acsami.0c19562 PMID: 33595277

Yan, K., Karthick Kannan, P., Doonyapisut, D., Wu, K., Chung, C., & Zhang, J. (2021). Advanced Functional Electroactive and Photoactive Materials for Monitoring the Environmental Pollutants. *Advanced Functional Materials*, *31*(12), 2008227. Advance online publication. DOI: 10.1002/adfm.202008227

Yang, C., Jiang, Y., Hao, S. H., Yan, X. Y., Hong, D. F., & Naranmandura, H. (2022). Aptamers: An emerging navigation tool of therapeutic agents for targeted cancer therapy. *Journal of Materials Chemistry. B*, *10*(1), 20–33. DOI: 10.1039/D1TB02098F PMID: 34881767

Yao, J., Li, P., Li, L., & Yang, M. (2018). Biochemistry and biomedicine of quantum dots: From biodetection to bioimaging, drug discovery, diagnostics, and therapy. *Acta Biomaterialia*, *74*, 36–55. DOI: 10.1016/j.actbio.2018.05.004 PMID: 29734008

Zhan, Z., Mao, H., Xue, M., Han, G., Zhou, G., & Zhang, Y. (2024). Ratiometric fluorescence detection of the angiotensin-converting enzyme via single-excitation and double-emission biomass-derived carbon quantum dots. *Methods and Applications in Fluorescence*, *12*(1), 015004. DOI: 10.1088/2050-6120/ad02dd PMID: 37827162

Chapter 5
Material Procurement Policy via Blockchain and Logistics 4.0

Ourssoula Sikal

Laboratory of Modelling and Simulation of Intelligent Industrial Systems, ENSET of Mohammedia, Hassan II University of Casablanca, Casablanca, Morocco

Hasna Hajji

Laboratory of Modelling and Simulation of Intelligent Industrial Systems, ENSET of Mohammedia, Hassan II University of Casablanca, Casablanca, Morocco

Abdellah Zamma

Laboratory of Modelling and Simulation of Intelligent Industrial Systems, ENSET of Mohammedia, Hassan II University of Casablanca, Casablanca, Morocco

Mohamed El Khaili

http://orcid.org/0000-0001-8730-2981

Laboratory of Modelling and Simulation of Intelligent Industrial Systems, ENSET of Mohammedia, Hassan II University of Casablanca, Casablanca, Morocco

ABSTRACT

While procurement can be considered a critical process in supply chain management, since it can provide new opportunities for supply chains to improve their efficiency and effectiveness. Most of companies have been thinking about how logistics 4.0 and blockchain could help them be more efficient and minimizing risks of error and disputes. Therefore, the applications of Industry 4.0 technologies such as blockchain, the Internet of Things and Artificial Intelligence in procurement processes have assisted organizations to improve their performance and to deal with SCM's environmental uncertainties. Implementing a materials supply policy via logistics

DOI: 10.4018/979-8-3693-6627-1.ch005

2027, IGI Global Scientific Publishing. Copying or distributing in print or electronic forms without written permission of IGI Global Scientific Publishing is prohibited. Use of this chapter to
rative artificial intelligence (AI) technologies is expressly prohibited. The publisher reserves all rights to license its use for generative AI training and machine learning model development.

4.0 involves leveraging advanced and cutting-edge technologies to optimize the sourcing process and to revolutionize SC management and the way materials are sourced, managed, and distributed. What can offer several advantages such as enhanced transparency, traceability and efficiency in the SC. The adoption of blockchain technology into the materials procurement process has huddled a crucial consideration within industry in recent years.

I. INTRODUCTION

1. Introduction to the Significance of Integrating Blockchain and Logistics 4.0 in Materials Procurement Policy

Procurement Policy refers to the set of guidelines and procedures governing the acquisition of goods and services by an organization, or resources needed for its operations. It outlines the principles, rules, and decision-making processes involved in the procurement process to ensure transparency, efficiency and accountability, fairness, and compliance with relevant regulations.

Demand management, category and sourcing management, contract management, supplier relationship management, compliance and risk management are the areas of a typical procurement function, as illustrated in Figure 1.

Figure 1. Procurement elements

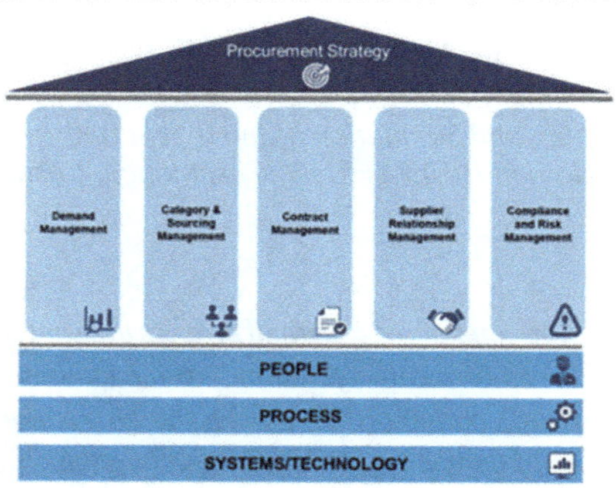

Source: Motaung & Sifolo (2023)

Strategic Sourcing is a systematic process of identifying, evaluating and analyzing a company's spending and supplier relationships to find ways to reduce costs, improve quality and overall value of the goods and services procured, and align procurement with overall business strategy.

Figure 2.

(Source: stock.adobe)

There are several definitions of procurement. The more comprehensive definition adopted is about "business process of selecting a source, ordering and acquiring goods and services" (Bakar et al., 2016). According to the Chartered Institute of Procurement and Supply (CIPS), procurement "ensures identification, sourcing, access and management of external resources that an organization requires or may need to meet its strategic objectives" (Motaung & Sifolo, 2023). The strategic management of procurement operations is one of the most significant SC processes that has a direct impact on its performance (Jahani et al., 2021). All purchasing procedures, such as buying, leasing, renting, or obtaining goods, services, or building materials from outside vendors, are considered procurement (Glas & Kleemann, 2016 ; Tan, 2001). This concept encompasses all of the management duties that procurement managers ought to perform, such as locating, sourcing, gaining access to, and overseeing resources that are external to the business and required to meet the strategic goals of assisting organizations in creating substantial business value and generating profit (Glas & Kleemann, 2016 ; Tan, 2001). As a result, it is important for the procurement department to continuously improve, as procurement is a crucial component of a supply chain. Purchasing has become easier as a result of the recent expansion of I4.0 application use in procurement. One of I4.0's components,

"Procurement 4.0," connects all SC upstream partners and allows for swift and flexible cooperation across organizational boundaries (Glas & Kleemann, 2016). Stated differently, the industrial revolution has led to the emergence of a new field of digitization and the revelation of the idea of smart manufacturing. As a result, concepts like "smart things"—that is, "smart factories," "smart logistics," "smart products," "smart contracts," etc.—are created, and businesses have employed digital technologies to change their supply chain operating model (SCs) (Hahn, 2020). These advancements in technology are a component of Industry 4.0, the designation given to the Fourth Industrial Revolution. A connection between smart and physical assets, such as smart products and machines that operate autonomously and produce self-coordinating systems, is depicted by I4.0, which has been employed in several SC processes (Hahn, 2020). To further clarify, the term "I4.0," which was initially used in Germany in 2011, refers to the linking of all components of integrated data chains and manufacturing elements via the Internet, which has a major impact on the quick competition among manufacturers, suppliers, and retailers (Glas & Kleemann, 2016 ; Tupa et al., 2017). A few examples of I4.0 technologies that can assist procurement managers in addressing purchasing challenges are the Internet of Things (IoT), Cyber-Physical Systems (CPS), Cloud Computing (CC), Big Data Analytics (BDA), Robotics, Blockchain, Artificial Intelligence (AI), Smart Manufacturing, ERP systems, simulation, Internet, procure-to-pay systems, Contract Management Systems (CMS), third-party providers, etc. (Ghadimi et al., 2019; Glas & Kleemann, 2016; Handfield et al., 2019).

Material procurement policy via Blockchain and Logistics 4.0 refers to the use of blockchain technology and the principles of Logistics 4.0 to streamline and optimize the process of acquiring raw materials and supplies. Blockchain provides a secure, transparent, and decentralized platform to track the provenance and movement of materials throughout the supply chain. Logistics 4.0, driven by the Internet of Things (IoT) and sensor integration, enables real-time monitoring and data-driven decision-making in the procurement process.

Material procurement policy via Logistics 4.0 refers to the use of advanced digital technologies, such as the Internet of Things, big data analytics, and automation, to optimize the procurement of materials and supplies throughout the supply chain. This approach aims to enhance efficiency, visibility, and responsiveness in the procurement process by leveraging real-time data, predictive analytics, and automated decision-making.

2. Brief Overview of the Importance of Integrating Blockchain and Logistics 4.0 in Materials Sourcing Policy

The procurement function plays a pivotal role in supply chain risk management. Its main activities—selecting vendors, sourcing the company's inputs, and managing the upstream supply chain—directly affect the company's risk exposure. They engage major global forces, including geopolitical tensions, climate change, economic uncertainty, and social equity. An effective procurement function must closely collaborate with the company's risk management and business units, often taking the lead in risk mitigation efforts. To do these well, it must use up-to-date digital technology, including logistics 4.0. Establishing transparency, prioritizing risks, developing real-time prediction algorithms, and integrating risk management into the organization are critical to procurement's value proposition provided by Logistics 4.0. Companies must aggressively address supply chain risks by integrating people and technology. Procurement professionals must collaborate with colleagues and use digital tools to enhance human capabilities. For example, Traditional AI tools support supplier risk management through data structuring, predictive analytics, real-time supply monitoring, and supply chain optimization strategies. Generative AI can assist with supplier evaluation and risk assessment by automating document generation, enhancing communication, and supporting predictive scenarios. Collaboration and digital technology under the name of Industry 4.0 are key to overcoming the main challenges of supply chain risk management. Utilizing customizable tools and enhancing collaboration can help companies detect and mitigate risks in global, complex supply chains (Boston Consulting Group, 2023).

Porter's value chain model illustrates how procurement functions as an organizational function that supports the main activities such as operations, marketing, service, etc.. The research also emphasizes how crucial procurement is in influencing the degree of cooperation and performance amongst supply chain participants. In this sense, procurement activities play a key role in supply chain management and encourage practitioners to create P4.0, which suggests values for supply chains (Jahani et al., 2021). As an example, artificial intelligence (AI) can assist in the analysis of various patterns that result in a deeper comprehension of certain supplier behaviors, hence assisting procurement managers in identifying suitable potential suppliers (Angkiriwang et al., 2014). Another example relates to blockchain technology, which can combine elements of a decentralized "trustless" database and facilitate international trade as well as process disintermediation. In the end, this can assist procurement managers in accessing and monitoring supplier transaction archives (You et al., 2008). These kinds of examples might highlight the importance of integrating I4.0 into technology to strengthen the procurement process.

Ultimately, a number of findings also show that the I4.0 technologies enhance the efficiency of the supply chain procedures.

II. ANALYSIS OF CURRENT CHALLENGES IN MATERIALS SUPPLY POLICY

1. Overview and Statistics of Materials Procurement Process Between Past and Today

Procurement is crucial since supplier selection, strategic relationships, green certifications, and green process adoption are all actions that allow suppliers to support a company's sustainability objectives

Everything that we use or eat is the outcome of an intricate process that goes through several phases, such as design, production, and distribution. Material procurement is the acquisition of supplies and raw materials required to make completed items, and it is an important step in the process. In the absence of appropriate material procurement practices, organizations risk delays, overspending, poor quality, and even legal troubles. We will discuss what material procurement comprises in this blog article and why it is crucial for companies to do it correctly Understanding the intricacies of material acquisition is imperative for achieving success in the competitive landscape of contemporary commerce, irrespective of whether one is a seasoned expert seeking to enhance supply chain protocols or a fledgling business owner embarking on a new venture.

The procurement system described in the manuscript includes several key players, including vendors, industry/customers, and delivery executives. Vendors play a crucial role in supplying goods or services within the procurement system, thereby guaranteeing the availability of essential products for acquisition

Another important group of stakeholders is industry/customers, who interact with the system to obtain goods or services for their organizations or individual need. Delivery executives are responsible for guaranteeing the timely and safe delivery of acquired items to the designated destinations, thereby completing the transactional process within the system. These stakeholders collaborate in the procurement process, which is aided by the single platform that uses blockchain technology, improving the system's security, effectiveness, and transparency.

Procurement and importation of materials play pivotal roles in the Moroccan and global economies, serving as critical components of supply chain management and economic development strategies. In Morocco, the process of procurement necessitates the adept navigation of regulatory frameworks, encompassing customs procedures and import tariffs, which possess the potential to influence the expenses and accessibility of materials. In figure 1 showing recent statistics of importations of many materials in Morocco as per exchange office data, which reflecting its position as a key player in regional trade dynamics Within the intricate realm of global trade, blockchain technology has emerged as a disruptive tool for enhancing the visibility and efficiency of acquisition procedures.

Figure 3. statistics of materials importation in Morocco

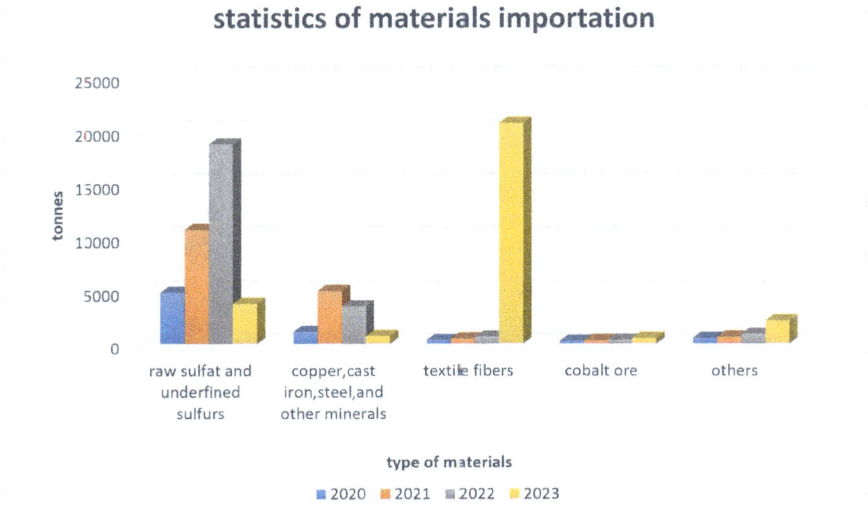

(change office, report 2020-2023)

By leveraging blockchain and logistics 4.0, Moroccan businesses can streamline documentation, verify transactions, and track shipments in real-time, thereby reducing administrative costs and minimizing fraud risks. This innovative approach not only fosters greater trust among stakeholders but also strengthens Morocco's position in global supply chains by improving the reliability and traceability of imported materials. As Morocco continues to integrate blockchain solutions into its procurement strategies, the potential for enhancing operational efficiency and maintaining competitive advantage in international markets remains promising.

2. Identification of the Principal Challenges Encountered in Traditional Materials Procurement Policies

A. Impact of Environmental Uncertainties

The impact of environmental uncertainties on procurement refers to how unpredictable external factors can affect the acquisition of goods and services by an organization. Factors like supply chain disruptions, market volatility, and regulatory changes can introduce risks and challenges into the procurement process, requiring flexibility and adaptive strategies to mitigate their effects.

Supply chains frequently deal with a variety of uncertainties. These can originate from different sources, such as supplier performance, market demand, or even natural disasters. It has been suggested that a key contributing cause to the requirement for supply chain resilience, agility, and flexibility is uncertainty. Supply chain uncertainty can be classified into three main categories: downstream (demand) uncertainty, internal (process) uncertainty, and upstream (supply) uncertainty (Angkiriwang et al., 2014).

a. Supply Uncertainty

Supply Uncertainty. This has to do with the unpredictability of material availability. Uncertainty about material pricing, lead time, alternative sourcing, availability, and capacity can all be signs of supply uncertainty. To provide better customer service in an unpredictable supply, more flexibility is needed. The flexibility to adapt the production schedule to make substitute materials or create other goods in the event of a sudden increase in the cost of a particular material, for instance, is important for preserving the efficacy and efficiency of the supply chain (Angkiriwang et al., 2014).

Impact of supply uncertainties on materials procurement refers to the challenges and risks that organizations face when procuring materials and resources due to unpredictable or unstable supply conditions. Supply uncertainties can lead to delays, shortages, price fluctuations, and disruptions in the production or delivery of materials, which can in turn impact an organization's operations, costs, and ability to meet customer demands. Effective materials procurement strategies that account for and mitigate supply uncertainties are crucial for organizations to maintain reliable supply chains and minimize the negative impacts on their business.

Here are some external factors that can affect materials procurement.

Supply Chain Disruptions refer to unexpected events that disrupt the normal flow of materials, goods, or services through the supply chain, creating challenges and uncertainties for organizations in materials procurement.

Political instability refers to the uncertain and unpredictable nature of a government or political system, which can disrupt the normal functioning of a supply chain.

Trade wars refer to the retaliatory tariffs and trade barriers imposed between nations in an attempt to protect domestic industries and markets. Unlike other supply chain disruptions, trade wars are driven by intentional political and economic policies between governments.

Figure 4. faster capital

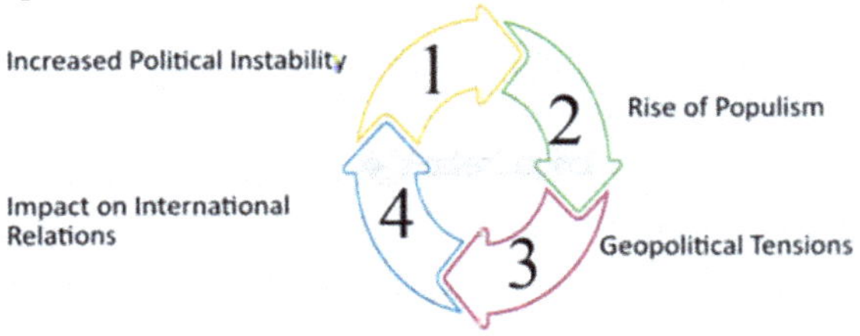

Here are some key insights into the ripple effect of trade wars on global supply chains:

Figure 5.

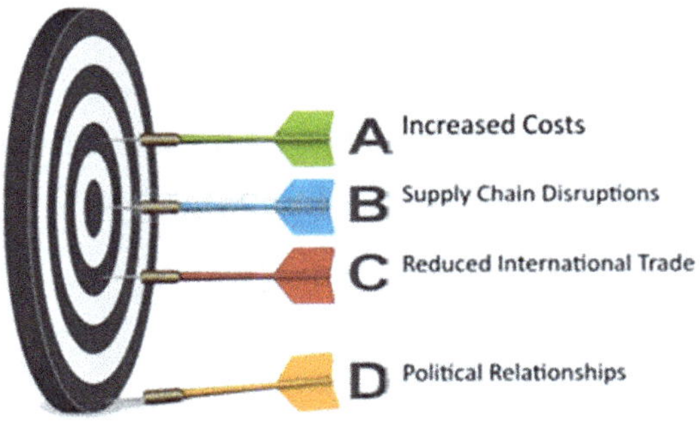

(Source: Globe Explorer)

Sanctions refer to economic and political penalties imposed by governments or international organizations on other countries, entities, or individuals, typically in response to actions deemed unacceptable. They aim to restrict access to resources, markets, or financial systems as a means of exerting pressure.

Supplier Failures refer to disruptions in the supply chain due to issues originating directly from the supplier, such as production problems, quality control issues, or financial difficulties. These types of failures differ from other supply chain disruptions in that they are often more localized and may be addressable through improved supplier monitoring and risk mitigation strategies.

Quality issues refer to problems with the products or services provided by a supplier, such as defects, non-compliance with specifications, or inconsistent performance. These issues can significantly disrupt the supply chain and must be addressed promptly (You et al., 2008).

Raw Material Availability refers to the fluctuations in the supply and accessibility of the physical materials required for production, which can be impacted by factors like resource depletion, geopolitics, and extreme weather events (You et al., 2008).

Raw Material Scarcity refers to the limited availability of natural resources that are used as the basic inputs for manufacturing processes, constraining the supply of certain materials.

Mineral Depletion refers to the declining availability of certain minerals due to extraction and mining surpassing natural replenishment rates, leading to resource scarcity concerns (Liotine & Ginocchio, 2019).

Economic Factors refer to the fluctuations in market prices, exchange rates, and other financial variables that can impact the procurement and supply of materials. These factors introduce uncertainty and risk into the materials procurement process.

Commodity Price Volatility refers to the significant and unpredictable fluctuations in the market prices of raw materials and goods, which can create economic uncertainty and challenges for businesses in their materials procurement.

Figure 6.

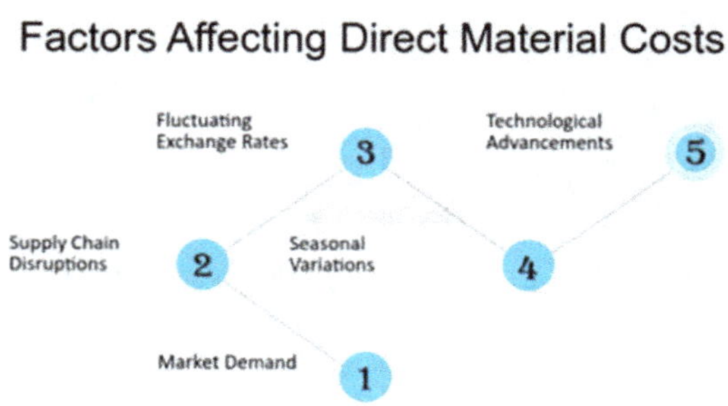

(Source Globe Explorer)

Currency exchange rate fluctuations refer to unpredictable changes in the relative value of currencies, which can significantly impact the cost of procuring materials from international suppliers (Motaung & Sifolo, 2023).

Trade policies and tariffs can significantly impact the supply and cost of materials by introducing barriers to international trade, which differentiates this factor from other economic uncertainties like commodity price volatility or currency exchange rate fluctuations (Radcliffe, n.d.).

Demand Fluctuations refer to the unpredictable and variable changes in the quantity of a material or product that customers demand over time, which can make procurement planning challenging.

Unpredictable market trends refer to the irregular, sudden, and often unexplained fluctuations in market conditions and consumer demand that make it challenging for materials procurement teams to effectively plan and manage their supply chains.

Sudden Demand Spikes refer to unexpected, rapid increases in the demand for a product or material, often caused by unforeseen events or market changes, which can disrupt supply chains and procurement planning.

Figure 7. Sudden Demand Spikes

(Source Globe Explorer)

Changing consumer preferences refer to the evolving and unpredictable shifts in the demands and behaviors of consumers, which can significantly impact material procurement and supply chain management

B. Impact of Sanitary, Social, Geopolitical and Economic Crisis on the Material Supply Process, Case Study

a. Pandemic Covid 19

The literature reveals that COVID-19 presented internal and external shocks that had caused massive disruption to the supply chain. Worldwide activities were impacted by the coronavirus pandemic, with supply chain (SC) disruptions being one of the most notable. Businesses are anticipated to be impacted permanently, therefore it is doubtful that the SC will return to its pre-COVID-19 state (Motaung & Sifolo, 2023; Korucuk et al., 2023). A devastating effect on global supply chains has been caused by the Covid-19 pandemic. In the meantime, a number of businesses have closed due to restrictions in developed market economies, resulting in worldwide instability and global chaos. As a result, thousands of factories across producing nations have closed entirely or partially. Millions of factory workers were consequently sent back to their home countries, many without receiving the required notice period or compensation. following a study by Mark Anner, Ph.D., Director, Center for Global Workers' Rights in Association with the Worker Rights Consortium on 2020; As the Covid-19 outbreak started to negatively impact customers' cash flow, suppliers saw a rise in payment lateness. Our poll validates these reports. In comparison to the contractually specified timeframes, 10.9% of suppliers reported experiencing payment delays of one to ten days, while 68.8% reported delays beyond ten days. In fact, Mostafiz Uddin said that some purchasers were delaying payments by up to thirty days. The sector was hit by an even more severe crisis by the middle of March 2020. Orders that were already in process as well as upcoming orders were suddenly canceled (or placed on hold) by buyers. Even though some of these orders were completely completed and prepared for shipping, the buyers would not accept the delivery of these orders or fulfill their contractual responsibilities to pay for them. At the time of the survey, 23.4% of suppliers said that “a lot” of their in-process orders had been canceled, followed by “most” (22.3%) and “all” (5.9%) of their in-process orders (Anner, 2020). There are three stages to the current crisis that Bangladesh's garment industry is experiencing. The purchase of “raw materials” was the first difficulty. The second issue was the late payments by buyers. Thirdly, there has been an issue with buyers canceling purchases that are still being processed. Over a million people and companies have suffered greatly as a result of these three periods coming to an end. These three stages are covered in the survey.

Lockdown on Wuhan's 11 million residents. Not long after, other cities in the province adopted a similar approach. This had a significant impact on Chinese exports of finished goods to the global apparel supply chains, but it also had an influence on Chinese raw material exports (particularly fabric) that were required by exporters in countries ranging from Bangladesh and Cambodia to Honduras. 93% of Bangladeshi suppliers experienced delays in the shipping of raw materials from China, based on the results of our poll. 53.4% of them stated that they were later fined by purchasers for the ensuing shipment delays. Along with the delays in raw material delivery, 38.3% of respondents said their raw material's price increased “a lot,” and 47.7% said prices increased “some” as a result of the Chinese issue. Ninety-one percent of respondents denied that they had changed their pricing in reaction to a significant rise in the cost of raw materials (Anner, 2020).

3. Importance of Facing These Challenges Through Blockchain and Logistics 4.0

Industry 4.0 practices have a significant impact on supply chain management by enabling greater flexibility, efficiency, and responsiveness in the supply chain operations. The integration of Industry 4.0 technologies such as IoT, AI, big data analytics, and automation allows for real-time data collection, analysis, and decision-making, leading to improved visibility and transparency across the supply chain. This enhanced visibility helps in optimizing processes, reducing lead times, improving inventory management, and enhancing overall supply chain performance. Additionally, Industry 4.0 practices facilitate the adoption of agile supply chain management strategies, enabling companies to quickly adapt to changing market conditions and customer demands. According to Jamwal et al. (2021) and Piyathanavong et al. (2024), the COVID-19 pandemic has recently caused an intensification of the utility owing to technical improvements in production and operation management brought about by Industry 4.0 and digital transformation (Korucuk et al., 2023).

Global supply networks faced substantial hurdles following the COVID-19 epidemic. This has rekindled conversations among several organizations on supply-chain resiliency. Consequently, many organizations are using digital technologies to increase resilience in order to deal with the production delays and distribution slowdowns brought on by interruptions in the labor and material supply chains (Motaung & Sifolo, 2023).

Using Technological Advancements in procurement which refers to the rapid evolution of digital tools, automation, and data analytics that are transforming the way organizations manage their supply chains and procurement processes, for example; Flexible supply contract: One tactic to promote supply chain flexibility is to proactively negotiate supply contracts to lower minimum order amounts or to get suppliers to promise to deliver goods or services in the event of a large spike in demand. According to Tang and Tomlin (2008), flexible procurement contracts can guarantee supplier stability, enable the buyer to adapt to changes in demand, and give supply flexibility (Motaung & Sifolo, 2023). As well, dynamic Pricing which refers to the practice of continually adjusting product or service prices in real-time based on factors such as supply, demand, and market conditions, rather than using static or predetermined pricing.

One illustration would be A multinational brewery keeps an eye out for threats in other critical material categories as well. For example, the situation in Ukraine is driving up the cost of natural gas and malted barley. Long-term resilience of the organization is strengthened by the timely mitigation of supply chain disruption consequences through the use of AI-enhanced planning tools and scenario modeling (Ghadimi et al., 2019).

III. THE ROLE OF LOGISTICS 4.0 AND BLOCKCHAIN IN MATERIALS PROCUREMENT POLICY

1. Definition of Blockchain Technology and its Potential Applications

Blockchain was developed to act as the cryptocurrency Bitcoin's public transaction log. This innovation could have a good effect on many different industries. These days, a wide range of industries that depend on intermediaries, including banks, insurance providers, notaries, law firms, and others, can innovate and create systems that almost do away with them completely. Offering advantages by enhancing existing business and operational frameworks. Unlike a conventional centralized database system, blockchain operates on a shared decentralized database model, enabling a social source of truth rather than a single source by integrating networking and cryptological elements, blockchain serves as a trust protocol that can establish trust without depending on intermediaries or central authorities. Through the creation of a dependable distributed digital ledger, blockchain records ownership and transactions reliably" (Liotine & Ginocchio, 2019). Blockchain is primarily dependent on hashing and encryption methods, which need to be updated frequently to fend against changing hacker tactics. It is necessary to take precautions in a distributed ledger environment to stop hackers from changing hash values. Although such changes can be automatically detected by blockchain protocols, preventive measures should be implemented to discourage hacker efforts (Liotine & Ginocchio, 2019). Table 1 lists blockchain usage in various global sectors. Blockchain technology is a "digital transcript" that eliminates fraud, improves efficiency and performance, and allows third parties to access information as needed. Blockchain technology is becoming increasingly widely adopted in a variety of areas and will undoubtedly become a significant tool in practically every business. (Mapping the potentials of blockchain in improving supply chain performance)

Table 1 lists blockchain usage in various global sectors.

Table 1. Blockchain usage in various global sectors

Number	Sector	Usage of Blockchain
1	Medical sector	implementation of blockchain enhance manufacturing process, transparency and overall efficiency (Ada et al., 2021)
2	Food supply chain sector	Implementing blockchain in food supply chain management allows for more dependable traceability and ensuring the safety and quality of the food (Katsikouli et al., 2021)
3	Finance and bank sector	The application of Blockchain technology improves the security of banking transactions by reducing fraud and cyberattack vulnerabilities, ultimately boosting customer trust in the safeguarding of their financial data (Garg et al., 2021)
4	Maritime industry sector	Blockchain technology has been proposed in order to boost the efficiency and transparency of shipping containers and freight in the shipping sector (Elmay et al., 2021).
5	Insurance sector	Blockchain technology is being used in the insurance business to deal with issues including trust, data abuse, pricing distinctions efficiency, regulatory compliance, challenges of interest, mediation, and transparency (Karmakar et al., 2023).
6	Textile and fashion sector	Using blockchain, textile suppliers can track the process of products from raw materials to finished output (Cuc, 2023).
7	Energy sector	Blockchain offers online energy trade, providing users to purchase and trade energy directly from each other, without the need of intermediaries. This promotes energy efficiency and cost savings (He et al., 2024).

2. Overview of Logistics 4.0 and its Fundamental Principles

Logistics 4.0 refers to the integration of advanced digital technologies, such as the Internet of Things, big data analytics, and automation, to optimize logistics operations and supply chain management. The aspects of Material Procurement Policy via Logistics 4.0 focus on optimizing supply chain efficiency and responsiveness through data-driven decision making and digital integration across the procurement lifecycle. The Principles of Logistics 4.0 outline the key tenets that guide the strategic application of advanced technologies to optimize material procurement processes. These principles can be summarized in:

Automation in the context of Logistics 4.0 refers to the use of advanced technologies to self-execute operational processes, with little to no human intervention, optimizing efficiency and reducing errors, in contrast to traditional manual processes (Elmay et al., 2021).

Figure 8. Industry 4.0 technology

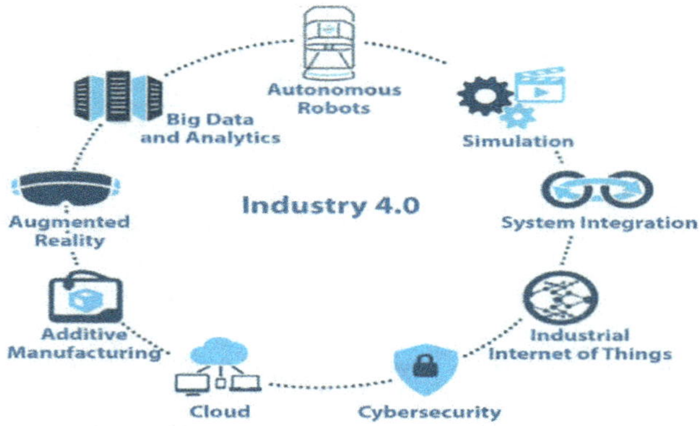

(Elmay et al., 2021).

Digital Transformation in Logistics 4.0 refers to the integration of digital technologies to fundamentally change how logistics operations are conducted, rather than simply automating existing processes. Enabling new capabilities and business models.

Figure 9. Digital transformation

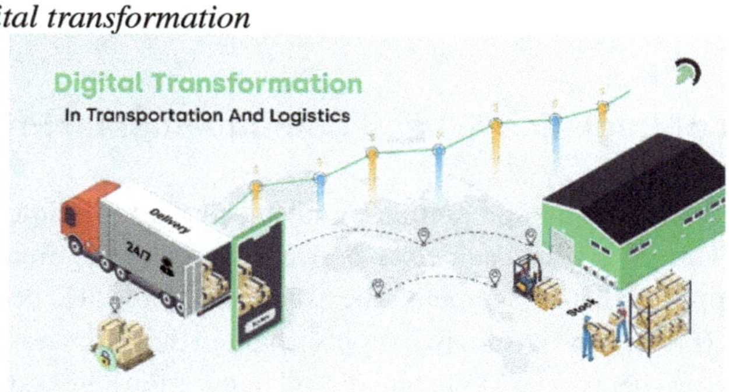

(Source: slitigenz.io)

Interconnectivity in Logistics 4.0 refers to the seamless integration, communication and data exchange between various systems, devices, and stakeholders across the supply chain, enabling real-time data sharing and visibility, and coordinated decision-making (Institute of Electrical and Electronics Engineers, 2019).

Figure 10. The main elements of logistics 4.0

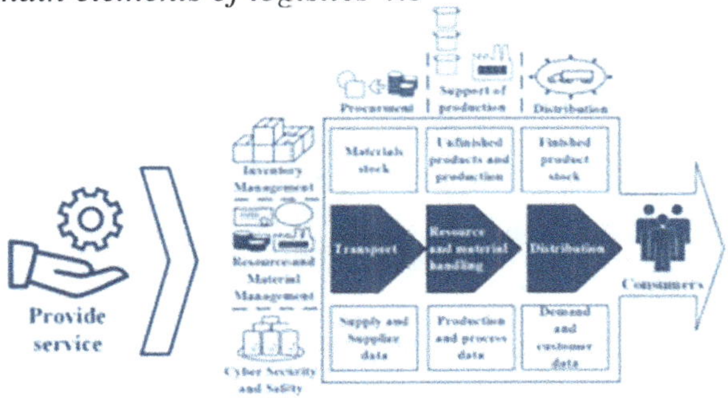

(Institute of Electrical and Electronics Engineers, 2019)

A. Distribution of the Most Expected Values Proposed in Procurement Using I4.0 and the Most Applied I4.0 Applications

A careful examination of the literature reveals a breakdown of the most frequently proposed values in purchasing using I4.0, and thus extracts the added (value value-added benefit) of procurement for companies. Purchasing performance management is the most important value and alongside supplier performance, management and supplier selection and evaluation enhancement are allocated the most key I4.0 applications, as illustrated in the figure below.

Following a study by Jahani et all of the literature in their article review, the number of articles dealing with different supply technologies is illustrated in the pie chart below. The most important I4.0 applications used to add value to procurement processes are summarized in figure 7. Of which BDA and IoT are two important elements that can have a significant impact on the value proposition in the area of procurement (Jahani et al., 2021).

Figure 11. Distribution of the most attended maritime transportation risks

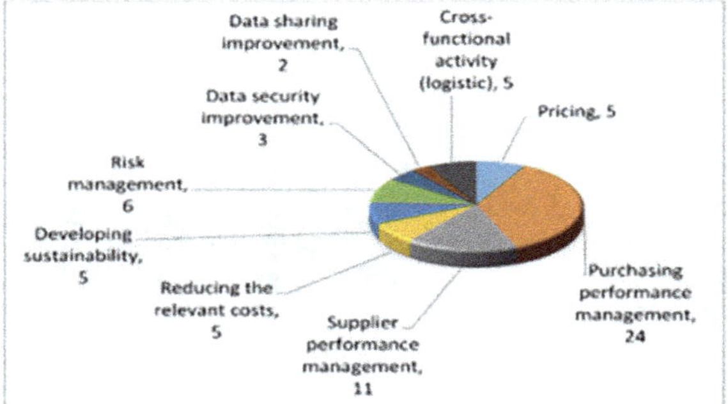

(Jahani et al., 2021).

Figure 12. Distribution of the most applied I4.0 applications

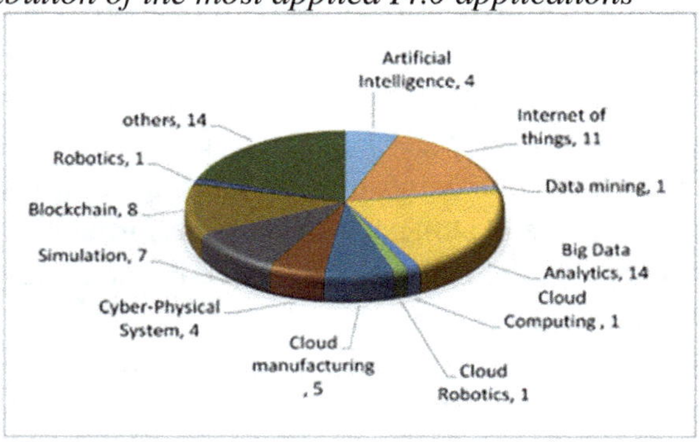

(Jahani et al., 2021).

B. Digital Technologies Associated With Procurement

As P4.0 is one of the most important concepts in the smart SC, the proliferation of digital technologies has significantly transformed the landscape of procurement, supporting daily commercial tasks enabling organizations to streamline their operations, enhance decision-making, and elevate their strategic capabilities.

Procurement professionals must adhere to the strategic goals and the company's procurement objectives in order to help the businesses succeed, since procurement can be defined as a combination of purchasing with inventory control, transportation, receiving, inspection, storekeeping, and salvage and disposal operations (Jumas et al., 2023). in this sense we have identified the most important logistics 4.0 applications and the major added values in the field of procurement which can be offered using L4.0 applications.

3. Application of Logistics 4.0 and Blockchain Principles in Materials Procurement Policy

A. Illustration of How Logistics 4.0 Technologies Help Manage Uncertainties and Optimize Collaboration in the Materials Supply Chain

According to Jahani et al. (2021) in their Systematic Literature Review; Under the term P4.0 which refers to applicable technologies that have contributed significantly to supply chains and represented substantial values in procurement, such as CPSs, BDAs, and IoT. Improvements in performance in the areas of price, supplier selection and assessment, cost-effectiveness, data collection and analysis, and buying management are among the additional advantages that come with procurement management in supply chains. In order to improve price, supplier management, and buying management decisions, or to elaborate sustainability and data-sharing security, the adoption of P4.0 leads to operating procedures related to procurement management becoming more autonomous, flexible, and transparent (Fatorachian & Kazemi, 2021). The first-rate values that the purchasing managers can offer with the aid of L4.0 are categorized as tariffication, performance management of purchases, performance management of suppliers, reduction of relevant costs, sustainable development, risk management, improvement of data security, improvement of data sharing, and transversal activity (logistic), added Jahani et al. (2021).

Industry 4.0 has a significant impact on procurement processes. With the integration of advanced technologies such as artificial intelligence, big data analytics, Internet of Things (IoT), and automation, procurement in the era of Industry 4.0 becomes more efficient, transparent, and data-driven. Some of the key impacts of Industry 4.0 on procurement include: **Increased Efficiency** (Automation and digitalization of procurement processes lead to increased efficiency by reducing manual tasks, streamlining workflows, and enabling real-time data analysis). **Improved Decision-Making** (Advanced analytics and AI algorithms help in making data-driven decisions, optimizing supplier selection, and identifying cost-saving opportunities). **Enhanced Transparency** (Blockchain technology can be used to create transparent and secure supply chains, providing visibility into every step of the procurement process). **Supply Chain Resilience** (Industry 4.0 technologies enable better risk management and mitigation strategies, making supply chains more resilient to disruptions). **Supplier Relationship Management** (Digital tools facilitate better communication and collaboration with suppliers, leading to stronger relationships and improved performance). **Cost Reduction** (By optimizing processes, identifying inefficiencies, and leveraging data analytics, Industry 4.0 can help in reducing procurement costs). Overall, Industry 4.0 revolutionizes procurement by leveraging technology to drive uncertainty management, efficiency, transparency, and strategic decision-making in the supply chain (Korucuk et al., 2023).

B. Explication of how Blockchain can Enhance Transparency, Traceability in the Materials Procurement Process

The growing scope of blockchain technology has garnered considerable interest in recent years, primarily due to its potential applications across various sectors. A particular focus area that is gaining prominence is the integration of blockchain in supply chain management, particularly in relation to materials procurement (Mahyuni et al., 2020; Guerpinar et al., 2021; Song et al., 2019).

The intrinsic characteristics of blockchain, which encompass transparency, traceability, and immutability, position it as an exceptionally promising technology for enhancing the efficacy and dependability of materials procurement processes (Mahyuni et al., 2020; Guerpinar et al., 2021). Through the implementation of a communal, distributed ledger that logs all transactions and physical movements within the supply chain, blockchain has the capability to deliver an unchangeable and verifiable account of the acquisition procedure (Song et al., 2019; Ada et al., 2021).

This enhanced visibility and traceability has the potential to create more trust among supply chain stakeholders, minimize incidents of fraud or errors and optimize inventory management (Ada et al., 2021 ; Mahyuni et al., 2020). Furthermore, the decentralized structure of blockchain may eliminate the require for intermediaries, thereby reducing costs and optimizing the purchase process (Mahyuni et al., 2020; Guerpinar et al., 2021).

While the integration of blockchain technology in materials procurement offers promising advantages, its widespread adoption remains in its nascent stages. Current research, including practical case studies, explores the implementation of blockchain-based solutions across various supply chain applications. The automotive sector is one such example (Ada et al., 2021). As technological advancements occur and blockchain becomes more widely accepted, it is expected to assume a more significant role in shaping future materials procurement policies, thereby improving transparency, traceability, and overall efficiency within the supply chain (Mahyuni et al., 2020).

C. Benefits of Blockchain Regarding Traditional ERP

Blockchain technology has received substantial interest in recent years, notably in the domain of supply chain management While the potential benefits of blockchain, such as improved data security, enhanced traceability, and increased transparency, are well-documented, the regulatory and legal considerations surrounding its implementation in the procurement materials field are less well-understood

Blockchain technology has the potential to significantly reduce transaction costs and enhance transparency without the need for a centralized intermediary. When integrated with distributed ledger systems, it effectively synchronizes data across multiple parties, ensuring uniformity, additionally offers a clear and unchangeable record of transactions, enabling all stakeholders in procurement to monitor and authenticate every stage of the procedure. This level of transparency aids in the mitigation of fraud and inaccuracies in procurement operations. These capabilities offer compelling prospects for adoption within supply chain operations. Despite ongoing pilot projects to explore these benefits, many organizations exhibit a cautious "wait and see" stance before integrating blockchain solutions into their workflows. As with all emerging technologies, a prudent, low-risk strategy is typically advisable. Therefore, the acquisition of materials through the utilization of blockchain technology presents distinct differences from conventional Enterprise Resource Planning (ERP) systems in various fundamental aspects. The utilization of blockchain technology in procurement systems, as examined in various scholarly works, provides heightened levels of security, genuineness, and integrity by means of decentralized networks, cryptographic techniques, and smart contracts These systems eradicate the requirement for intermediaries such as payment gateways, thereby ensuring a more effective and economical process. Moreover, blockchain technology enables the transparent and immutable recording of transactions, thereby enhancing the traceability and answerability in material procurement, as emphasized in prior studies. Conversely, ERP systems lack the decentralized and transparent characteristics of blockchain, often necessitating manual interventions, numerous approvals, and extensive documentation, resulting in inefficiencies and inaccuracies, as observed in academic literature Therefore, the incorporation of blockchain technology in the domain of material procurement has the capability to enhance processes, reduce expenses, and improve overall efficiency in contrast to traditional ERP systems.

IV. INTEGRATION OF LOGISTICS 4.0 AND BLOCKCHAIN IN MATERIALS SUPPLY POLICY

1. Risk Management and Security Related to the Use of Blockchain and Logistics 4.0 in Materials Procurement

The National Association of State Procurement (NASPO) (Lee & Lee, 2015) states that security is a key worry when integrating digital technologies like the Internet of Things, which serves as a barrier to the adoption of digital procurement. The adoption of cloud computing raises serious questions for both individuals and organizations about data privacy and security. Glas & Kleemann (2016) also agree that there are hazards associated with the digitalization trend, including those related to data security and transparency. Therefore, it is essential that companies develop adaptable strategies for reducing security risks related to the refusal to implement digital procurement.

The context of Logistics 4.0 in materials procurement, risk management and security play crucial roles in ensuring the smooth and secure flow of materials throughout the supply chain. Here are some key considerations related to risk management and security in Logistics 4.0 for materials procurement:

Data Security: With the digitization of logistics processes in Logistics 4.0, there is a significant amount of data generated and exchanged. It is essential to implement robust data security measures to protect sensitive information related to procurement activities, supplier details, inventory levels, and shipment details.

Cybersecurity: As logistics processes become more interconnected through IoT devices and cloud-based systems, the risk of cyber threats increases. Implementing cybersecurity measures such as encryption, firewalls, and regular security audits is essential to protect against cyberattacks and data breaches.

Supply Chain Visibility: Logistics 4.0 provides enhanced visibility into the supply chain, allowing real-time tracking of materials and shipments. However, this increased visibility also poses risks in terms of data privacy and confidentiality. Proper access controls and data encryption can help mitigate these risks.

Supplier Risk Management: Effective risk management in materials procurement involves assessing and managing risks associated with suppliers. Logistics 4.0 enables real-time monitoring of supplier performance, quality, and delivery times, allowing for proactive risk mitigation strategies.

Compliance and Regulations: With the use of advanced technologies in Logistics 4.0, it is important to ensure compliance with data protection regulations, industry standards, and international trade laws. Non-compliance can lead to legal risks and penalties.

Physical Security: In addition to cybersecurity measures, physical security of materials during transportation and storage is critical. Implementing measures such as GPS tracking, tamper-evident packaging, and secure warehousing facilities can help prevent theft and damage to materials.

Risk Assessment: Conducting regular risk assessments and scenario planning exercises can help identify potential risks in the procurement process and develop mitigation strategies. This proactive approach is essential in managing uncertainties in the supply chain.

By addressing these considerations and implementing robust risk management and security measures, organizations can leverage the benefits of Logistics 4.0 in materials procurement while safeguarding their operations against potential risks and threats (Korucuk et al., 2023).

Jiri Tupa and All discussed in their article (Aspects of risk management implementation for Industry 4.0), the risk management implementation for Industry 4.0, which includes aspects related to materials procurement. In the context of Industry 4.0, new risks may arise due to increased data volumes, real-time data availability, and the integration of information and communication technologies (ICT) in manufacturing processes. Risks such as cyber-attacks, data integrity issues, and information availability problems are highlighted. Of which they suggest using an information security management system (ISMS) to address these risks and protect manufacturing systems. Additionally, it proposes integrating risk management principles with business process management to ensure the effectiveness of corporate performance in the procurement of materials within the industry 4.0 framework (Tupa et al., 2017).

Blockchain technology is crucial in strengthening risk management and security in materials procurement within Logistics 4.0, Utilizing the decentralized ledger technology of blockchain enables supply chain operations to reach heightened levels of transparency, traceability, and security. The integration of blockchain in managing risks within the supply chain has demonstrated favorable results in ensuring the reliability of financial transactions and information, granting a competitive edge to organizations. Moreover, blockchain guarantees secure and transparent transactions, thereby improving data integrity, traceability, and availability while minimizing vulnerabilities associated with single-point failures. Within the domain of supply chain finance, blockchain transforms the flow of information, enhances credibility and transparency, and broadens the scope of credit via smart contracts and distributed ledger technologies, thus mitigating risks and ensuring the authenticity and security of transactions. In essence, the application of blockchain technology in Logistics 4.0 transforms materials procurement by addressing issues such as counterfeiting, fostering trust, and securing transactions, ultimately optimizing supply chain security and risk management

2. Innovation and Competitive Advantages of Integrating These Technologies in Materials Procurement

A common definition of e-procurement or P4.0 among many researchers is a digital platform that centralizes, optimizes, simplifies and streamlines the procurement process (Motaung & Sifolo, 2023). It is the opinion of Lindskog, Vegas, and Brege (n.d.) that electronic bidding, ordering, and rendering through markets, extranets, private platforms, portals, and/or electronic data exchange constitute electronic procurement. Electronic marketplaces (e-marketplace), electronic auctions (e-auction), and electronic catalogues (e-catalogue) are just a few examples of end-to-end solutions that are used in e-procurement, according to Costa, Arantes, and Tavares (2013). The web-based ERP system, electronic sourcing, electronic reverse auction, and electronic informing are among the e-procurement features and capacities that Oh et al. (2014) describe as being particularly useful.

The use of new digital technologies to digitalize procurement procedures is a significant next step in the evolution of procurement (Motaung & Sifolo, 2023). According to Gottenberg (as cited in Motaung & Sifolo, 2023), businesses must be digitally aware given the current working environment. This is necessary because to the massive volume of data and amount of information flow that needs to be collected, examined, and deciphered in order to assist procurement organizations in making decisions about their company in real time. According to Burnson (as cited in Motaung & Sifolo, 2023), "leading companies are making the necessary preparations to create a true digital procurement organization through their actions."

Procurement organizations are undergoing a transformation thanks to digital technology, which also facilitates information access and enhances service delivery. In order to conduct a thorough examination of "procurement practices, consumption models, and supplier transactions," the Oliver Wyman Group contends that digitizing procurement procedures "enables the generation of new data" (Motaung & Sifolo, 2023). Digital technology adoption will have a long-lasting and really revolutionary influence on corporate operations, particularly public procurement, according to Ashwell (2017). Organizations that have completely digitalized their procurement processes are using robotically or automatically enabled procedures to improve internal stakeholder and supplier collaboration (Motaung & Sifolo, 2023).

Burnson (n.d.) states that "deep and rich data" and "rigorous process" are the cornerstones of digital procurement and are necessary for creating "(artificial intelligence) AI-enabled predictive models." According to Motaung & Sifolo (2023), digital procurement entails using cloud solutions, artificial intelligence (AI), big data, block chain, and Internet of Things technologies to accomplish transactions. Adopting digital procurement solutions, according to Pereira et al. (2014), "involves both organizational changes and reengineering of processes." Big data, digital technology, and people are the three primary components of digital transformation capabilities, according to Ashwell (2017). Ashwell goes on to say that these qualities, along with connections, will transform procurement (Ashwell, 2017). Conversely, big data was identified by Dremel, Wulf, Herterich, Waizmann, and Brenner (2017) as the essential skill for starting the digital transformation process. business processes, as well as procurement processes have also been affected by digital transformation (Motaung & Sifolo, 2023). Government agencies have been impacted by these interruptions, and public procurement mechanisms urgently need to be reformed (Kramer, 2016).

It is clear from the aforementioned advancements that procurement technologies are developing and will keep developing in the future (Sali et al., 2023). As on Brandyberry's (2017) findings, procurement technology is expected to undergo substantial advancements that would boost process efficiency, conduct automatic spend analysis, provide small and incremental savings, and facilitate efficient control and contracting of crucial expenses.

According to Accenture's research report "Procurement Next Frontier," procurement organizations and their professionals will undergo a revolution during the next five to seven years. According to the report, procurement's relationship with internal and external stakeholders would undergo a profound transformation as a result of the fourth industrial revolution. According to KPMG, digital breakthroughs will bring about "the big procurement transformation," which will fundamentally alter procurement organizations. Predictive technologies will need to be included "as integral elements in the procurement strategy." ATKearney forecasts in the research "The Future of Procurement" that businesses will use "intelligent technologies" to automate their procurement processes. ATKearney goes on to say that highly automated procurement in the future would lead to automated buying via AI, AI sourcing managers, and a supplier help desk chatbot. From the perspective of GEP, companies that use digital technologies like artificial intelligence (AI), big data, robotic process automation (RPA), and three-dimensional (3D) printing to revolutionize their procurement will be well-positioned going forward. According to Accenture, as seen in Figure 12, future procurement will be fully integrated with a completely different structure and made possible by digital technology (Motaung & Sifolo, 2023).

Figure 13. Future procurement framework. Source: (Motaung & Sifolo, 2023).

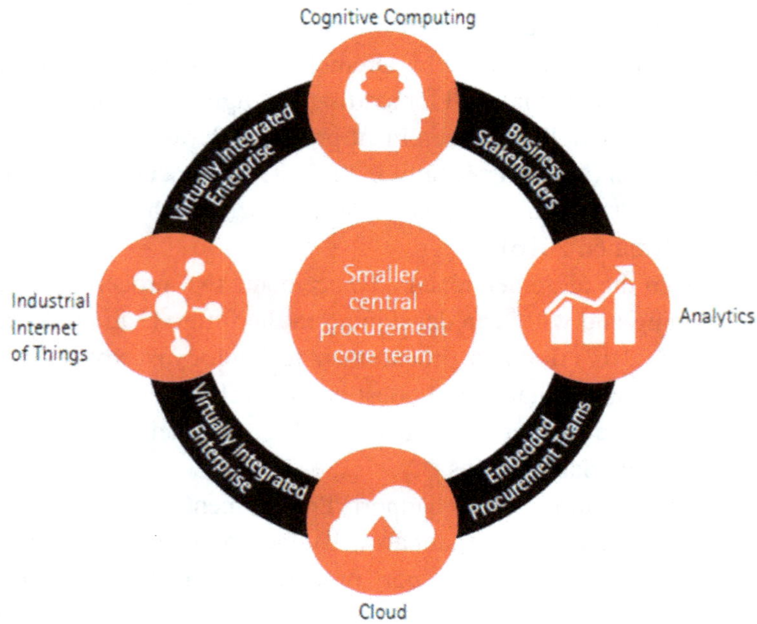

Organizations that use digital procurement experience digital transformation as a result of the fundamental shift in the services provided by procurement. Among the advantages are practical strategies for boosting supply chain efficiency by integrating big data analytics with procurement. Mishra, Devaraj, and Vaidyanathan (2013) claim that companies who have adopted digital procurement competency have the potential to reduce procurement expenses. This might potentially affect an organization's competitive advantage and thereby reduce the risk to the organization's supply chain (Motaung & Sifolo, 2023).

V. CASE STUDIES AND BEST PRACTICES

1. Concrete Cases Where Companies Have Successfully Implemented Blockchain and Logistics 4.0 in the Field of Material Procurement

One concrete case where a company successfully implemented blockchain and Logistics 4.0 in the field of material procurement is IBM's Food Trust platform. IBM's Food Trust leverages blockchain technology to enhance transparency and traceability in the food supply chain, ensuring the authenticity and quality of food products. Here's how IBM's Food Trust platform utilizes blockchain and Logistics 4.0 in material procurement: Supply Chain Transparency (By using blockchain technology, IBM's Food Trust provides end-to-end visibility into the food supply chain, allowing stakeholders to track the journey of food products from farm to table. This transparency helps in verifying the origin of ingredients and ensuring compliance with quality standards). Traceability (Blockchain enables the recording of every transaction and movement of food products on an immutable ledger. This traceability feature allows quick identification of the source of any issues or contamination, facilitating targeted recalls and reducing the impact of food safety incidents). Smart Contracts (Smart contracts embedded in the blockchain automate various processes in the supply chain, such as payment settlements, quality checks, and compliance verification. This automation streamlines procurement operations and reduces manual intervention). Data Security (Blockchain technology ensures the security and integrity of data shared among supply chain participant. The decentralized nature of blockchain prevents unauthorized access and tampering of sensitive information, enhancing data security in material procurement). Collaboration (IBM's Food Trust platform fosters collaboration among supply chain partners by providing a shared platform for data exchange and communication. This collaborative approach improves coordination, reduces delays, and enhances efficiency in material procurement processes).

Real-time Monitoring (Logistics 4.0 technologies, such as IoT sensors and RFID tags, integrated with blockchain enable real-time monitoring of temperature, humidity, and other environmental conditions during transportation and storage. This real-time data helps in ensuring the quality and integrity of materials throughout the supply chain). Quality Assurance (By leveraging blockchain and Logistics 4.0, IBM's Food Trust enables quality assurance mechanisms, such as digital certifications, product authenticity verification, and compliance tracking. These features enhance trust among stakeholders and ensure the procurement of high-quality materials). Overall, IBM's Food Trust platform serves as a successful example of how blockchain and Logistics 4.0 can revolutionize material procurement processes by enhancing transparency, traceability, efficiency, and security in the supply chain (Korucuk et al., 2023).

Another example; ACSA, which is a public company that owns and manages a network of nine airports in South Africa. According to a study conducted by Joel R. Motaung in his article, it was the consensus of all panelists that ACSA would benefit from digital procurement. Using digital procurement at ACSA was seen to have several advantages, one of them being the enhancement of tender process efficiency. Following are the aspects that the results showed: Efficiency and efficacy of the process: According to participant 2, "efficiencies would be greatly improved...

especially when it comes to sourcing." "It would definitely increase productivity and efficiencies within the procurement space, especially the RFQ process," said participant number 4. Put differently, digital procurement improves ACSA's process efficacy and efficiency. "Efficiencies can be obtained by digitizing some of the processes," stated participant number 6. Gardenal makes the case that implementing digital procurement can have favorable effects on "efficiency, effectiveness, dematerialization, competitiveness, and transparency." Stated differently, airport management firms should anticipate enhancements in the caliber and velocity of procurement procedures. Benefits for competitiveness: Participant 9 concurred, saying that "everything is going to be easier done and it's going to be computerized the more you put into place your digital procurement." Mishra, Devaraj, and Vaidyanathan (2013) support this by pointing out that the use of e-procurement systems has benefited numerous organizations in a variety of ways. "Opportunities are there, for example automating your RFPs and RFQs," stated participant 1. Furthermore, the TAM's emphasis on user technology adoption and behavioral intention may have an impact on e-procurement potential. Improved compliance and lower procurement costs: At ACSA, there are financial advantages to digital procurement solutions. "ACSA can benefit from the improvement in efficiencies where we can do more with less," participant number seven clarified. Among the advantages mentioned by Mishra et al. (2013) are "reduced cycle time, better coordination with partners, and enhanced financial performance." The performance of the organization is positively impacted when IT technologies are used to facilitate procurement procedures and the value chain. As stated by Participant 3, "We can shorten the cycle time. You know why, when I can utilize tools to do it for me, I have to spend time designing a spend cube? "So obviously you are shortening procurement cycle time, say from... let's say from eight days to cut it down to four days," observes Participant 4. One participant said, "The tendering process is quite time-consuming." Thus, turnaround times can be shortened. Digitalization has a significant impact on day-to-day procurement processes, according to Gartner. Participants also highlighted cost reduction and concentrating on initiatives that would provide value to the business (Motaung & Sifolo, 2023).

A case in point A multinational brewery used digital tools, such as artificial intelligence (AI), to analyze risk exposure data and forecast a glass shortage. The risk management team examined and projected supply and demand patterns for the key inputs in the molded glass vial value chain. Significant supply threats were identified by the investigation as a result of rising costs for raw materials including limestone, soda ash, and industrial sand—which together account for about one-third of manufacturing expenses. Equipped with this knowledge of potential risks and pricing projections, the company's purchasing department created and implemented strategies to lessen the ensuing increase in glass bottle costs. These precautions included gathering goods in advance, signing long-term agreements, and locating substitute providers who posed less of a danger (Boston Consulting Group, 2023).

When most of the logistics parties, such as customs authorities, shipping companies, port authorities, and other logistics departments, have appended the shipment data, the blockchain logistics enterprise can accomplish all the relevant data and serve as a temple information tool for common logistics processes. Then importers and exporters, with a QR code or other information, can obtain digitized customs information about the cargo in advance. Enterprises will then benefit from customs declarations quickest (if not sooner). Goods can easily flow. The safety, accuracy, and reliability of the blockchain are the features that the logistics industry is most worried about, regardless of the product's aspects. Such features make it an absolute necessity in the transport of goods industry. Before or after the physical transit of goods, the blockchain is the specialist tool to store the data relevant to the material. With blockchain technology, the issuer can secure the digital goods receipt to approve the acquisition. When the carrier receives the product and the invoice, the blockchain distributes and imputes the receipt of the goods to the shipper, carrier, and insurance company, thus allowing the contract among the four parties to be implemented and validated.

The traditional methods of procurement in business organizations still handle a multitude of problems that nullify the auspiciousness of technology, thereby making the business processes more sluggish. Therefore, recent technological advancements have provided an alternative to such traditional inefficient business processes. The material procurement function is no exception in the company because the online method of procurement is now receiving more attention than the traditional method. Blockchain, recognized as a budding technological advancement, operates as a decentralized and secure network for facilitating direct exchanges between participants, thus obviating the need for intermediaries such as trust organizations. The use of the blockchain in the logistics process has been most economically justified. Blockchain provides a highly secure and multi-party management solution for transporting data and contains the data as a distributed file method.

2. Comparative Analysis Between Traditional Approaches to Material Procurement via Blockchain and Logistics 4.0 Based Approach

Supply chain management, a more comprehensive strategic approach to materials and distribution management, is the result of the evolution of the conventional buying and logistics roles over the past decade. Because of developing technology, the supply chain's traditional functions have transformed into more comprehensive and contemporary strategic approaches (Tan, 2001).

According to a study provided by Motaung & Sifolo (2023), there is a discussion about the evolution of procurement towards Industry 4.0 and the benefits of digital procurement technologies. While a direct comparative analysis between traditional approaches to material procurement and Industry 4.0-based approaches is not explicitly outlined, we can draw insights from the content to highlight the differences and advantages of Industry 4.0 in material procurement compared to traditional methods:

Efficiency and Automation: Industry 4.0-based approaches leverage technologies like artificial intelligence, big data analytics, and robotic process automation to automate and streamline procurement processes. This leads to increased efficiency, reduced cycle times, and improved accuracy compared to manual and paper-based traditional methods. Data-Driven Decision Making: Industry 4.0 enables data-driven decision-making by utilizing advanced analytics to analyze large volumes of data. This allows organizations to make informed decisions, optimize procurement strategies, and identify cost-saving opportunities based on real-time insights, which may not be possible with traditional approaches. Supply Chain Visibility: Industry 4.0 technologies provide enhanced visibility and transparency across the supply chain. With features like the Internet of Things (IoT) and blockchain, organizations can track and trace materials in real-time, monitor supplier performance, and ensure compliance with regulations, leading to improved supply chain management compared to traditional methods. Innovation and Collaboration: Industry 4.0 fosters innovation and collaboration through digital platforms and tools. Companies can engage with suppliers, partners, and stakeholders in a more integrated and connected manner, (Motaung & Sifolo, 2023) facilitating communication, knowledge-sharing, and strategic decision-making, which may be limited in traditional procurement approaches. Cost Reduction and Risk Mitigation: Industry 4.0 technologies enable cost reduction through process optimization, predictive analytics, and automation. By leveraging digital tools, organizations can identify cost-saving opportunities, mitigate risks, and enhance compliance, resulting in improved financial performance compared to traditional procurement methods. Sustainability and Compliance: Industry 4.0-based approaches support sustainability initiatives and compliance requirements by enabling organizations to track environmental impact, ensure ethical sourcing practices, and meet regulatory standards. This focus on sustainability and compliance is more robust and integrated in Industry 4.0 compared to traditional procurement practices. Overall, Industry 4.0-based approaches to material procurement offer significant advantages in terms of efficiency, data-driven decision-making, supply chain visibility, innovation, cost reduction, risk mitigation, sustainability, and compliance compared to traditional methods. By embracing digital technologies and automation, organizations can transform their procurement processes and achieve competitive advantages in today's rapidly evolving business landscape (Motaung & Sifolo, 2023).

The findings are consistent with the research conducted by Bienhaus and Haddud (2018), which found that the following advantages of digitizing procurement procedures are possible: (1) Increasing productivity in routine administrative work; (2) streamlining the intricate decision-making process; and (3) helping procurement concentrate more on strategic endeavors.

Considering the current level of advancement and utilization of mathematical tools in addressing logistical and control issues, a holistic strategy was proposed for creating decision support services to oversee and manage procurement activities. This strategy involves merging conventional optimization techniques with contemporary blockchain technology. The fundamental principles of conventional and blockchain-driven decision-making processes have been elucidated, each with its own specific objectives. The detailed examination of the four-phase procurement process life cycle includes Substance Request, Tender, Purchase, and Confirmation stages. The research successfully accomplishes its objectives by presenting a viable solution to procurement requests for mining companies. Various traditional optimization methods, such as Small Lot, Periodic Batch, and Just-in-Time, were employed at each of the four procurement stages

The advent and extensive growth of blockchain technology have signaled a significant influence on the global economy due to its ability to enhance service quality through attributes such as transparency, immutability, confidentiality, swift transaction processing, and quantifiable relationships. In the last decade, technology has extended its array of uses, particularly in the spheres of management and finance This article aims to explore and evaluate conventional and blockchain-driven decision-making processes to assess their practical implementation potential. The study aims to conduct a thorough comparative analysis of two alternative information systems for optimizing spot markets - the conventional methods commonly used in supply chain management and the parametric as well as blockchain-based approaches throughout the life cycle of reclaimed resources in open pit and underground mines. An overview of the existing information systems for enterprise procurement functions is provided, along with the unique procurement challenges faced by mining companies. Proposing an integrated approach to system development that combines blockchain technology with optimization techniques tailored to mining enterprises, the article outlines the core principles of traditional and blockchain-based optimization in the mining sector, along with the functionalities of conventional decision-making processes.

All things considered, the industry 4.0-based approach to material procurement offers significant advantages over traditional methods by leveraging digital technologies, data-driven insights, automation, and connectivity to enhance efficiency, transparency, agility, and decision-making in the supply chain (Korucuk et al., 2023).

The summary makes it clear that digitalization reduces the vulnerability of procurement activities by enabling real-time risk mitigation. The summary concludes that there are several opportunities associated with the transition from traditional to digital procurement (Motaung & Sifolo, 2023).

VI. ADOPTION AND IMPLEMENTATION STRATEGIES

1. Challenges of Adopting These Technologies

The dismal organizational acceptance of technology, which has an adverse effect on the performance outcomes, implementation, and adoption of it, is acknowledged by proponents of e-procurement systems (Kademaunga & Phiri, 2019). Proponents of electronic procurement systems recognize that organizations have not adopted technology as expected, which has a detrimental effect on the technology's adoption, deployment, and performance outcomes (Motaung & Sifolo, 2023). For instance, Kademaunga and Phiri (2019) agree that selling the e-procurement concept internally to organizational stakeholders, such as senior management and end-users, is one of the Challenges preventing the successful implementation of electronic procurement in an African context.

According to semi-structured interviews carried out by Motaung & Sifolo (2023), based on the transcripts, participants expressed worries about job security as a result of the widespread use of data and the digital transition. The implementation of digital procurement has been hindered by several organizational challenges, including job losses, cyber security, lack of interoperability, lack of expertise, and system downtimes, according to the participants. Participants believed that as a result of automated procedures, fewer workers would be needed in the procurement sector, which would prevent job losses. Participants acknowledged that ACSA's vulnerability to cybercrime would make its systems and data available to other parties. These results indicate that, with regard to the obstacles brought on by digital procurement, participant opinions are consistent with existing research. Digitalization has been linked to a number of problems, including ownership, security, and openness of data (Glas & Kleemann, 2016). Cybersecurity and system outages brought on by network congestion were two further obstacles mentioned by the attendees.

At ACSA, the function of digital procurement adds value and has the potential to lower supply chain costs. The participants did, however, identify some organizational and practical challenges to the implementation of digital procurement, including job losses, cyber security, a lack of interoperability, a lack of expertise, and system outages. Digital technology integration continues to be a complicated, dynamic, fluid, and unstable phenomena that affects many organizational processes and results. When it comes to the implementation of digital procurement or digitalization, a framework is offered that highlights important priority areas. Literature has also added an in-depth survey on the main obstacles to the adoption of digital procurement (Motaung & Sifolo, 2023), Costs as a barrier to adopting digital procurement: The costs of transforming procurement functions are crucial in making digitalization decisions. Digitalization offers cost reduction, cost savings, and allows resources to focus on value-adding tasks. However, transitioning from conventional to digital procurement comes with heavy costs, emphasizing the importance of understanding both monetary and unforeseen costs (Riley-Reid, 2015). Even though utilizing I4.0 technologies in procurement is crucial, investing in these technologies might be difficult for businesses. The primary issue at hand is the expenses associated with implementing Industry 4.0 technology. One example of a high-performing and high-maintenance technology that can cost up to $250,000 per hour is AI-based techniques (Wanner et al., 2020). This has been one of the primary causes of the firms' continued hesitancy to adopt I4.0 technologies, raising concerns about whether these pricy applications would enable them to achieve the necessary advantages. As a result, giving firms a defined set of values that the I4.0 application may suggest improves their ability to execute I4.0. Stated differently, businesses must understand how their technology investment can yield the necessary returns to meet any forthcoming obstacles. The use of I4.0 in procurement (P4.0) is the subject of several publications in the relevant literature. Nonetheless, the development of a thorough framework outlining how I4.0 might benefit the procurement department remains imperative (Lee & Lee, 2015). Security as a barrier to adopting digital procurement: Security is a key worry when implementing digital technologies like the Internet of Things, according to the National Association of State Procurement (NASPO) (Lee & Lee, 2015), which is one of the barriers to embracing digital procurement. The adoption of cloud computing raises serious questions for both individuals and organizations about data privacy and security. Glas and Kleemann (2016 also agree that there are hazards associated with the digitalization trend, including those related to data security and transparency. Therefore, it is essential that companies develop adaptable strategies for reducing security risks related to the refusal to implement digital procurement. Talent capabilities as a barrier to adopting digital procurement: The study “Trends in Procurement Evolution” highlights the importance of talent capabilities for digital procurement strategies. However, the knowledge gap regarding procurement transformation is the biggest barrier. People lack necessary knowledge and skills, and internal capabilities may hinder adoption. Despite increasing demand for digital skills, only 25% of organizations have the required resources and competencies. Therefore, organizations must adopt training plans to address the digital knowledge gap (Motaung & Sifolo, 2023 ; Kramer, 2016). Rigid procurement operating models as a barrier to adoption of digital procurement: Digital procurement proponents argue that its non-adoption is due to factors such as unsuitable software platforms, organizational resistance, lack of strategic systems integration, and failure to involve procurement professionals, as well as ineffective digital business strategies and lack of an accepted technology standard (Roman & Thai, 2013 ; Panda & Sahu, 2012; Remane et al., 2017). Immature processes as a barrier to adoption of digital procurement: Digital procurement proponents argue that its non-adoption is due to factors such as unsuitable software platforms, organizational resistance, lack of strategic systems integration, and failure to involve procurement professionals, as well as ineffective digital business strategies and lack of an accepted technology standard (Motaung & Sifolo, 2023 ; Remane et al., 2017). Resistance to change as a barrier to the adoption of digital procurement: The adoption of digital technologies in organizations is hindered by a lack of change management capability, with only 13% possessing mature technology change management abilities. Resistance to change is a significant barrier to digital procurement adoption, and change management is crucial for preparing employees for the changes and preparing them for the impact on their jobs. The Oliver Wyman Group recommends intensive change management with stakeholders and should be included in the digitalization implementation roadmap. There is no consensus among researchers on the main factors affecting digital procurement adoption, but most scholars argue that the benefits outweigh the effects of non-adoption, making digital procurement beneficial for organizations (Motaung & Sifolo, 2023 ; Glas & Kleemann, 2016).

A research carried out by Jahani et al. (2021), highlights some challenges associated with adopting Industry 4.0 technologies in procurement processes within the supply chain. Some of the challenges include Cost of Implementation: One of the significant challenges is the high cost associated with implementing Industry 4.0 technologies. For example, technologies like Artificial Intelligence can be expensive to implement and maintain, which may deter companies from investing in these technologies. Uncertainty of Benefits: Companies may be hesitant to invest in Industry 4.0 technologies due to uncertainty about the expected benefits. It can be challenging for companies to determine whether the expensive applications will meet their expectations and provide the desired value. Lack of Comprehensive Frameworks: There is a lack of comprehensive frameworks and systematic approaches for the implementation of Industry 4.0 in procurement processes. This lack of guidance can make it difficult for companies to effectively adopt and integrate these technologies. Digital Transformation Readiness: Companies may face challenges in preparing their supply chain environment for digital transformation. This includes ensuring that the necessary infrastructure, resources, and capabilities are in place to support the adoption of Industry 4.0 technologies. Complexity of Supply Chain

Activities: The involvement of complex supply chain activities and multi-criteria decision-making in procurement processes can pose challenges when integrating Industry 4.0 technologies. Companies may struggle to navigate the complexities of these processes. Addressing these challenges requires careful planning, strategic decision-making, and a clear understanding of how Industry 4.0 technologies can bring value to procurement processes within the supply chain (Glas & Kleemann, 2016 ; Jahani et al., 2021).

2. Strategies of Implementation

It is a complicated occurrence for any organization to attain its strategic goals by putting in place a standard procurement system and adopting well-integrated processes. A challenge that persists is navigating digital procurement processes to achieve supply-chain resiliency. included within the Technology Acceptance Model (TAM), a crucial concept for comprehending the factors that influence people's decisions to accept or reject technology. Overall, the study discusses the implementation of digital procurement strategies within the context of Industry 4.0. It highlights the importance of leveraging digital technologies such as artificial intelligence, blockchain, Internet of Things, cognitive computing, and robotic process automation in the procurement process. The strategies involve embracing digital transformation, generating economies for stakeholders, making timely decisions, leveraging digital disruption activities, predicting demand, creating value for users, strengthening supplier relationships, and focusing on strategic activities. Additionally, as well emphasizes the need to address barriers such as talent capabilities, rigid procurement operating models, immature processes, resistance to change, costs, and security concerns to successfully (Motaung & Sifolo, 2023).

A suitable Technology Acceptance Model (TAM) in the procurement sector is necessary for the selection of various I4.0 technologies. This TAM specifies many indicators to assess the acceptance and implementation of each technology (Brandon-Jones & Kauppi, 2018).

According to Jiri Tupa's and All article, the implementation of Industry 4.0 in the context of risk management. It highlights the need for new approaches to manage risks in the era of Industry 4.0, where the integration of IT and key infrastructure for digitalization of manufacturing poses new potential dangers. The study suggests a framework for implementing an integrated management system that combines elements of Enterprise Risk Management (ERM) and Information Security Management System (ISMS) to address risks associated with Industry 4.0. The framework is based on the Deming PDCA cycle (Plan-Do-Check-Act) and emphasizes the importance of identifying, analyzing, and responding to risks to achieve project objectives. Additionally, it discusses the integration of performance and risk management, emphasizing the need to connect Key Performance Indicators (KPIs) and Key Risk Indicators (KRIs) to enhance risk management in relation to company performance (Tupa et al., 2017).

A study conducted by Niloofar Jahani and All, provides insights into the implementation of Industry 4.0 technologies in procurement processes within the supply chain. Some key strategies for implementing Industry 4.0 include: Identifying Appropriate Technologies: Selecting different Industry 4.0 technologies such as Artificial Intelligence, Big Data Analytics, Internet of Things, Blockchain, and others based on the specific needs and challenges faced in procurement processes. Technology Acceptance Model (TAM): Utilizing a Technology Acceptance Model to determine the adoption and application of each technology in the procurement domain. Focus on Specific Functions and Processes: Emphasizing functions and processes like e-sourcing, contract management, and e-procurement to enhance the efficiency and effectiveness of procurement operations. Building Technological Resources: Managers should focus on building technological resources within their companies to support the adoption of Industry 4.0 technologies. Aligning Values with Supply Chain Performance Measures: Ensuring that the proposed values from Industry 4.0 technologies align with supply chain performance measures and strategies. Developing a Strategic Management Framework: Developing a strategic management framework to identify the most important values to be developed in the supply chain and determining which technologies can be implemented based on the supply chain's infrastructure. These strategies aim to leverage Industry 4.0 technologies to enhance procurement processes, improve visibility, and resilience within the supply chain (Jahani et al., 2021).

3. Keys Considerations for Implementing Blockchain in Materials Procurement

The manufacturing sector has undergone a significant transformation in recent years, primarily influenced by the emergence of pioneering technologies like Blockchain, known as a distributed ledger technology, presents a promising resolution to tackle the challenges encountered in materials procurement within the supply chain (Li et al., 2020).

An essential advantage of integrating blockchain into materials procurement is the heightened transparency and traceability it delivers (Abeyratne & Monfared, 2016). By utilizing blockchain, the monitoring of materials and products across the supply chain is enabled, thereby ensuring the genuineness and origin of raw materials (Song et al., 2019). This aspect is especially critical in sectors where quality assurance and regulatory adherence hold significant importance.

Furthermore, systems based on blockchain can aid in secure and tamper-proof transactions among supply chain participants, thus diminishing the risk of fraud and enhancing the overall efficiency of the procurement process (Pulmano et al., 2023). The decentralized structure of blockchain permits real-time visibility and coordination among various stakeholders, fostering improved decision-making and proactive responses to disruptions (Song et al., 2019).

Although the potential advantages of employing blockchain in materials procurement are considerable, there exist numerous challenges that necessitate attention (Pulmano et al., 2023). For example, the integration of blockchain with current enterprise resource planning (ERP) systems and legacy infrastructure can present a significant obstacle, demanding extensive technical know-how and organizational preparedness (Zhang, 2020). Additionally, the scalability and compatibility of blockchain solutions with the expanding volumes of data and transactions in the supply chain need to be thoughtfully evaluated (Zhang, 2020).

VII. PERSPECTIVES AND FUTURE DIRECTIONS

This section provides insights into the perspectives and future directions related to the adoption of Industry 4.0 technologies in procurement processes within the supply chain. Some key perspectives and future directions include:

1. Value Propositions: The upstream studies emphasize the importance of understanding the value propositions offered by Industry 4.0 technologies in procurement. Future research can focus on further exploring and defining these value propositions to help procurement managers better understand the benefits of implementing these technologies.

2.Change management: In terms of future guidance, the importance of understanding the predictors of human behavior in terms of acceptance or rejection of technologies, such as electronic procurement systems, in the context of organizational success is emphasized, as not translating the initial adoption decision into individual acceptance of electronic public procurements can lead to some complexity in the workplace.

3. Risk management: As well, in terms of future directions, the literature suggests the need for a tool to link key performance indicators (KPIs) and key risk indicator (KRIs) to improve risk management in relation to the measurement of company performance. It also emphasizes the importance of adapting existing risk management tools and measures to manage the increasing volumes of data and real-time data availability that accompany Industry 4.0. The authors indicate that a future research objective is to develop a new framework based on relevant data analysis impacted by Industry 4.0. This suggests a continued focus on improving risk management strategies to align with the changing landscape of digital manufacturing processes.

4.Integration of Technologies: Future studies can explore the integration of different Industry 4.0 technologies, such as Artificial Intelligence, Big Data Analytics, Internet of Things, and Blockchain, to optimize procurement processes and enhance supply chain performance. There are now more opportunities thanks to digital technology.

5.Measurement and Evaluation: Future studies can focus on developing metrics and evaluation methods to measure the effectiveness of Industry 4.0 technologies in procurement. This can help companies assess the ROI and performance improvements resulting from the adoption of these technologies.

6.Strategic Management Frameworks: There is a need for the development of strategic management frameworks to guide companies in implementing Industry 4.0 technologies in procurement processes. Future research can focus on creating comprehensive frameworks that provide step-by-step guidance for successful adoption. Companies should consider the future role of procurement and supply chain network nodes when aligning their procedures and processes toward the smart supply chain, since smart development has a favorable influence on procurement. The Accenture study goes on to say that digital technologies will compel businesses to build relationships with strategic suppliers in order to become "virtually integrated enterprises." It also claims that the Internet of Things will improve procurement procedures and that "cognitive systems" will replace routine and manual labor, enabling "better" and faster decision-making.

7.Supply Chain Resilience: Exploring how Industry 4.0 technologies can enhance supply chain resilience and agility in the face of disruptions and uncertainties is another important future direction. Understanding how these technologies can improve supply chain responsiveness and adaptability is crucial for companies operating in dynamic environments.

By addressing these perspectives and future directions, companies can better leverage Industry 4.0 technologies to enhance procurement processes, improve supply chain performance, and drive sustainable growth in the digital era (Tan, 2001 ; Kramer, 2016; Jahani et al., 2021 ; Motaung & Sifolo, 2023; Tupa et al., 2017).

Blockchain technology is swiftly reshaping various sectors, with procurement being no exemption. Its capacity to revolutionize the sourcing and management of materials holds significant importance. This article delves into the existing viewpoints and upcoming trends in leveraging blockchain for material procurement.

Current Perspectives

Augmented Transparency and Traceability: Blockchain provides an unchangeable log of transactions, granting full visibility into the entire supply chain. This enables the tracking of materials from their source to their destination, ensuring genuineness and pinpointing potential counterfeits or ethically questionable origins.

Efficient procedures: Smart contracts, defined as self-executing agreements recorded on the blockchain, possess the capability to automate standard procurement activities. This results in a reduction of manual validation, acceleration of endorsement processes, and minimization of inaccuracies.

Enhanced Data Security: The decentralized nature of blockchain renders data manipulation nearly unfeasible. This promotes trust and secure information exchange among all involved parties.

Cost Reduction: The optimization of processes, automation of tasks, and prevention of fraud can result in substantial cost savings in procurement.

Future Directions

Standardization and Interoperability: Establishing universal standards for integrating blockchain across diverse procurement platforms is imperative for broader acceptance.

Scalability and Efficiency: Blockchain technology must advance to manage the vast data volumes linked with intricate supply chains. Enhanced scalability and transaction processing speed are fundamental.

Applications for Sustainability: Blockchain can oversee the origins of materials and ensure ethical sourcing methods. This aligns with the increasing emphasis on sustainable procurement practices.

Integration with Artificial Intelligence (AI): Merging blockchain with AI can give rise to intelligent procurement systems that scrutinize past data to optimize future acquisitions and identify potential risks.

Regulatory Frameworks: With the expansion of blockchain adoption, explicit regulations tailored to procurement must be implemented to ensure legal certainty and safeguard consumers.

VIII. CONCLUSION

This study highlights the need for digital transformation and the adoption of digital technologies to improve operational efficiency and customer experience while protecting information and systems, providing insight into the advantages and obstacles of digital purchases.

The integration of digital technologies into various aspects of procurement processes in general and material procurements in particular and organizational outcomes is not an easy task. Given the conclusions of this chapter, it is proposed that there be a concerted effort to ensure that institutional structures are resilient, adaptive and durable in order to overcome obstacles and maximize the benefits of digital purchases.

Blockchain technology presents vast potential for reshaping material procurement. By promoting transparency, streamlining processes, and bolstering data security, it can establish a more efficient, dependable, and sustainable procurement environment. Nevertheless, tackling obstacles like scalability and formulating explicit regulations are pivotal for its widespread triumph. With the evolution of technology and advancing industry cooperation, blockchain stands to become the cornerstone of future procurement methodologies.

REFERENCES

Abeyratne, S. A., & Monfared, R. P. (2016). Blockchain ready manufacturing supply chain using distributed ledger. *International Journal of Research in Engineering and Technology*, *5*(9), 1–10.

Ada, N.. (2021). Blockchain technology for enhancing traceability and efficiency in automobile supply chain—A case study. *Sustainability*, *13*(24), 13667. Advance online publication. DOI: 10.3390/su132413667

Angkiriwang, R., Pujawan, I. N., & Santosa, B. (2014). Managing uncertainty through supply chain flexibility: Reactive vs. proactive approaches. *Production & Manufacturing Research*, *2*(1), 50–70. DOI: 10.1080/21693277.2014.882804

Anner, M. (2020). *Abandoned? The impact of Covid-19 on workers and businesses at the bottom of global garment supply chains*. Center for Global Workers' Rights. https://www.workersrights.org/wp-content/uploads/2020/03/Abandoned-Penn-State-WRC-Report.pdf

Ashwell, M. L. (2017). The digital transformation of intelligence analysis. *Journal of Financial Crime*, *24*(3), 393–411. DOI: 10.1108/JFC-03-2017-0020

Ashwell, M. L. (2017). The digital transformation of intelligence analysis. *Journal of Financial Crime*, *24*(3), 393–411. DOI: 10.1108/JFC-03-2017-0020

Bakar, N. A., Peszynski, K., & Sundram, V. P. K. (2016). Abridgment of traditional procurement and e-procurement: Definitions, tools and benefits. *Journal of Emerging Economies and Islamic Research*, *4*(1). 74–91. DOI: 10.24191/jeeir.v4i1.6284

Bienhaus, F., & Haddud, A. (2018). Procurement 4.0: Factors influencing the digitisation of procurement and supply chains. *Business Process Management Journal*, *24*(4), 965–984. DOI: 10.1108/BPMJ-06-2017-0139

Boston Consulting Group. (2023). *How to mitigate supply chain risk*. https://www.bcg.com/publications/2023/how-to-mitigate-supply-chain-risk

Brandon-Jones, A., & Kauppi, K. (2018). Examining the antecedents of the technology acceptance model within e-procurement. *International Journal of Operations & Production Management*, *38*(1), 22–42. DOI: 10.1108/IJOPM-06-2015-0346

Costa, A. A., Arantes, A., & Tavares, L. V. (2013). Evidence of the impacts of public e-procurement: The Portuguese experience. *Journal of Purchasing and Supply Management*, *19*(4), 238–246. DOI: 10.1016/j.pursup.2013.07.004

Cuc, S. (2023). Unlocking the potential of blockchain technology in the textile and fashion industry. *FinTech*, *2*(2), 311–326. DOI: 10.3390/fintech2020018

[] Elmay, F. K., Salah, K., Jayaraman, R., & Omar, I. A. (2021). Using NFTs and blockchain for traceability and auctioning of shipping containers and cargo in maritime industry. *IEEE Access*. https://doi.org/ (Note: Complete DOI may vary; verify latest version)DOI: 10.1109/ACCESS.2021.XXXX

Fatorachian, H., & Kazemi, H. (2021). Impact of Industry 4.0 on supply chain performance. *Production Planning and Control*, *32*(1), 63–81. DOI: 10.1080/09537287.2020.1712487

Garg, P., Gupta, B., Chauhan, A. K., Sivarajah, U., Gupta, S., & Modgil, S. (2021). Measuring the perceived benefits of implementing blockchain technology in the banking sector. *Technological Forecasting and Social Change*, *163*, 120407. Advance online publication. DOI: 10.1016/j.techfore.2020.120407

Ghadimi, P., Wang, C., Lim, M. K., & Heavey, C. (2019). Intelligent sustainable supplier selection using multi-agent technology: Theory and application for Industry 4.0 supply chains. *Computers & Industrial Engineering*, *127*, 588–600. DOI: 10.1016/j.cie.2018.10.050

Glas, A. H., & Kleemann, F. C. (2016). The impact of Industry 4.0 on procurement and supply management: A conceptual and qualitative analysis. *International Journal of Business Management and Invention*, *5*(6), 55–66. http://www.ijbmi.org

Glas, A. H., & Kleemann, F. C. (2016). The impact of Industry 4.0 on procurement and supply management: A conceptual and qualitative analysis. *International Journal of Business Management and Invention*, *5*(6), 55–66. http://www.ijbmi.org

Guerpinar, T., Guadiana, G., Ioannidis, P. A., Straub, N., & Henke, M. (2021). The current state of blockchain applications in supply chain management. In *ACM International Conference Proceeding Series* (pp. 168–175). Association for Computing Machinery. https://doi.org/DOI: 10.1145/3460537.3460568

Hahn, G. J. (2020). Industry 4.0: A supply chain innovation perspective. *International Journal of Production Research*, *58*(5), 1425–1441. DOI: 10.1080/00207543.2019.1641642

Handfield, R., Jeong, S., & Choi, T. (2019). Emerging procurement technology: Data analytics and cognitive analytics. *International Journal of Physical Distribution & Logistics Management*, *49*(10), 972–1002. DOI: 10.1108/IJPDLM-11-2017-0348

He, Y., Zhou, Z., Pan, Y., Chong, F., Wu, B., Xiao, K., & Li, H. (2024). Review of data security within energy blockchain: A comprehensive analysis of storage, management, and utilization. *High-Confidence Computing. High-Confidence Computing, 100233*(3), 100233. Advance online publication. DOI: 10.1016/j.hcc.2024.100233

Institute of Electrical and Electronics Engineers. (2019). *Proceedings of 2019 8th International Conference on Industrial Technology and Management: ICITM 2019, March 2-4, 2019, Cambridge, UK.*

Jahani, N., Sepehri, A., Vandchali, H. R., & Tirkolaee, E. B. (2021). Application of industry 4.0 in the procurement processes of supply chains: A systematic literature review. *Sustainability (Basel), 13*(14), 7520. Advance online publication. DOI: 10.3390/su13147520

Jahani, N., Sepehri, A., Vandchali, H. R., & Tirkolaee, E. B. (2021). Application of industry 4.0 in the procurement processes of supply chains: A systematic literature review. *Sustainability (Basel), 13*(14), 7520. Advance online publication. DOI: 10.3390/su13147520

Jahani, N., Sepehri, A., Vandchali, H. R., & Tirkolaee, E. B. (2021). Application of industry 4.0 in the procurement processes of supply chains: A systematic literature review. *Sustainability (Basel), 13*(14), 7520. Advance online publication. DOI: 10.3390/su13147520

Jamwal, A., Agrawal, R., Sharma, M., & Giallanza, A. (2021). Industry 4.0 technologies for manufacturing sustainability: A systematic review and future research directions. *Applied Sciences (Basel, Switzerland), 11*(12), 5725. DOI: 10.3390/app11125725

Jumas, D., Mohd-Rahim, F. A., Ariani, V., & Florisa, A. N. (2023). Obtaining professional skills in the public construction procurement system: A challenge from the viewpoint of Indonesian government worker. *Journal of Design and Built Environment, 23*(1), 48–58. DOI: 10.22452/jdbe.vol23no1.4

Kademaunga, C. K., & Phiri, J. (2019). Factors affecting successful implementation of electronic procurement in government institutions based on the technology acceptance model. *Open Journal of Business and Management, 7*(4), 1705–1714. DOI: 10.4236/ojbm.2019.74118

Karmakar, A., Ghosh, P., Banerjee, P. S., & De, D. (2023). ChainSure: Agent free insurance system using blockchain for Healthcare 4.0. *Intelligent Systems with Applications, 17*, 200177. Advance online publication. DOI: 10.1016/j.iswa.2023.200177

Katsikouli, P., Wilde, A. S., Dragoni, N., & Høgh-Jensen, H. (2021). On the benefits and challenges of blockchains for managing food supply chains. *Journal of the Science of Food and Agriculture*, *101*(6), 2175–2181. DOI: 10.1002/jsfa.10883 PMID: 33063342

Korucuk, S., Tirkolaee, E. B., Aytekin, A., Karabasevic, D., & Karamaşa, Ç. (2023). Agile supply chain management based on critical success factors and most ideal risk reduction strategy in the era of Industry 4.0: Application to plastic industry. *Operations Management Research : Advancing Practice Through Research*, *16*(4), 1698–1719. DOI: 10.1007/s12063-023-00360-5

Korucuk, S., Tirkolaee, E. B., Aytekin, A., Karabasevic, D., & Karamaşa, Ç. (2023). Agile supply chain management based on critical success factors and most ideal risk reduction strategy in the era of Industry 4.0: Application to plastic industry. *Operations Management Research : Advancing Practice Through Research*, *16*(4), 1698–1719. DOI: 10.1007/s12063-023-00360-5

Kramer, G. (2016). Electronic public procurement as a tool of reform in South African public procurement. *African Public Procurement Law Journal*, *3*(1), 1–43. DOI: 10.14803/3-1-16

Lee, I., & Lee, K. (2015). The Internet of Things (IoT): Applications, investments, and challenges for enterprises. *Business Horizons*, *58*(4), 431–440. DOI: 10.1016/j.bushor.2015.03.008

Lee, I., & Lee, K. (2015). The Internet of Things (IoT): Applications, investments, and challenges for enterprises. *Business Horizons*, *58*(4), 431–440. DOI: 10.1016/j.bushor.2015.03.008

Li, J., Maiti, A., Springer, M., & Gray, T. (2020). Blockchain for supply chain quality management: Challenges and opportunities in context of open manufacturing and industrial internet of things. *International Journal of Computer Integrated Manufacturing*, *33*(12), 1321–1355. DOI: 10.1080/0951192X.2020.1815853

Lindskog, H., Vegas, N., Babb, J., & Brege, S. (n.d.). Will electronic procurement change the public sector's purchasing behaviour? In *9th Annual ISOneWorld Conference: Issues in Information Systems*.

Liotine, M., & Ginocchio, D. (2019). The supply blockchain: Integrating blockchain technology within supply chain operations. In *Technology in supply chain management and logistics: Current practice and future applications* (pp. 57–89). Elsevier., DOI: 10.1016/B978-0-12-815956-9.00004-1

Mahyuni, L. P., Adrian, R., Darma, G. S., Krisnawijaya, N. N. K., Dewi, I. G. A. A. P., & Permana, G. P. L. (2020). Mapping the potentials of blockchain in improving supply chain performance. *Cogent Business & Management*, *7*(1), 1788329. Advance online publication. DOI: 10.1080/23311975.2020.1788329

Mishra, A. N., Devaraj, S., & Vaidyanathan, G. (2013). Capability hierarchy in electronic procurement and procurement process performance: An empirical analysis. *Journal of Operations Management*, *31*(6), 376–390. DOI: 10.1016/j.jom.2013.07.011

Motaung, J. R., & Sifolo, P. P. S. (2023). Benefits and barriers of digital procurement: Lessons from an airport company. *Sustainability (Basel)*, *15*(5), 4610. Advance online publication. DOI: 10.3390/su15054610

Oh, S., Yang, H., & Kim, S. W. (2014). Managerial capabilities of information technology and firm performance: Role of e-procurement system type. *International Journal of Production Research*, *52*(15), 4488–4506. DOI: 10.1080/00207543.2013.867084

Panda, P., & Sahu, G. P. (2012). *E-procurement implementation: Critical analysis of success factors' impact on project outcome. SSRN*. https://ssrn.com/abstract=2019575 DOI: 10.2139/ssrn.2019575

Pereira, C. R., Christopher, M., & Lago Da Silva, A. (2014). Achieving supply chain resilience: The role of procurement. *Supply Chain Management*, *19*(5/6), 626–642. DOI: 10.1108/SCM-09-2013-0346

Piyathanavong, V., Huynh, V. N., Karnjana, J., & Olapiriyakul, S. (2024). Role of project management on Sustainable Supply Chain development through Industry 4.0 technologies and Circular Economy during the COVID-19 pandemic: A multiple case study of Thai metals industry. *Operations Management Research : Advancing Practice Through Research*, *17*(1), 13–37. DOI: 10.1007/s12063-022-00283-7

Pulmano, C. E., Estuar, M. R. J. E., De Leon, M. M., Tan, H. C. L., Co, N. A. S., & Tamayo, L. P. V. (2023). Towards the development of a blockchain-based decentralized digital credential system using hyperledger fabric for participatory governance. *Procedia Computer Science*, *219*, 99–106. DOI: 10.1016/j.procs.2023.01.269

Radcliffe, B. (n.d.). *The basics of tariffs and trade barriers*. Investopedia. https://www.investopedia.com/articles/economics/08/tariff-trade-barrier-basics.asp

Remane, G., Hanelt, A., Nickerson, R. C., & Kolbe, L. M. (2017). Discovering digital business models in traditional industries. *The Journal of Business Strategy*, *38*(2), 41–51. DOI: 10.1108/JBS-10-2016-0127

Riley-Reid, T. D. (2015). The hidden cost of digitization – Things to consider. *Collection Building*, *34*(3), 89–93. DOI: 10.1108/CB-01-2015-0001

Roman, A. V., & Thai, K. V. (2013). Public procurement in the United States. In *Exzellente öffentliche Beschaffung* (pp. 245–258). Springer., DOI: 10.1007/978-3-658-00567-2_14

Sali, N., Biwalkar, S., Swamy, A., Khadilkar, N., & Sahu, S. (2023). Blockchain in food supply chain: Benefits, challenges and review. *International Journal for Research in Applied Science and Engineering Technology*, *11*(4), 4654–4661. DOI: 10.22214/ijraset.2023.50661

Song, J. M., Sung, J., & Park, T. (2019). Applications of blockchain to improve supply chain traceability. *Procedia Computer Science*, *162*, 119–122. DOI: 10.1016/j.procs.2019.11.266

Tan, K. C. (2001). A framework of supply chain management literature. *European Journal of Purchasing & Supply Management*, *7*(1), 39–48. DOI: 10.1016/S0969-7012(00)00020-4

Tan, K. C. (2001). A framework of supply chain management literature. *European Journal of Purchasing & Supply Management*, *7*(1), 39–48. DOI: 10.1016/S0969-7012(00)00020-4

Tupa, J., Simota, J., & Steiner, F. (2017). Aspects of risk management implementation for Industry 4.0. *Procedia Manufacturing*, *11*, 1223–1230. DOI: 10.1016/j.promfg.2017.07.248

Tupa, J., Simota, J., & Steiner, F. (2017). Aspects of risk management implementation for Industry 4.0. *Procedia Manufacturing*, *11*, 1223–1230. DOI: 10.1016/j.promfg.2017.07.248

Wanner, J., Heinrich, K., & Zschech, P. (2020). *How much AI do you require? Decision factors for adopting AI technology*. https://www.researchgate.net/publication/344350604

Wanner, J., Heinrich, K., & Zschech, P. (2020). *How much AI do you require? Decision factors for adopting AI technology*. https://www.researchgate.net/publication/344350604

You, F., Wassick, J. M., & Grossmann, I. E. (2008). Risk management for a global supply chain planning under uncertainty: Models and algorithms. *AIChE Journal. American Institute of Chemical Engineers*, *55*(4), 931–946. DOI: 10.1002/aic.11721

Zhang, J. (2020). Deploying blockchain technology in the supply chain. In *Computer security threats*. IntechOpen., DOI: 10.5772/intechopen.86530

Chapter 6
Magnetic Materials: Advances and Applications in Modern Industry

Dipan Kumar Das
http://orcid.org/0000-0003-0224-3295
Centurion University of Technology and Management, India

Padmaja Patnaik
http://orcid.org/0000-0003-4468-7871
Centurion University of Technology and Management, India

Nibedita Nayak
Centurion University of Technology and Management, India

Siddharth Parthasarathy
http://orcid.org/0000-0001-8319-3239
Centurion University of Technology and Management, India

Sudip Kumar Das
http://orcid.org/0009-0003-0814-0884
Centurion University of Technology and Management, India

Niharika Das
Centurion University of Technology and Management, India

Shalom Akhai
http://orcid.org/0000-0002-7533-457X
Maharishi Markandeshwar Engineering College, Maharishi Markandeshwar (Deemed to be) University, Mullana, India

ABSTRACT

Magnetic materials have revolutionized modern industry with their versatile applications across various sectors. This chapter explores the comprehensive landscape of these materials, starting with their historical evolution and fundamental principles. It delves into traditional and modern synthesis methods, highlighting techniques like sol-gel, sputtering, and vapor deposition. Detailed characterization methods such as X-ray diffraction, SEM, and VSM are discussed to elucidate the properties of magnetic materials. The chapter also examines their pivotal role in

DOI: 10.4018/979-8-3693-6627-1.ch006

© 2027, IGI Global Scientific Publishing. Copying or distributing in print or electronic forms without written permission of IGI Global Scientific Publishing is prohibited. Use of this chapter to [generative] artificial intelligence (AI) technologies is expressly prohibited. The publisher reserves all rights to license its use for generative AI training and machine learning model development.

energy applications, including renewable energy systems and superconducting magnets, and their integration into electronics and sensing technologies. Biomedical applications, such as drug delivery and MRI contrast agents, are explored with a focus on biocompatibility. Emerging trends like quantum materials and spintronics are considered, providing a forward-looking perspective. The concluding part summarizes key advancements, addresses industry challenges, and predicts future research and application trends.

1. INTRODUCTION TO MAGNETIC MATERIALS

The history and evolution of magnetic materials date back over 3000 years, from the discovery of the first natural magnet, lodestone, known as magnetite iron ore (Coey,2010). Magnetism fascinated ancient civilizations and inspired many legends. Around 2500 BC, the discovery of magnetite, or lodestone, in Greece led to stories about its mysterious power, including a tale of a shepherd in Magnesia whose iron shoe nails were pulled toward a rock. Such unexplained attractions made magnetism seem magical. Similar myths, like iron parts being drawn from ship hulls by magnetic rocks in the South Pacific, further deepened its aura of wonder and mystery in early storytelling. Stories of knights in iron armor being disarmed and immobilized were attributed to the magnetic forces of lodestone, highlighting how magnetism was intertwined with legends and myths of ancient times. The magnetic properties of Fe3O4 influenced the naming of magnetite ore in different regions, such as Magnesia stone, lodestone, l'aimant, chumbak, and ts'u she, reflecting the cultural significance and impact of magnetism on language and folklore. The allure of magnetism in ancient stories not only entertained people but also sparked curiosity and exploration, leading to further discoveries and advancements in understanding the natural world. These ancient tales of magnetism laid the foundation for the development of the modern concept of magnets and their applications in various fields, showcasing how ancient stories influenced the scientific understanding and technological progress related to magnetic materials over time. Over the centuries, the study of magnetism has progressed, leading to the development of sophisticated magnetic materials crucial in modern technology, with applications in power generation, communications, and information storage (Zhao et al.,2024). Recent advances have uncovered new classes of magnetic materials exhibiting magneto-thermal, magneto-elastic, magneto-optic, and magneto-electrical couplings, fueling innovation in information technology, wireless communication, and biotechnology. The rise of magnetic topological materials further enables next-generation electronics, particularly memory devices with controllable band topology (Tang and Wan, 2023). The history of magnetic materials spans over 3000 years, from lodestones to modern rare-earth magnets (Polytechnic, 1991). Soft magnetic materials, crucial for transformers and electric machines, have evolved from soft iron to silicon-iron, nickel-iron alloys, and advanced materials like amorphous metallic glasses (Sowter, 1987; Krings et al., 2017). Permanent magnets have seen significant improvements, with the development of metal, ceramic, and rare-earth magnet families in the 20th century, increasing performance two-hundredfold. Measurement techniques for magnetic properties have progressed alongside material development, addressing challenges in traceability, reproducibility, and standardization (Fiorillo, 2010).

Magnetic materials are classified based on their response to external magnetic fields. Soft magnetic materials exhibit high magnetic permeability and low coercivity, making them ideal for applications like transformers and magnetic shielding (Hajalilou et al., 2023). Magnetic materials in the polycrystalline state are made of many tiny crystals and are classified based on their type and crystal structure, which determines their properties and uses. They are mainly divided into soft and hard magnetic materials. Soft magnetic materials are easily magnetized and demagnetized, making them suitable for applications like transformers and electromagnetic devices. Hard magnetic materials have high coercivity and remanence, allowing them to retain magnetization without an external field, which makes them ideal for permanent magnets. Magnetic behaviour is studied using hysteresis loops, which show how magnetization changes with an applied magnetic field. A Vibrating Sample Magnetometer (VSM) is commonly used to measure these properties and analyse magnetic response and energy losses under external magnetic fields (Mushnikov,2022). The paper reviews the fundamentals of magnetism in solid materials, including the origin of magnetic moments and magnetic ordering in localized and itinerant electron systems. It explains key techniques such as electron paramagnetic resonance (EPR) and nuclear magnetic resonance (NMR) for studying magnetic properties, along with methods of generating magnetic fields. It also discusses soft, hard, and semi-hard magnetic materials and their applications, especially in medical technologies. Applications include magneto-caloric systems, magneto-impedance sensors, and magnetoresistance devices. Overall, the paper highlights the vital role of magnetic materials in medicine, biology, and modern technology (Narayanan et al., 2024). Hard magnetic soft materials are smart composites made of a soft polymer matrix mixed with hard magnetic particles. They can deform reversibly under external magnetic fields, making them valuable in soft robotics, adaptive structures, and biomedical devices. Studying hysteresis loops is important to understand their magnetization behaviour, while atomic-level interactions influence key properties such as magnetic domains and susceptibility (Florio, 2022).

Key properties include magnetic field (H), magnetic induction (B), and magnetization, which can be measured using various methods such as open sample measurements and vibrating sample magnetometers. Reliable measurements must meet criteria of uniformity, correspondence, and direction (Abderahmane & Daniel, 2023). Challenges in measurement arise from factors like sample shape, magnetic circuit configuration, and material anisotropy (Petrescu, 2009). Modern digital techniques and the development of novel materials have driven advancements in measurement methodologies. Standards play a crucial role in ensuring measurement reproducibility and traceability (Fiorillo, 2010). The characterization of both soft and hard magnetic materials is essential for understanding their properties and applications, with recent focus on high-performance materials like rare-earth magnets and nanostructures (Fiorillo, 2004). Magnetic nanoparticles with their applications, properties, synthesis methods described in Table 1.

Table 1. Magnetic nanoparticles with their applications, properties, synthesis methods

Magnetic Nanoparticles	Applications	Key Properties	Common Synthesis Methods
Iron Oxide (Fe_3O_4)	Biomedical (drug delivery, MRI), catalysis, data storage	Super-paramagnetism, biocompatibility, high magnetization	Co-precipitation, thermal decomposition, solvothermal
Cobalt Ferrite ($CoFe_2O_4$)	Magnetic recording, microwave devices, catalysis	High coercivity, high saturation magnetization	Sol-gel, hydrothermal, combustion
Nickel Ferrite ($NiFe_2O_4$)	Magnetic fluids, microwave devices, catalysis	High Curie temperature, good chemical stability	Co-precipitation, microemulsion, hydrothermal

Figure 1. Classification of magnetic materials

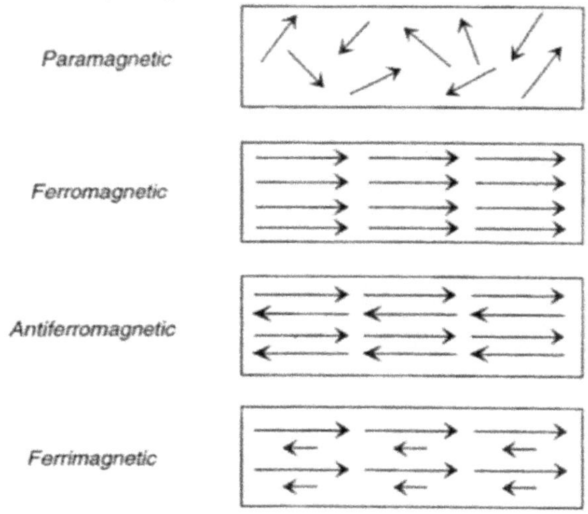

(https://www.researchgate.net/figure/Figure-A3-Classification-of-magnetic-materials-10_fig118_310770506)

2. SYNTHESIS OF MAGNETIC MATERIALS

Magnetic nanoparticles are synthesized using various physical, chemical, and biological methods, each offering distinct advantages. Common techniques include sol-gel, co-precipitation, hydrothermal, sputtering, laser ablation, chemical vapor deposition (CVD), thermal decomposition, microwave-assisted synthesis, mechanical alloying, electrochemical, biological, pyrolysis, aerosol, and emulsion methods. The table 2 provides a comparative overview of various magnetic material synthesis methods.

Table 2. Comparative Analysis of Magnetic Nanoparticle Synthesis Techniques

Synthesis Method	Description	Advantages	Disadvantages	Particle Size	Morphology	Applications
Sol-Gel	Versatile technique for synthesizing magnetic materials	Controlled properties, room-temperature synthesis	Limited scalability	5-50 nm	Spherical, nanoparticles	Catalysis, drug delivery, neurodegenerative diseases
Co-Precipitation	Widely used method for synthesizing magnetic nanomaterials	Simple, low-cost, controllable conditions	Limited control over particle size	10-100 nm	Spherical, nanoparticles	Biosynthesis, catalysis, electronics, environmental treatment
Hydrothermal	Simple, cost-effective, and environmentally friendly method	High specific surface area, strong magnetic responsivity, biocompatibility	Limited control over particle size	20-200 nm	Rod-shaped, nanoparticles	Biosensors, drug delivery, data storage, hyperthermia treatment
Sputtering	High-rate deposition of magnetic materials	High deposition rates, uniform film deposition	Substrate heating, high-energy ion bombardment	10-100 nm	Thin films, coatings	Soft and hard magnetic coatings
Laser Ablation	Innovative method for fabricating and manipulating magnetic nanostructures	Oriented nanowires, nanocomposites, epitaxial growth	Limited control over particle size, complex equipment	10-100 nm	Nanowires, nanoparticles	Biotechnology, catalysis, magneto-optics
Chemical Vapor Deposition (CVD)	Versatile method for fabricating magnetic thin films and structures	Large arrays of micron-scale magnetic features, controllable magnetic vector orientation	Complex equipment, limited scalability	10-100 nm	Thin films, coatings	High-density recording media, magnetic materials with customized properties
Thermal Decomposition	Synthesis of magnetic nanoparticles through decomposition of precursors	High crystallinity, uniform particle size	Limited control over particle shape	5-50 nm	Spherical, nanoparticles	Catalysis, drug delivery, biomedical applications
Microwave-Assisted Synthesis	Rapid synthesis of magnetic materials using microwave radiation	Fast reaction times, high yields	Limited control over particle size, non-uniform heating	10-100 nm	Spherical, nanoparticles	Biomedical applications, catalysis, energy storage
Mechanochemical Synthesis	Synthesis of magnetic materials through mechanical alloying	High-energy ball milling, uniform particle size	Limited control over particle shape, contamination risk	10-100 nm	Spherical, nanoparticles	Permanent magnets, magnetic sensors, biomedical applications
Electrochemical Synthesis	Synthesis of magnetic materials through electrochemical reactions	High purity, controlled particle size	Limited scalability, complex equipment	10-100 nm	Thin films, coatings	Energy storage, biomedical applications, sensors
Biological Synthesis	Synthesis of magnetic materials using biological systems	Biocompatibility, environmentally friendly	Limited control over particle size, slow reaction times	10-100 nm	Spherical, nanoparticles	Biomedical applications, biosensors, drug delivery

continued on following page

Table 2. Continued

Synthesis Method	Description	Advantages	Disadvantages	Particle Size	Morphology	Applications
Pyrolysis	Synthesis of magnetic materials through thermal decomposition of precursors	High crystallinity, uniform particle size	Limited control over particle shape, high temperatures required	5-50 nm	Spherical, nanoparticles	Catalysis, drug delivery, biomedical applications
Aerosol Synthesis	Synthesis of magnetic nanoparticles through aerosol processing	High purity, uniform particle size	Limited control over particle shape, complex equipment	10-100 nm	Spherical, nanoparticles	Biomedical applications, catalysis, energy storage
Emulsion Synthesis	Synthesis of magnetic nanoparticles through emulsion processing	High yields, uniform particle size	Limited control over particle shape, surfactant requirements	10-100 nm	Spherical, nanoparticles	Biomedical applications, drug delivery, biosensors
Microwave-Assisted Synthesis	Rapid synthesis of magnetic materials using microwave radiation	Fast reaction times, high yields	Limited control over particle size, non-uniform heating	10-100 nm	Spherical, nanoparticles	Biomedical applications, catalysis, energy storage
Mechanochemical Synthesis	Synthesis of magnetic materials through mechanical alloying	High-energy ball milling, uniform particle size	Limited control over particle shape, contamination risk	10-100 nm	Spherical, nanoparticles	Permanent magnets, magnetic sensors, biomedical applications

2.1. Overview of Traditional and Modern Synthesis Methods

Magnetic nanoparticles (MNPs) have garnered significant attention due to their unique properties and diverse applications. Various synthesis methods have been developed to produce MNPs with controlled size, shape, and stability. These include co-precipitation, thermal decomposition, solvo-thermal, and microwave-assisted techniques, among others (Majidi et al., 2016). Magnetic fields have emerged as a novel parameter in the synthesis and assembly of nanomaterials, influencing their structure and properties (Lin Hu et al., 2014). The integration of MNPs into functional nanocomposites has been achieved through various solution-based methods, resulting in multicomponent nanostructures and matrix-dispersed composites (Behrens, 2011). Magnetic oxide nanomaterials, such as iron, nickel, and cobalt oxides, exhibit unique magnetic properties at the nanoscale, with each particle acting as a single magnetic domain (Tripathi et al., 2015; Das et al., 2024).

2.2. Chemical Synthesis Routes: Sol-Gel, Co-Precipitation, Hydrothermal

The sol-gel method is a versatile technique for synthesizing magnetic materials with controlled properties and applications. It allows for the production of nanoscale ferri-, para-, and superparamagnetic materials with unique characteristics (J. Masi & S. Masi, 1997). This method has been used to synthesize zinc-substituted cobalt ferrite magnetic nanoparticles, demonstrating the ability to tailor magnetic properties through composition (Ranjani et al., 2016). The sol-gel approach enables the encapsulation of various materials, including metallic and magnetic nanoparticles, within bulk, nano-, and mesostructured materials, opening up possibilities for applications in catalysis, drug delivery, and treatment of neurodegenerative diseases (Aguilar, 2019). Furthermore, the sol-gel technique can be used to produce uniform silica spheres of submicron size doped with magnetic particles, offering potential for biological applications due to their biocompatibility and non-toxicity (Kornak et al., 2005).

Chemical co-precipitation is a widely used method for synthesizing magnetic nanomaterials, particularly iron oxide (Fe3O4) and cobalt ferrite nanoparticles, due to its simplicity, low cost, and controllable conditions (Li et al., 2023). The process typically involves mixing metal salts in alkaline solutions, with parameters such as temperature, base addition method, and inert gas use affecting the final product characteristics (Al-Alawy et al., 2018). X-ray diffraction (XRD), scanning electron microscopy (SEM), and vibrating sample magnetometry (VSM) are commonly employed to characterize the resulting nanoparticles' structure, morphology, and magnetic properties (Rathore et al., 2012). Particle sizes ranging from 6.8 to 29 nm have been reported, with crystallite sizes influenced by reaction temperature (Khanam et al., 2022).

Hydrothermal synthesis has emerged as a popular method for producing nanoscale magnetic materials due to its simplicity, cost-effectiveness, and environmental friendliness (Li et al., 2013). This technique has been successfully employed to synthesize various magnetic nanoparticles, including cobalt ferrite (Zalite et al., 2015), rare earth iron garnets, and iron oxides (Islam et al., 2011).

2.3. Physical Methods: Sputtering, Laser Ablation, Vapor Deposition

Various sputtering methods have been developed for high-rate deposition of magnetic materials. The "Gap Type target" (GT target) utilizes narrow gaps in the target to allow magnetic flux leakage, enabling high-rate sputtering of magnetic materials with common permanent magnets (Nakamura et al., 1982). The facing-targets method employs parallel targets with a perpendicular magnetic field, achieving high deposition rates without substrate heating (Hoshi et al., 1982). This technique prevents high-energy ion bombardment and allows uniform film deposition over large areas (Hoshi et al., 1982). Gas flow sputtering has demonstrated high potential for preparing soft and hard magnetic coatings, with deposition rates exceeding 50 μm/h for NiFe and 85 μm/h for CoSm (Bandorf et al., 2007). These methods have successfully produced magnetic films with desirable properties, such as high permeability and low coercivity for soft magnetic materials, and high coercivity for hard magnetic materials, while maintaining low substrate temperatures during deposition (Naoe et al., 1980).

Laser ablation in the presence of magnetic fields offers innovative methods for fabricating and manipulating magnetic nanostructures. This technique can produce oriented nanowires composed of nanoparticles when performed in air. By simultaneously ablating multiple targets, nanocomposites of iron oxide with noble or transition metals can be created (Nikov et al., 2020). In liquid environments, the method allows for separation of micro- and nanoparticles, with the possibility of depositing ordered one-dimensional microstructures on substrates. The application of magnetic fields during ablation can enhance ionization of growth species near the substrate, enabling epitaxial growth of films at lower temperatures and increasing deposition rates when combined with eclipse methods (Tachiki & Kobayashi, 1999).

Chemical vapor deposition (CVD) is a versatile method for fabricating magnetic thin films and structures. It allows for the creation of large arrays of micron-scale magnetic features with well-defined spatial and structural properties (Welipitiya et al., 1996; Das et al., 2023). CVD can produce single-crystal films of ferrites and garnets on various substrates, with controllable magnetic vector orientation, composition, and surface conditions (Pulliam, 1967). The technique enables the growth of high-quality homogeneous thin metal films with properties comparable to those produced by other methods (Zhou et al., 1993). Continuous vapor deposition, a variation of CVD, can be used to create films with excellent characteristics for high-density recording media. The deposition direction (HIN or LIN) influences the magnetic properties and microstructures of the resulting films (Nakamura et al., 1982).

Figure 2. Magnetic hysteresis loop of a ferromagnetic material

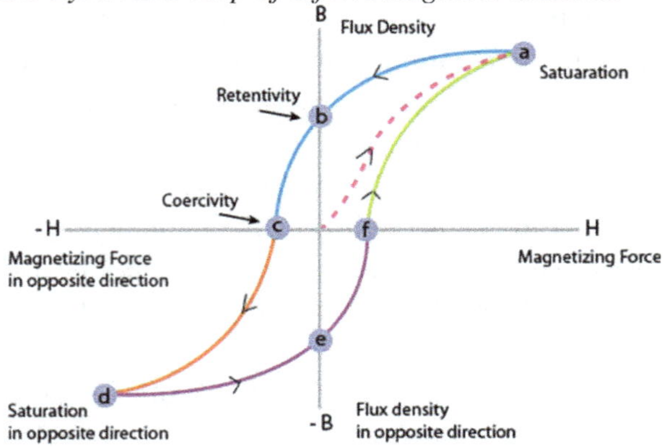

3. CHARACTERIZATION TECHNIQUES FOR MAGNETIC MATERIALS

SEM analyzes magnetic materials and domain structures, providing high spatial resolution and magnetic contrast. TEM analyzes structure, composition, and magnetic properties on nanoscale magnetic materials, providing comprehensive characterization. VSM measures magnetic properties with good sensitivity and precise temperature control, while SQUID measures magnetic properties with high sensitivity, speed, resolution, accuracy, and stability. MOKE measures magnetic properties of ferromagnetic thin films, providing high angular accuracy and contactless characterization. Eddy-Current Measurements determine electrical conductivity and magnetic permeability, while TR-MOKE ultrafast thermal characterization of magnetic thin films is advanced. DSC identifies Curie and Néel transition temperatures, while MFM imaging magnetic domain structures provides high spatial resolution and magnetic contrast. X-ray Magnetic Circular Dichroism (XMCD) probes magnetic properties of materials with high sensitivity and element-specificity. Table 3 provides a comprehensive overview of the characterization techniques for magnetic materials.

Table 3. Characterization Techniques for Magnetic Materials

Technique	Description	Advantages	Applications	Resolution	Sample Requirements
X-ray Diffraction	Structural analysis of magnetic materials	High-resolution structural information	Magnetic material characterization	Atomic scale	Powder or single crystal
SEM	Analyzing magnetic materials and domain structures	High spatial resolution, magnetic contrast	Recording tapes, cobalt, magnetic oxides	1-10 nm	Bulk or thin film
TEM	Analyzing structure, composition, and magnetic properties	Comprehensive characterization, atomic-scale analysis	Nanoscale magnetic materials	0.1-10 nm	Thin film or nanoparticle
VSM	Measuring magnetic properties of materials	Good sensitivity, precise temperature control	Materials science laboratories, educational settings	10^-3 emu	Bulk or thin film
SQUID	Measuring magnetic properties of materials	High sensitivity, speed, resolution, accuracy, and stability	Geology, materials science	10^-6 emu	Bulk or thin film
MOKE	Measuring magnetic properties of ferromagnetic thin films	High angular accuracy, contactless characterization	Ferromagnetic thin films, exchange-bias structures	0.1-10°	Thin film
Eddy-Current Measurements	Determining electrical conductivity and magnetic permeability	Accurate, non-destructive	Electrical machine applications	1-10%	Bulk or thin film
TR-MOKE	Ultrafast thermal characterization of magnetic thin films	Advanced technique for thermal characterization	Magnetic thin films	10^-6 s	Thin film
DSC	Identifying Curie and Néel transition temperatures	Commonly used, thermal analysis	Magnetic material characterization	0.1-10°C	Powder or bulk
MFM	Imaging magnetic domain structures	High spatial resolution, magnetic contrast	Magnetic material characterization	10-100 nm	Thin film or bulk
Kerr Microscopy	Imaging magnetic domain structures	High spatial resolution, magnetic contrast	Magnetic material characterization	1-10 μm	Thin film or bulk

continued on following page

Table 3. Continued

Technique	Description	Advantages	Applications	Resolution	Sample Requirements
X-ray Magnetic Circular Dichroism (XMCD)	Probing magnetic properties of materials	Element-specific, high sensitivity	Magnetic material characterization	Atomic scale	Thin film or bulk
Mössbauer Spectroscopy	Probing magnetic properties of materials	High sensitivity, element-specific	Magnetic material characterization	Atomic scale	Powder or bulk

3.1. Structural Analysis Using X-ray Diffraction, SEM, and TEM

X-ray diffraction techniques offer powerful tools for structural analysis of magnetic materials. Magnetic X-ray microdiffraction enables probing complex magnetic structures at domain-relevant scales, combining reciprocal and real space resolution (Evans & Isaacs, 2006). Synchrotron X-ray beams with tunable energy and polarization have expanded the capabilities of magnetic X-ray diffraction for ferromagnetic materials (Vettier & Mcwhan, 1989). In situ studies of magnetic field-induced structural changes can be performed using specialized diffractometers coupled with cryostats and superconducting magnets, allowing for high-precision measurements at various temperatures and magnetic fields (Holm et al., 2004). For materials available only as microcrystals, magnetic orientation techniques can enhance powder XRD by aligning crystals uniaxially or biaxially, effectively "decompressing" 1D information into 2D or 3D. This approach facilitates structure determination and extraction of unit cell parameters and space group information from powder samples (Kimura & Kimura, 2018).

Scanning electron microscopy (SEM) has emerged as a powerful tool for analyzing magnetic materials and their domain structures. Various techniques have been developed to enhance magnetic contrast in SEM, including secondary electron energy discrimination and spatial aperturing (Wardly, 1971). Spin-polarized SEM (spin SEM) offers unique capabilities such as full magnetization orientation detection and high spatial resolution (Kohashi, 2018). Another technique, scanning electron microscopy with polarization analysis (SEMPA), combines SEM with secondary electron spin polarization analysis to provide high-resolution images of surface magnetic microstructures (Unguris et al., 1991). These methods have been applied to study a wide range of magnetic materials, including recording tapes, cobalt, magnetic oxides, and silicon-iron transformers (Jones, 1978).

Transmission electron microscopy (TEM) is a powerful tool for analyzing the structure, composition, and magnetic properties of nanoscale magnetic materials

(Smith et al., 2006). Various TEM techniques, including imaging, diffraction, and spectroscopy, can be combined to provide comprehensive characterization. Proper sample preparation is crucial, with methods varying for bulk samples, thin films, nanowires, and powders (Liu, 2002). Quantitative TEM analysis often requires image simulations and can be applied to semiconductor islands, nanowires, quantum dots, and soft magnetic materials. Advanced techniques like electron holography enable three-dimensional shape reconstruction, while quantitative high-resolution TEM (qHRTEM) allows for atomic-scale analysis of structure and composition (Neumann et al., 2009). Lorentz microscopy and electron holography can be used to correlate structure and magnetic properties in soft magnetic materials (Neumann et al., 2009).

3.2. Magnetic Properties Measurement: VSM, SQUID, and MOKE

Vibrating Sample Magnetometers (VSMs) are versatile instruments for measuring magnetic properties of materials. Several researchers have developed low-cost, high-quality VSMs suitable for both educational and research purposes (Niazi et al.,2000). These instruments offer good sensitivity, precise temperature control, and ease of sample change. Innovations in VSM design include a parallel motion system, which facilitates faster sample replacement and positioning (Mészáros, 2007). Advanced analytical techniques, such as VSM depth profiling and Differential Vibrating Sample Magnetometry (DVSM), have been developed to study magnetic properties as a function of depth in thin films (Haines, 1984). VSMs have been successfully used to measure magnetic properties of various materials, including Ni wires and floppy disk media, with precision limits as low as 5×10−3 emu (Burgei et al., 2003). These developments have made VSMs valuable tools for materials science laboratories and educational settings.

SQUID magnetometers have emerged as powerful tools for measuring magnetic properties of materials. They offer high sensitivity, speed, resolution, accuracy, and stability (Deaver et al., 1978). Room-temperature scanning SQUID microscopes can image magnetic thin films, providing quantifiable results on magnetization, total dipole moment, and demagnetizing fields (Fleet et al., 2001). Advanced systems incorporate in-situ magnetization/demagnetization fields, enabling the measurement of remanent curves at specific points on a sample (Du et al., 2018). To optimize measurement efficiency, a method has been developed to select characteristic points based on the magnetic field distribution, balancing time savings with information retention (Du et al., 2018). SQUID magnetometers are increasingly integrated with automated data acquisition systems and closed-cycle helium refrigerators, enhancing their convenience and applicability (Deaver et al., 1978). The magneto-optical Kerr effect (MOKE) is a versatile technique for measuring magnetic properties of ferromagnetic thin films. MOKE systems can be designed for both polar and longitudinal configurations, offering high angular accuracy (Fu et al., 2018). These setups typically include lasers, polarizers, and photoelastic modulators, enabling contactless characterization of various magnetic materials like FeNi and SmCo (Allocca et al., 2014). MOKE measurements can determine key magnetic features such as hysteresis loops and anisotropy fields. An alternative method using magnetic susceptibility measurements with MOKE has been demonstrated for measuring anisotropy fields in thin film ferromagnets (Bartlett et al., 2019).

3.3. Electrical and Thermal Property Characterization

Characterizing magnetic materials is crucial for electrical machine applications. While Epstein frame measurements are commonly used, ring core measurements are more suitable for complex magnetic flux paths in electrical machines (Krings & Soulard, 2015). Various methods exist to determine electrical conductivity and magnetic permeability, including eddy-current measurements, which can achieve accuracy within 9% for conductivity and 14% for effective permeability (Karpen et al., 1998; Nayak et al.,2024). A methodology for core loss prediction using quasi-static B-H loop measurements and dynamic loss measurements on non-sinusoidal flux density has been developed, eliminating the need for close loop control (Rodrigues & Jewell, 2014). Characterization techniques can be applied to different materials, such as structural steel used in large AC machines, to explore B-H curves at various frequencies and field intensities (Muhit & Hoque, 2011). These methods provide valuable insights into the electromagnetic properties of materials used in electrical machines.

Thermal characterization of magnetic materials is crucial for understanding their behavior and optimizing their performance in various applications. Time-resolved magneto-optical Kerr effect (TR-MOKE) has emerged as an advanced technique for ultrafast thermal characterization of magnetic thin films (Chen et al., 2016). For integrated magnetic components, experimental methods using multiple sensors have been developed to measure operating temperatures, which can significantly affect material properties (Basma et al., 2024). Thermal analysis techniques, such as differential scanning calorimetry (DSC), are commonly used to identify Curie and Néel transition temperatures in magnetic materials, although their efficacy and interpretation require further research (Reddy et al., 2023). When selecting permanent magnets for various applications, it is essential to consider their thermal properties correctly, as these characteristics play a crucial role in material performance and design optimization (Trout, 2001).

4. MAGNETIC MATERIALS IN ENERGY APPLICATIONS

Magnetic materials play a crucial role in the energy sector, with applications in wind turbines, electric vehicles, refrigeration, heat pumping, nanocomposites, high-performance superconductors (HTS), low-temperature superconductors (LTS), amorphous magnetic materials, and magnetic shape memory alloys (MSMAs). Permanent magnets improve motor and generator efficiency, while soft magnetic materials enhance electricity transmission and utilization. Magnetocaloric materials offer energy-efficient refrigeration and heat pumping, while nanocomposite magnetic materials enhance energy storage capacity. High-performance superconductors (HTS) increase energy efficiency, reduce size, and offer flexible transmission in power applications. Low-temperature superconductors (LTS) improve performance and reliability of LTS systems. Amorphous magnetic materials offer high permeability, low coercivity, and low magnetostriction, but are high in cost compared to crystalline materials. Magnetic shape memory alloys (MSMAs) are used in actuators, sensors, and microfluidics due to their shape memory effect and high strength. Developing new MSMAs with improved properties and reduced production costs is a focus of research. Table 4 provides an overview of the applications of magnetic materials in the energy sector.

Table 4. Magnetic Materials and Their Applications in Energy sector

Magnetic Materials	Application	Property Requirements	Challenges	Focus of Research
Permanent magnets	Improve efficiency of motors and generators in wind turbines and electric vehicles	High Remanence (Br), High Coercivity (Hc), Maximum Energy Product (BHmax)	Limited rare-earth element supply	Develop alternative materials with good performance and abundant elements
Soft magnetic materials	Improve efficiency of electricity transmission and utilization	High permeability (μ), Low coercivity (Hc), Low loss (eddy current and hysteresis)	High energy consumption	New materials with improved efficiency and lower energy loss
Magnetocaloric materials	Energy-efficient refrigeration and heat pumping	Large magnetocaloric effect (MCE) at operating temperature, good thermal properties	Finding materials with optimal properties for room temperature applications	Develop new materials (rare-earth and rare-earth-free) with high energy efficiency and suitable operating temperatures

continued on following page

Table 4. Continued

Magnetic Materials	Application	Property Requirements	Challenges	Focus of Research
Nanocomposite magnetic materials	Enhance energy storage capacity	High saturation magnetization (Ms), good cycling stability	-	Optimize properties of nanocomposites for improved performance
High-performance superconductors (HTS)	Increased energy efficiency, reduced size, and flexible transmission in power applications	High critical temperature (Tc), high critical current density (Jc)	High cost and complexity	Develop more affordable HTS with improved performance
Low-temperature superconductors (LTS)	Energy storage and transportation	High critical current density (Jc), good mechanical properties	-	Improve performance and reliability of LTS systems
Amorphous magnetic materials	Transformers, magnetic cores	High permeability (μ), Low coercivity (Hc), Low magnetostriction	High cost compared to crystalline materials	Develop cost-effective amorphous magnetic materials with good performance
Magnetic shape memory alloys (MSMAs)	Actuators, sensors, microfluidics	Shape memory effect in presence of magnetic field, high strength, good ductility	High cost, limited material selection	Develop new MSMAs with improved properties and reduce production costs

4.1. Wind Turbines and Renewable Energy Systems

Magnetic materials play a crucial role in renewable energy systems such as wind turbines and electric vehicles. Research emphasizes the importance of finding materials based on abundant elements to ensure sustainability and efficiency in energy applications (Gutfleisch et al., 2011). Magnetic materials play an important role in improving the efficiency and performance of devices in electric power generation, conditioning, conversion, transportation, and other energy-use sectors. Permanent magnets, essential for renewable energy technologies, face challenges due to limited rare-earth metal supplies, prompting the need for alternative materials (Matizamhuka, 2018). The increased reliance on renewable energy sources has accelerated research in energy-related technologies worldwide. The use of rare-earth (RE) metals in permanent magnets continues to be a source of greater concern due to the limited RE supply and dwindling reserves globally. Efforts to enhance energy efficiency and reduce environmental impacts highlight the significance of magnetic materials in electric motors, generators, and transformers (Franco & Gutfleisch, 2012). Magnetic materials can contribute to improving the efficiency of electricity transmission, utilization, and transportation through electric motors and generators. The development of new magnetic materials with improved performance is essential for achieving energy sustainability goals and mitigating climate change impacts (Naima et al., 2013).

Figure 3. magnetization loops for soft and hard magnets

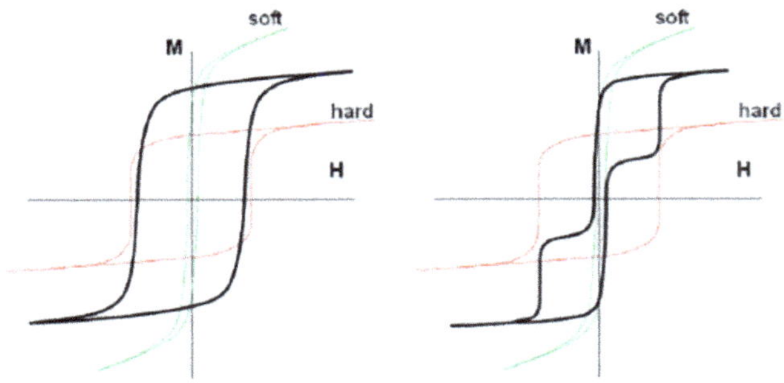

4.2. Magnetic Refrigeration and Energy Storage Systems

Magnetic materials play a crucial role in energy-efficient technologies, including magnetic refrigeration and energy storage systems. Magnetocaloric materials offer promising alternatives for environmentally friendly cooling and heating applications around room temperature (Kitanovski, 2020). Magnetocaloric materials have potential for energy-efficient refrigeration, heat pumping, and thermal energy harvesting applications. These materials exhibit significant entropy changes in response to applied magnetic fields, making them suitable for magnetic cooling processes (Vasile et al., 2008). Magnetic refrigeration is an energy-efficient and environmentally-friendly solid-state cooling technology based on the magnetocaloric effect in certain materials. In the field of energy storage, magnetic materials with high energy products (BH) max are essential. Nanotechnology has significantly influenced the development of magnetic energy storage composites, enhancing their performance through exchange coupling between soft and hard magnetic phases (Sreenivasulu & Srikanth, 2016; Mallick et al.,2024). Nanotechnology enhances the energy storage capacity of magnetic materials through exchange coupling in soft-hard magnetic composites. Both rare-earth and rare-earth-free magnetic materials are being explored for these applications, with ongoing research focusing on optimizing their properties for various energy-related technologies (Gutfleisch et al., 2011).

4.3. Superconducting Magnets for Power Applications

Superconducting magnets offer significant advantages in power applications, including increased energy efficiency, reduced equipment size, and flexible transmission (Mikkonen, 2002). There are ongoing efforts and consortia in Europe focused on the development and application of superconducting power technology. These magnets utilize high-performance superconductors to produce strong magnetic fields with lower power consumption (Zlobin, 2014). Superconducting magnets have a wide range of applications in scientific research, industry, medicine, transportation, and other fields. The widespread adoption of superconducting magnets has been enabled by progress in supporting technologies, such as cryogenics and superconducting composite manufacturing. Applications range from energy storage and transportation to medical imaging and particle accelerators (Yao & Ma, 2021). Superconducting materials have great potential for a wide range of applications, including electric power generation and transmission, transportation, and scientific instrumentation. The performance, economy, and operating parameters of these applications depend on the electromagnetic, mechanical, and cost-related properties of superconductors. While low-temperature superconductors (LTS) have been commercialized in superconducting magnetic energy storage (SMES) systems, high-temperature superconductors (HTS) show promise for broader adoption due to simplified cryogenics and increased stability. HTS materials operate at higher temperatures (30-80 K) and offer economic advantages over LTS systems (Malozemoff et al., 2002).

5. MAGNETIC MATERIALS IN ELECTRONICS AND SENSING

5.1. Sensors and Actuators Based on Magnetic Materials

Magnetic materials play a crucial role in developing sensors and actuators for various applications, including soft robotics. These materials enable the creation of strain gauges, force sensors, and displacement sensors (Pasquale, 2003). Amorphous ferromagnetic materials are particularly suitable for constructing magnetic sensors due to their availability and robustness, allowing for contactless measurements in harsh environments (Morón et al., 2015). Amorphous ferromagnetic materials are widely used in magnetic sensors because they are cost-effective, readily available, and offer high reliability. Along with nanocrystalline materials, they exhibit high sensitivity and low coercivity, making them ideal for precise magnetic field detection. Sensors made from these materials are robust and suitable for diverse applications (Kwon et al., 2024). The directional properties of magnetic nanomaterials support selective and efficient sensing and actuation. Recent advances highlight how these anisotropic materials enhance soft robotics in terms of safer human interaction, improved navigation, and miniaturization. Overall, smart magnetic materials have played a key role in shifting robotics from rigid structures to adaptive, flexible systems (Hao et al., 2022).

5.2. RF and Microwave Applications

Magnetic materials play a crucial role in RF and microwave applications, offering opportunities for device miniaturization and performance enhancement. These materials are utilized in various wireless communication systems, including base stations and point-to-point radio links (Cruickshank, 2011). Recent advancements have focused on developing self-biased materials, such as cubic and hexagonal ferrites, to improve high-frequency performance (Harris, 2010). Multiferroic materials, which exhibit both electric polarization and magnetization, have shown promise in creating compact and power-efficient RF and microwave devices, including tunable inductors and non-reciprocal filters (Lin et al., 2016). Integrated magnetics and multiferroics enable compact and power-efficient RF and microwave electronics. Yttrium Iron Garnet (YIG) and related composites have been studied for their potential in ferrite resonator antennas and microstrip antennas, demonstrating effectiveness in the microwave frequency range and the ability to control radiation characteristics through external magnetic fields (Mallmann et al., 2013). These developments in magnetic materials are driving innovations in wireless technology, enabling more efficient and versatile RF and microwave devices. Yttrium iron garnet (YIG) and gadolinium iron garnet (GdIG) composites have applications as ferrite resonator antennas and microstrip antennas in the microwave frequency range. The YIG and GdIG composites were effective as ferrite resonator antennas and microstrip antennas in the microwave frequency range.

5.3. Magnetic Materials in Data Storage Technologies

Traditional materials like Ni80Fe20 are being replaced by high-iron nickel-iron alloys and nanoengineered composite materials to meet the demands of high-coercivity media and high-frequency writing (Andricacos & Robertson, 1998; Coufal et al., 2006). High-iron nickel-iron alloys (Ni45Fe55) have higher magnetic moment and resistivity than low-iron nickel-iron alloys and are increasingly used in head fabrication. Controlled addition of impurities can produce drastic improvements in the properties of electroplated materials. Magnetic data storage on disks and tape has provided constantly increased storage density, requiring innovation and breakthroughs in magnetic materials. Today's disk and tape drives take advantage of novel Nano engineered composite magnetic materials and quantum mechanical processes. Technology has been developed to fabricate laminated materials by electroplating from a single plating solution. Perpendicular recording is expected to be superseded by heat-assisted and bit-patterned magnetic recording to achieve areal densities above 1 Tbit/in^2 (Tannous & Comstock, 2017). A wide range of magnetic materials, including soft magnetic materials, antiferromagnetic alloys, and ferromagnetic alloys, are essential for advances in magnetic recording technology. Magnetic recording media has seen advances in high-magnetization metal alloys with high coercivity. Perpendicular recording will be replaced by heat-assisted and bit-patterned magnetic recording to achieve areal densities above 1 Tbit/in^2. The development of giant magnetoresistance (GMR) sensors and magnetic tunneling junctions has led to advancements in read heads and magnetic random-access memory (MRAM) technology (Tannous & Comstock, 2017).

6. BIOMEDICAL APPLICATIONS OF MAGNETIC MATERIALS

6.1. Magnetic Nanoparticles in Drug Delivery and Hyperthermia

Magnetic nanoparticles (MNPs) have emerged as promising tools in cancer diagnosis, drug delivery, and treatment. Their superparamagnetic properties make them ideal for magnetic resonance imaging (MRI) contrast enhancement and targeted drug delivery (Wu & Huang, 2017). MNPs can be used as contrast agents for various imaging modalities to enable early cancer detection. MNPs can be used as drug delivery vehicles to target chemotherapeutic drugs and antibodies to tumors. MNPs can be used for various cancer treatment modalities like magnetic hyperthermia, photodynamic therapy, and photothermal therapy, often in combination for improved efficacy. MNPs can be functionalized for various biomedical applications, including protein and DNA separation, and hyperthermia-based therapy (Namdeo et al., 2008). Magnetic nanoparticles have been explored for various biomedical applications due to their superparamagnetic properties and high magnetic saturation. Surface functionalization of magnetic nanoparticles enables their use in protein/DNA separation, imaging, drug delivery, and hyperthermia applications. Recent advancements have expanded the concept of hyperthermia to include thermotherapy and magnetically modulated controlled drug delivery. Two approaches for hyperthermia-based drug delivery have been identified: bond breaking (DBB) and enhanced permeability (DEP) (Kumar & Mohammad, 2011). Magnetic nanoparticles can be used for hyperthermia-based therapy and controlled drug delivery. Core-shell magnetic nanomaterials, particularly nanoshells prepared using layer-by-layer self-assembly, show potential for combined hyperthermia therapy and controlled drug release (Kumar & Mohammad, 2011).

6.2. MRI Contrast Agents Based on Magnetic Materials

Magnetic nanomaterials have emerged as promising contrast agents for MRI, offering enhanced sensitivity and specificity compared to conventional agents (Caspani et al., 2020). Magnetic nanoparticles, such as SPIONs, nanodiscs, SAFs, and nanowires, have shown promising potential as contrast agents for MRI with improved performance compared to traditional gadolinium-based agents. These magnetic nanoparticles can be functionalized and tuned for targeted imaging and theranostic applications, where they can serve as both contrast agents and therapeutic agents. These agents can be classified based on their composition, magnetic properties, and biodistribution (Geraldes & Laurent, 2009). Gadolinium-based complexes and superparamagnetic iron oxide nanoparticles are commonly used as T1 and T2 contrast agents, respectively. Recent research has focused on improving their targeting capabilities and developing novel materials with higher relaxivity values (Cheng et al., 2013). Several types of Gd3+-based and SPIO nanoparticle contrast agents have been used clinically for tumor diagnosis. New chelates, carriers, and coating materials have been explored to improve the sensitivity and targeting of MRI contrast agents. MRI contrast agents for detecting tumor metabolism have also been developed. Superparamagnetic nanoparticles, typically synthesized from iron oxide and coated with polymers, offer multifunctional applications beyond imaging, including drug delivery and hyperthermia for cancer treatment. The effectiveness of these contrast agents depends on various chemical and physical properties, with different models describing their behavior in different motional regimes (Issa & Obaidat, 2019).

6.3. Biocompatibility Issues and Solutions

Magnetic materials show promise for biomedical applications, but biocompatibility remains a crucial concern. Iron oxide nanoparticles require comprehensive evaluation at integrated, cellular, and molecular levels to ensure safety (Lin et al., 2011). Biocompatibility of magnetic iron oxide nanoparticles is evaluated at three levels: integrated, cellular, and molecular. Magnetoelastic materials, such as iron-gallium alloys, demonstrate potential as biodegradable implants with controllable degradation rates and no adverse effects on cell viability (Holmes et al., 2014). The degradation rates of magnetoelastic materials can be controlled by their composition. The degradation rates of magnetoelastic materials can be remotely controlled by applying a magnetic field, which causes vibrations that accelerate degradation. While one type of magnetoelastic material (Metglas) was found to be non-biocompatible, another type (iron-gallium alloy) showed no adverse effects on cell viability from its degradation products. Coating magnetic nanoparticles with alumina can improve biocompatibility, as demonstrated by in vitro studies using human WBC cell lines (Giri et al., 2003). Magnetic nanoparticles with a Curie temperature between 42-50°C and good biocompatibility are the best candidates for effective hyperthermia treatment, as they can act as an in vivo temperature control switch to prevent overheating. The coated magnetic nanoparticles were found to be biocompatible based on in vitro cell culture experiments using human white blood cell lines. However, some magnetic materials, like Fe-Co alloys, exhibit toxicity within 24 hours, while others, such as Terfenol-D, maintain 90% cell viability after 72 hours (P. Pouponneau et al., 2006).

Summary of Applications

Magnetic materials have various applications presented in Table 5 like in renewable energy, electricity transmission, refrigeration, superconducting magnets, soft robotics, RF and microwave, data storage, and spintronics. Permanent magnets are crucial for wind turbines and electric vehicles due to limited rare-earth metal supplies. Magnetic refrigeration uses magneto-caloric materials for environmentally friendly cooling and high-efficiency solid-state cooling. Superconducting magnets increase energy efficiency and reduce equipment size. Soft robotics uses amorphous ferromagnetic materials for robust measurements and sensors.

Table 5. Magnetic Materials and Their Applications: A Review

Application	Magnetic Material	Properties/Advantages	Challenges/Limitations
Renewable Energy	Permanent magnets	Crucial for wind turbines, electric vehicles, high efficiency	Limited rare-earth metal supplies, cost
	Magnetic materials	Improved efficiency in electricity transmission, utilization, and transportation	Material development, cost
Magnetic Refrigeration	Magnetocaloric materials	Environmentally friendly cooling, high efficiency, solid-state cooling	Material development, long-term reliability, cost
Superconducting Magnets	High-temperature superconductors (HTS)	Increased energy efficiency, reduced equipment size, flexible transmission	Material development, cost, reliability, cryogenics
Soft Robotics	Amorphous ferromagnetic materials	Robust, contactless measurements, flexible sensors, soft actuators	-
RF and Microwave	Ferrites (cubic and hexagonal)	Miniaturization, performance enhancement, compact devices, high-frequency performance	Material development, high-frequency performance, cost

continued on following page

Table 5. Continued

Application	Magnetic Material	Properties/Advantages	Challenges/Limitations
	Multiferroic materials	Compact and power-efficient RF and microwave devices, tunable inductors, non-reciprocal filters	Material development, high-frequency performance, cost
	Yttrium Iron Garnet (YIG) and composites	Effective in microwave frequency range, controlled radiation characteristics	Material development, cost
Data Storage	High-iron nickel-iron alloys	Improved storage density, high-coercivity media, high magnetic moment and resistivity	Material development, writability, cost
	Heusler compounds	Exceptional magnetic properties, spintronics applications, GMR spin valves, magnetic tunnel junctions	Material development, writability, cost

7. FUTURE TRENDS AND EMERGING TECHNOLOGIES

7.1. Quantum Materials and Spintronics

Recent research highlights emerging trends in quantum materials and magnetic storage technologies. Quantum materials exhibit macroscopic quantum effects, reduced dimensionality, and strong interactions, with applications ranging from electronics to biomedical sensing (Lau et al., 2020). The field of quantum materials is an important and active area of research with applications across many fields. Recent research has focused on developing quantum materials with high-temperature properties and the ability to precisely tune their characteristics at the atomic scale. The field of quantum materials has seen major advancements and is now well-positioned to contribute to the emerging field of quantum information science and technology. Future developments in magnetic storage may include perpendicular recording, heat-assisted magnetic recording, and patterned media (Kryder, 2003). Microelectronics innovations encompass new physical principles like quantum computing, novel materials such as piezoelectric nanostructures, and emerging technologies including spin lasers and biocompatible microelectronics (Luryi et al., 2013). Quantum materials, including superconductors and topological quantum matter, demonstrate intertwined charge, spin, orbital, and lattice degrees of freedom, leading to exotic properties with potential applications in quantum sensing and information processing (Guo et al., 2022).

7.2. Nanostructured Magnetic Materials and Their Potential

Nanostructured magnetic materials are at the forefront of technological advancements, offering promising applications in data storage, spintronics, and biomedical devices (Chander, 2024). Nanotechnology has revolutionized magnetism, enabling manipulation of materials at the nanoscale with diverse applications. Recent developments include the creation of multilayer film structures, granular alloys, and nanowires with unique magnetic and magneto resistive properties (Fedosyuk, 2018). The integration of magnetic nanostructures in micro-electromechanical systems (MEMS) is expected to revolutionize sensor and actuator technologies (Azzerboni et al., 2007). The development of nano-structured magnetic materials has enabled the miniaturization of devices with improved performance. Magnetic random access memory (MRAM) is a key application of the advancements in nano-structured magnetic materials, enabling the development of non-volatile RAM with fast dynamics. Emerging trends focus on three-dimensional magnetic nanostructures, which exhibit complex spin textures and topologies, potentially leading to improved energy efficiency and processing speed in future spintronic devices (Fischer et al., 2020).

7.3. Predictions on the Future Scope of Magnetic Material Applications

Magnetic materials are poised for significant advancements and applications in the coming decades. Nanocrystalline materials show promise for both soft and hard magnetic applications, with microstructure playing a crucial role in their properties (McHenry & Laughlin, 2000). This study reviews recent advances in nanocrystalline magnetic materials, covering both soft and hard types, along with progress in magnetic recording materials. It highlights that microstructure plays a crucial role in determining magnetic performance. Key features such as grain size, grain shape, and crystallographic texture strongly influence properties, and precise microstructural control has significantly improved material efficiency. The study also notes that functional magnetic materials, which respond to external fields, hold great potential for applications in data storage, magnetic refrigeration, and medical technologies (Abdelbasir & Shalan, 2019). Functional magnetic materials have important physical properties that can be affected by external magnetic fields, enabling applications in data storage, refrigeration, and medical studies. Key developments in spintronic devices are anticipated, with a focus on achieving half-metallicity and perpendicular anisotropy at room temperature. Materials such as Heusler alloys, ferrites, and L10 alloys are being explored for these purposes (Hirohata et al., 2015). The authors evaluated five major magnetic materials for their potential to achieve half-metallicity at room temperature: Heusler alloys, ferrites, rutiles, perovskites, and dilute magnetic semiconductors. The authors also evaluated L10 alloys and D022-Mn alloys for their potential to achieve perpendicular anisotropy in nanoscale devices at room temperature. The future of magnetic materials is likely to involve improvements in ultrasonic transducer materials, storage and logic elements, and microwave materials (Alperin et al., 1970).

8. CONCLUSION

The study of magnetic materials has progressed from ancient lodestone to advanced materials central to modern technology. Today, magnetic materials play vital roles in power generation, communication, data storage, spintronics, and biomedical devices. Recent developments include multifunctional materials with magneto-thermal, magneto-elastic, magneto-optic, and magneto-electrical properties, as well as magnetic topological materials for next-generation electronics and memory devices. Advanced synthesis methods such as sol-gel, co-precipitation, hydrothermal, sputtering, and vapor deposition allow precise control over nanoparticles, thin films, and nanocomposites. Characterization techniques like XRD, SEM, TEM, VSM, SQUID, and MOKE ensure detailed understanding of structural and magnetic behaviour. The classification into soft, hard, and semi-hard materials, along with hysteresis analysis, remains fundamental. Overall, magnetic materials have evolved into essential components of modern industry and continue to drive innovation across electronics, communication, and healthcare.

REFERENCES

Abdelbasir, M. S., & Esmail Shalan, A. (2019). Intriguing Properties and Applications of Functional Magnetic Materials. Functional Materials.

Abderahmane, A., & Daniel, L. (2023). Measurement Criteria for the Magnetic Characterization of Magnetic Materials. *IEEE Transactions on Instrumentation and Measurement*, *72*, 1–18. DOI: 10.1109/TIM.2023.3322510

Aguilar, G. V. (2019). *Sol-Gel Method - Design and Synthesis of New Materials with Interesting Physical.* Chemical and Biological Properties.

Al-Alawy, A. F., Al-Abodi, E. E., & Kadhim, R. M. (2018). Synthesis and Characterization of Magnetic Iron Oxide Nanoparticles by Co-Precipitation Method at Different Conditions. *Journal of Engineering*, *24*(10), 60–72. DOI: 10.31026/j.eng.2018.10.05

Allocca, L., Baltador, C., Morone, A., Attanasio, C., Cirillo, C., & Gambardella, U. (2014). Magnetic properties of ferromagnetic thin films obtained by Magneto-Optic Kerr Effect measurements. *2014 Third Mediterranean Photonics Conference*, 1-3. DOI: 10.1109/MePhoCo.2014.6866477

Alperin, H.A., Belson, H.S., Clark, A.E., Krall, A.D., & Pickart, S.J. (1970). Forecast on Magnetic Materials.

Andricacos, P. C., & Robertson, N. (1998). Future directions in electroplated materials for thin-film recording heads. *IBM Journal of Research and Development*, *42*(5), 671–680. DOI: 10.1147/rd.425.0671

Azzerboni, B., Asti, G., Pareti, L., & Ghidini, M. (2007). *Magnetic Nanostructures in Modern Technology: Spintronics.* Magnetic MEMS and Recording.

Bandorf, R., Bloche, A., Ortner, K., Lüthje, H., & Jung, T. D. (2007). High Rate Deposition of Magnetic Material by Gas Flow Sputtering. *Plasma Processes and Polymers*, *4*(S1), 4. DOI: 10.1002/ppap.200730504

Bartlett, A. P., Belanger, R., Amman, M. S., & Venus, D. (2019). In situ determination of the anisotropy field in ferromagnetic films using magnetic susceptibility measurements by MOKE. *Journal of Magnetism and Magnetic Materials*, *477*, 49–54. DOI: 10.1016/j.jmmm.2019.01.037

Basma, B. M., Piétroy, D., Boukhari, M. L., Chahbi, Z., Blanchet, T., Chatelon, J. P., Capraro, S., & Rousseau, J. J. (2024). On thermal characterization method of integrated magnetic components. *Scientific Reports*, *14*(1), 14. DOI: 10.1038/s41598-024-54588-7 PMID: 38418493

Behrens, S. (2011). Preparation of functional magnetic nanocomposites and hybrid materials: Recent progress and future directions. *Nanoscale*, *3*(3), 877–892. DOI: 10.1039/C0NR00634C PMID: 21165500

Burgei, W. A., Pechan, M. J., & Jaeger, H. (2003). A simple vibrating sample magnetometer for use in a materials physics course. *American Journal of Physics*, *71*(8), 825–828. DOI: 10.1119/1.1572149

Caspani, S., Magalhães, R., Araújo, J. P., & Sousa, C. T. (2020). Magnetic Nanomaterials as Contrast Agents for MRI. *Materials (Basel)*, *13*(11), 13. DOI: 10.3390/ma13112586 PMID: 32517085

Chander, S. (2024). *Harnessing Nanotechnology for Advancements in Magnetism: Current Trends and Future Prospects. International Journal of Innovative Science and Research Technology*. IJISRT.

Chen, J., Zhu, J., Zhang, D., Lattery, D. M., Li, M., Wang, J., & Wang, X. (2016). Time-Resolved Magneto-Optical Kerr Effect of Magnetic Thin Films for Ultrafast Thermal Characterization. *The Journal of Physical Chemistry Letters*, *7*(13), 2328–2332. DOI: 10.1021/acs.jpclett.6b00945 PMID: 27269127

Cheng, W., Ping, Y., Zhang, Y., Chuang, K., & Liu, Y. (2013). Magnetic resonance imaging (MRI) contrast agents for tumor diagnosis. *Journal of Healthcare Engineering*, *4*(1), 23–45. DOI: 10.1260/2040-2295.4.1.23 PMID: 23502248

Coey, J. M. (2010). *Magnetism and magnetic materials*. Cambridge university press.

Coufal, H., Dhar, L., & Mee, C. D. (2006). Materials for Magnetic Data Storage: The Ongoing Quest for Superior Magnetic Materials. *MRS Bulletin*, *31*(5), 374–378. DOI: 10.1557/mrs2006.96

Cruickshank, D.B. (2011). Microwave Materials for Wireless Applications.

Das, D. K., Patnaik, P., Nayak, S. K., & Barala, M. (2023). Assessing the Effect of Electronic Pseudopotentials and Relativistic Treatments on the Structural and Electrical Properties of GaN: A DFT Study. *Physics and Chemistry of Solid State*, *24*(4), 714–721. DOI: 10.15330/pcss.24.4.714-721

Das, D. K., Patnaik, P., Nayak, S. K., & Barala, M. (2024). Electronic and half-metallic behavior of copper-doped silicon carbide armchair single-walled nanotubes using density functional theory. *International Journal of Computational Materials Science and Engineering*, •••, 2450011.

Deaver, B.S., Bucelot, T.J., & Finley, J.J. (1978). SQUIDs for measuring the magnetic properties of materials.

Du, J., Wang, Q., & Song, T. (2018). Improvement of the measurement method of rock magnetic properties based on the scanning SQUID microscope with an in-situ magnetization/demagnetization field. *2018 IEEE International Magnetic Conference (INTERMAG)*, 1-1. DOI: 10.1109/INTMAG.2018.8508800

Evans, P. G., & Isaacs, E. D. (2006). Magnetic x-ray microdiffraction. *Journal of Physics. D, Applied Physics*, *39*(15), R245–R263. DOI: 10.1088/0022-3727/39/15/R01

Fedosyuk, V. M. (2018). MAGNETIC NANOMATERIALS AND NANOSTRUCTURES. TRENDS OF DEVELOPMENT. Proceedings of the National Academy of Sciences of Belarus, Physical-Technical Series.

Fiorillo, F. (2004). Measurement and characterization of magnetic materials.

Fiorillo, F. (2010). Measurements of magnetic materials. *Metrologia*, *47*(2), S114–S142. DOI: 10.1088/0026-1394/47/2/S11

Fiorillo, F. (2010). Measurements of magnetic materials. *Metrologia*, *47*(2), S114–S142. DOI: 10.1088/0026-1394/47/2/S11

Fischer, P., Sanz-Hernández, D., Streubel, R., & Fernández-Pacheco, A. (2020). Launching a new dimension with 3D magnetic nanostructures. *APL Materials*, *8*(1), 010701. DOI: 10.1063/1.5134474

Fleet, E. F., Chatraphorn, S., Wellstood, F. C., & Eylem, C. (2001). Determination of magnetic properties using a room-temperature scanning SQUID microscope. *IEEE Transactions on Applied Superconductivity*, *11*(1), 1180–1183. DOI: 10.1109/77.919559

Florio, G. (2022). Structural features of magnetic materials.

Franco, V. R., & Gutfleisch, O. (2012). Magnetic Materials for Energy Applications. *Journal of the Minerals Metals & Materials Society*, *64*(7), 750–751. DOI: 10.1007/s11837-012-0348-7

Fu, X., Mol, W., Yuan, S., Xu, L., Song, J., Dong, K., & Jin, F. (2018). Measurement System of Ferromagnetic Film Magnetic Properties Based on Mazneto-optical Kerr Effect. 2018 37th Chinese Control Conference (CCC), 7370-7373.

Geraldes, C., & Laurent, S. (2009). Classification and basic properties of contrast agents for magnetic resonance imaging. *Contrast Media & Molecular Imaging*, *4*(1), 1–23. DOI: 10.1002/cmmi.265 PMID: 19156706

Giri, J., Ray, A., Dasgupta, S., Datta, D., & Bahadur, D. (2003). Investigation on Tc tuned nano particles of magnetic oxides for hyperthermia applications. *Bio-Medical Materials and Engineering*, *13*(4), 387–399. PMID: 14646053

Guo, S. J., Xu, Y. S., Cheng, R., Zhou, J., & Chen, X. (2022). Thermal Hall effect in insulating quantum materials. *Innovation (Cambridge (Mass.))*, *3*(5), 3. DOI: 10.1016/j.xinn.2022.100290 PMID: 36039089

Gutfleisch, O., Willard, M. A., Brück, E., Chen, C. H., Sankar, S. G., & Liu, J. P. (2011). Magnetic Materials and Devices for the 21st Century: Stronger, Lighter, and More Energy Efficient. *Advanced Materials*, *23*(7), 23. DOI: 10.1002/adma.201002180 PMID: 21294168

Gutfleisch, O., Willard, M. A., Brück, E., Chen, C. H., Sankar, S. G., & Liu, J. P. (2011). Magnetic Materials and Devices for the 21st Century: Stronger, Lighter, and More Energy Efficient. *Advanced Materials*, *23*(7), 23. DOI: 10.1002/adma.201002180 PMID: 21294168

Haines, W. (1984). VSM profiling of CoCr films: A new analytical technique. *IEEE Transactions on Magnetics*, *20*(5), 812–814. DOI: 10.1109/TMAG.1984.1063365

Hajalilou, A., Tavakoli, M., & Parvini, E. (2023). Type and Characteristics of Magnetic Materials.

Hao, Y., Zhang, S., Fang, B., Sun, F., Liu, H., & Li, H. (2022). A Review of Smart Materials for the Boost of Soft Actuators, Soft Sensors, and Robotics Applications. *Chinese Journal of Mechanical Engineering*, *35*(1), 1–16. DOI: 10.1186/s10033-022-00707-2

Harris, V. G. (2010). The role of magnetic materials in RF, microwave, and mm-wave devices: the quest for self-biased materials. *Proceedings of the IEEE 2010 National Aerospace & Electronics Conference*, 1-96. DOI: 10.1109/NAECON.2010.5712910

Hirohata, A., Sukegawa, H., Yanagihara, H., Žutić, I., Seki, T., Mizukami, S., & Swaminathan, R. (2015). Roadmap for Emerging Materials for Spintronic Device Applications. *IEEE Transactions on Magnetics*, *51*(10), 1–11. DOI: 10.1109/TMAG.2015.2457393

Holm, A. P., Pecharsky, V. K., Gschneidner, K.Jr, Rink, R., & Jirmanus, M. (2004). X-ray powder diffractometer for in situ structural studies in magnetic fields from 0 to 35 kOe between 2.2 and 315 K. *The Review of Scientific Instruments*, *75*(4), 1081–1088. DOI: 10.1063/1.1667253

Holmes, H. R., DeRouin, A. J., Wright, S., Riedemann, T. M., Lograsso, T. A., Rajachar, R. M., & Ghee Ong, K. (2014). Biodegradation and biocompatibility of mechanically active magnetoelastic materials. *Smart Materials and Structures*, *23*(9), 23. DOI: 10.1088/0964-1726/23/9/095036

Hoshi, Y., Naoe, M., & Yamanaka, S. (1982). High-rate, low-temperature sputtering method of facing-targets type and its application for deposition of magnetic films. *Electronics and Communications in Japan (Part I Communications)*, *65*(6), 106–112. DOI: 10.1002/ecja.4410650614

Hu, L., Zhang, R., & Chen, Q. (2014). Synthesis and assembly of nanomaterials under magnetic fields. *Nanoscale*, *6*(23), 14064–14105. DOI: 10.1039/C4NR05108D PMID: 25338267

Islam, M. S., Kurawaki, J., Kusumoto, Y., Abdulla-Al-Mamun, M., & Mukhlish, M. Z. (2011). Hydrothermal Novel Synthesis of Neck-structured Hyperthermia-suitable Magnetic (Fe 3 O 4, γ-Fe 2 O 3 and α-Fe 2 O 3) Nanoparticles. *Journal of Scientific Research*, *4*, 99. DOI: 10.3329/jsr.v4i1.8727

Issa, B.A., & M. Obaidat, I. (2019). Magnetic Nanoparticles as MRI Contrast Agents. Magnetic Resonance Imaging.

Jones, G. A. (1978). Magnetic contrast in the scanning electron microscope: An appraisal of techniques and their applications. *Journal of Magnetism and Magnetic Materials*, *8*(4), 263–285. DOI: 10.1016/0304-8853(78)90096-3

Karpen, W. L., Becker, R. J., & Arnold, W. (1998). Characterization of Electric and Magnetic Material Properties with Eddy-Current Measurements. *Nondestructive Testing and Evaluation*, *15*(2), 93–107. DOI: 10.1080/10589759908952867

Khanam, J., Ahmed, M.F., Zaman, S., Sharmin, N., & Ahmed, S. (2022). Synthesis of nano-sized magnetic iron oxide by a simple and facile co-precipitation method.

Kimura, F., & Kimura, T. (2018). Magnetically textured powders—An alternative to single-crystal and powder X-ray diffraction methods. *CrystEngComm*, *20*(7), 861–872. DOI: 10.1039/C7CE01305A

Kitanovski, A. (2020). Energy Applications of Magnetocaloric Materials. *Advanced Energy Materials*, •••, 10.

Kohashi, T. (2018). Magnetization Analysis by Spin-Polarized Scanning Electron Microscopy. *Scanning*, *2018*, 2018. DOI: 10.1155/2018/2420747 PMID: 29507649

Kornak, R., Niznansky, D., Haimann, K., Tylus, W., & Maruszewski, K. (2005). Synthesis of magnetic nanoparticles via the sol-gel technique. *Materials Science Poland.*

Krings, A., Boglietti, A., Cavagnino, A., & Sprague, S. (2017). Soft Magnetic Material Status and Trends in Electric Machines. *IEEE Transactions on Industrial Electronics*, *64*(3), 2405–2414. DOI: 10.1109/TIE.2016.2613844

Krings, A., & Soulard, J. (2015). Experimental characterization of magnetic materials for electrical machine applications. *2015 IEEE Workshop on Electrical Machines Design, Control and Diagnosis (WEMDCD)*, 85-89. DOI: 10.1109/WEMDCD.2015.7194514

Kryder, M. H. (2003). Future trends in magnetic storage technology. Joint NAPMRC 2003. *Digest of Technical Papers*, •••, 68.

Kumar, C. S., & Mohammad, F. (2011). Magnetic nanomaterials for hyperthermia-based therapy and controlled drug delivery. *Advanced Drug Delivery Reviews*, *63*(9), 789–808. DOI: 10.1016/j.addr.2011.03.008 PMID: 21447363

Kwon, H., Yang, Y., Kim, G., Gim, D., & Ha, M. (2024). Anisotropy in magnetic materials for sensors and actuators in soft robotic systems. *Nanoscale*, *16*(14), 6778–6819. DOI: 10.1039/D3NR05737B PMID: 38502047

Lau, C. N., Xia, F., & Cao, L. (2020). Emergent quantum materials. *MRS Bulletin*, *45*(5), 340–347. DOI: 10.1557/mrs.2020.125

Li, Y., Zhu, H. Y., Hon, C. J., Jiang, Y., & Li, Y. (2013). Hydrothermal & Solvothermal Synthesis of Nanoscale Magnetic Materials (SCI). *Huaxue Jinzhan*, *25*, 276.

Li, Z., Sun, Y., Ge, S., Zhu, F., Yin, F., Gu, L., Yang, F., Hu, P., Chen, G., Wang, K., & Volinsky, A. A. (2023). An Overview of Synthesis and Structural Regulation of Magnetic Nanomaterials Prepared by Chemical Coprecipitation. *Metals*, *13*(1), 152. DOI: 10.3390/met13010152

Lin, H., Gao, Y., Wang, X., Nan, T., Liu, M., Lou, J., Yang, G., Zhou, Z., Yang, X., Wu, J., Li, M., Hu, Z., & Sun, N. X. (2016). Integrated Magnetics and Multiferroics for Compact and Power-Efficient Sensing, Memory, Power, RF, and Microwave Electronics. *IEEE Transactions on Magnetics*, *52*(7), 1–8. DOI: 10.1109/TMAG.2016.2514982

Lin, X., Chen, A., & Wang, S. (2011). Research progress in biocompatibility of magnetic iron oxide nanoparticles. *Kexue Tongbao*, *56*(26), 2223–2228. DOI: 10.1360/972010-2264

Liu, Y. (2002). Solving the Structural Mysteries of Magnetic Materials in TEM. *Microscopy and Microanalysis*, *8*(S02), 402–403. DOI: 10.1017/S1431927602100778

Luryi, S., Xu, J., & Zaslavsky, A. (2013). Future Trends in Microelectronics: Frontiers and Innovations.

Majidi, S., Zeinali Sehrig, F., Farkhani, S. M., Soleymani Goloujeh, M., & Akbarzadeh, A. (2016). Current methods for synthesis of magnetic nanoparticles. *Artificial Cells, Nanomedicine, and Biotechnology*, *44*(2), 722–734. DOI: 10.3109/21691401.2014.982802 PMID: 25435409

Mallick, P., Das, D. K., Mohanty, D., Patnaik, P., Mohapatra, R. K., & Satpathy, S. K. (2024). Introduction to Smart Materials in Micro-and Nanoscale. In Smart Micro-and Nanomaterials for Drug Delivery (pp. 1-17). CRC Press. DOI: 10.1201/9781003468424-1

Mallmann, E. J., Sombra, A. S., Góes, J. C., & Fechine, P. B. (2013). Yttrium Iron Garnet: Properties and Applications Review. *Diffusion and Defect Data, Solid State Data. Part B, Solid State Phenomena*, *202*, 65–96. DOI: 10.4028/www.scientific.net/SSP.202.65

Malozemoff, A. P., Maguire, J. F., Gamble, B. B., & Kalsi, S. S. (2002). Power applications of high-temperature superconductors: Status and perspectives. *IEEE Transactions on Applied Superconductivity*, *12*(1), 778–781. DOI: 10.1109/TASC.2002.1018517

Masi, J.V., & Masi, S.D. (1997). Sol-gel magnetic materials. Proceedings: Electrical Insulation Conference and Electrical Manufacturing and Coil Winding Conference, 565-570.

Matizamhuka, W. R. (2018). The Impact of Magnetic Materials in Renewable Energy-Related Technologies in the 21st Century Industrial Revolution: The Case of South Africa. *Advances in Materials Science and Engineering*, *2018*(1), 3149412. DOI: 10.1155/2018/3149412

McHenry, M. E., & Laughlin, D. E. (2000). Nano-Scale Materials Development for Future Magnetic Applications. *ChemInform*, •••, 31.

Mészáros, I. (2007). Development of a Novel Vibrating Sample Magnetometer. *Materials Science Forum*, *537-538*, 413–418. DOI: 10.4028/www.scientific.net/MSF.537-538.413

Mikkonen, R. (2002). Highlights of SC power applications in Europe. *IEEE Transactions on Applied Superconductivity*, *12*(1), 782–787. DOI: 10.1109/TASC.2002.1018518

Morón, C., Cabrera, C., Morón, A., García, A., & González, M. (2015). Magnetic Sensors Based on Amorphous Ferromagnetic Materials: A Review. *Sensors (Basel)*, *15*(11), 28340–28366. DOI: 10.3390/s151128340 PMID: 26569244

Muhit, M.S., & Hoque, R. (2011). Magnetic and Electric Characterization of Materials for Electrical Machines.

Mushnikov, N. V. (2022). Basics of magnetism and magnetic materials. In *Magnetic Materials and Technologies for Medical Applications* (pp. 23–55). Woodhead Publishing. DOI: 10.1016/B978-0-12-822532-5.00010-8

Naima, S.G., Zohra, S.F., & Samira, S. (2013). Magnetic materials for clean environment energy applications.

Nakamura, K., Ohta, Y., Itoh, A., & Hayashi, C. (1982). Magnetic properties of thin films prepared by continuous vapor deposition. *IEEE Transactions on Magnetics*, *18*(6), 1077–1079. DOI: 10.1109/TMAG.1982.1062143

Nakamura, K., Yamada, T., Ohta, Y., & Itoh, A. (1982). "GT target", A new high rate sputtering target of magnetic materials. *IEEE Transactions on Magnetics*, *18*(6), 1080–1082. DOI: 10.1109/TMAG.1982.1062177

Namdeo, M., Saxena, S., Tankhiwale, R., Bajpai, M., Mohan, Y. M., & Bajpai, S. K. (2008). Magnetic nanoparticles for drug delivery applications. *Journal of Nanoscience and Nanotechnology*, *8*(7), 3247–3271. DOI: 10.1166/jnn.2008.399 PMID: 19051873

Naoe, M., Yamanaka, S., & Hoshi, Y. (1980). Facing targets type of sputtering method for deposition of magnetic metal films at low temperature and high rate. *IEEE Transactions on Magnetics*, *16*(5), 646–648. DOI: 10.1109/TMAG.1980.1060683

Narayanan, P., Pramanik, R., & Arockiarajan, A. (2024). Hard magnetics and soft materials—A synergy. *Smart Materials and Structures*, *33*(4), 043001. DOI: 10.1088/1361-665X/ad2bd8

Nayak, S. K., Patnaik, P., Das, D. K., & Barala, M. (2024). Electronic and half-metallic behaviour of manganese doped SiC Zigzag nanotubes using density functional theory. *Chemical Physics Impact*, *8*, 100417. DOI: 10.1016/j.chphi.2023.100417

Neumann, W., Kirmse, H., Haeusler, I., Zheng, C., & Mogilatenko, A. (2009). Quantitative Structure Analysis of Nano-sized Materials by Transmission Electron Microscopy. MRS Proceedings, 1184. DOI: 10.1557/PROC-1184-GG03-02

Niazi, A., Poddar, P., & Rastogi, A. K. (2000). A precision, low-cost vibrating sample magnetometer. *CURRENT SCIENCE-BANGALORE*, *79*(1), 99–109.

Nikov, R. G., Avdeev, G., Dikovska, A. O., Koleva, M. E., & Nedyalkov, N. (2020). Microstructural characterization of nanocomposites produced by laser ablation in a magnetic field. *Journal of Physics: Conference Series*, *1492*(1), 1492. DOI: 10.1088/1742-6596/1492/1/012057

Pasquale, M. (2003). Magnetic sensors and actuators. *Journal of Magnetics*, *8*(1), 60–69. DOI: 10.4283/JMAG.2003.8.1.060

Petrescu, L. (2009). MAGNETIC MATERIAL CHARACTERIZATION BY OPEN SAMPLE MEASUREMENTS.

Polytechnic, B. (1991). Magnetism: It is Permanent.

Pouponneau, P., Yahia, L., Merhi, Y., Epure, L. M., & Martel, S. (2006). Biocompatibility of Candidate Materials for the Realization of Medical Microdevices. 2006 International Conference of the IEEE Engineering in Medicine and Biology Society, 2362-2365.

Pulliam, G. R. (1967). Chemical Vapor Growth of Single-Crystal Magnetic Oxide Films. *Journal of Applied Physics*, *38*(3), 1120–1126. DOI: 10.1063/1.1709508

Ranjani, M., Jesurani, D., Priyadharshini, M., & Vennila, S. M. (2016). SOL-GEL SYNTHESIS AND CHARACTERIZATION OF ZINC SUBSTITUTED COBALT FERRITE MAGNETIC NANOPARTICLES. *International Journal of Approximate Reasoning*, *4*, 53–58.

Rathore, D., Kurchania, R., & Pandey, R. K. (2012). Synthesis and Characterization of Cobalt Ferrite Nanoparticles Grown by Chemical Co-precipitation Method. *Invertis Journal of Science & Technology*, *5*, 132–137.

Reddy, J., Lakshya, A.K., Raj, R., Kumar, L., & Chowdhury, A. (2023). Characterizing Curie and Néel Point Phase Transitions via Thermal Techniques. physica status solidi (b), 260.

Rodrigues, L., & Jewell, G. W. (2014). Model Specific Characterization of Soft Magnetic Materials for Core Loss Prediction in Electrical Machines. *IEEE Transactions on Magnetics*, *50*(11), 1–4. DOI: 10.1109/TMAG.2014.2328099

Smith, D.J., McCartney, M.R., & Dunin-Borkowski, R.E. (2006). Nanoscale Structural and Magnetic Characterization Using Electron Microscopy.

Sowter, G. A. (1987). Soft magnetic materials for audio transformers: History, production, and applications. *Journal of the Audio Engineering Society*, *35*, 760–777.

Sreenivasulu, K. V., & Srikanth, V. V. (2016). Fascinating Magnetic Energy Storage Nanomaterials: A Brief Review. *Recent Patents on Nanotechnology*, *11*(2), 116–122. PMID: 28286941

Tachiki, M., & Kobayashi, T. (1999). Manipulation of Laser Ablation Plume by Magnetic Field Application. *Japanese Journal of Applied Physics*, *38*(6R), 3642. DOI: 10.1143/JJAP.38.3642

Tang, F., & Wan, X. (2023). Evolution of topological states in magnetic materials by symmetry-breaking. arXiv preprint arXiv:2302.13622.

Tannous, C., & Comstock, R. (2017). Magnetic Information-Storage Materials.

Tripathi, S. K., Kaur, R., Bharti, S., Gawri, I., & Kaur, J. (2015). Recent Advances in Synthesis, Properties and Applications of Magnetic Oxide Nanomaterials. *Diffusion and Defect Data, Solid State Data. Part B, Solid State Phenomena*, *232*, 1–44. DOI: 10.4028/www.scientific.net/SSP.232.1

Trout, S.R. (2001). Material selection of permanent magnets, considering thermal properties correctly. Proceedings: Electrical Insulation Conference and Electrical Manufacturing and Coil Winding Conference (Cat. No.01CH37264), 365-370.

Unguris, J., Scheinfein, M. R., Celotta, R. J., & Pierce, D. T. (1991). Scanning Electron Microscopy with Polarization Analysis: Studies of Magnetic Microstructure. *Journal of Vacuum Science and Technology*, •••, 239–265.

Vasile, C., Engel, T., Risser, M., Muller, C., & Imbs, I.A. (2008). ENERGY EFFICIENT AND ENVIRONMENTAL SAFE MAGNETIC COOLING SYSTEM.

Vettier, C., & Mcwhan, D. B. (1989). Magnetic X-ray diffraction from ferromagnetic materials. *Physica B, Condensed Matter*, *159*(2), 106–110. DOI: 10.1016/0921-4526(89)90027-6

Wardly, G. A. (1971). Magnetic Contrast in the Scanning Electron Microscope. *Journal of Applied Physics*, *42*(1), 376–386. DOI: 10.1063/1.1659604

Welipitiya, D., He, Y. L., Zhang, J., Oden, P. I., Thundat, T., Warmack, R. J., Gobulukoglu, I., Shan, Z., Sellmyer, D. J., & Dowben, P. A. (1996). Fabrication of large arrays of micron-scale magnetic features by selective area organometallic chemical vapor deposition. *Journal of Applied Physics*, *80*(3), 1867–1871. DOI: 10.1063/1.362999

Wu, M., & Huang, S. (2017). Magnetic nanoparticles in cancer diagnosis, drug delivery and treatment. *Molecular and Clinical Oncology*, *7*(5), 738–746. DOI: 10.3892/mco.2017.1399 PMID: 29075487

Yao, C., & Ma, Y. (2021). Superconducting materials: Challenges and opportunities for large-scale applications. *iScience*, *24*(6), 24. DOI: 10.1016/j.isci.2021.102541 PMID: 34136765

Zalite, I., Heidemane, G., Kuznetsova, L. V., & Maiorov, M. (2015). Hydrothermal Synthesis of Cobalt Ferrite Nanosized Powders. *IOP Conference Series. Materials Science and Engineering*, *77*, 012011. DOI: 10.1088/1757-899X/77/1/012011

Zhao, P., Tang, D., Xie, J., & Zhang, C. (2024). Magnetism and Magnetic Materials. In *Magnetic Levitation: Innovation of Density-Based Applications* (pp. 1–15). Springer Nature Singapore. DOI: 10.1007/978-981-99-8314-8_1

Zhou, L., Boyd, E. P., Dengb, H., Wigen, P. E., & Shoreb, S. G. (1993). Synthesis and evaluation of magnetic thin films. *IEEE Transactions on Magnetics*, *29*(6), 3052–3054. DOI: 10.1109/20.281114

Zlobin, A. V. (2014). *Superconducting Magnets - Principles*. Operation, and Applications.

Chapter 7

The Intersection of Machine Learning and Structural Health Monitoring:
A Bibliometric Analysis

Houssine Laaz
http://orcid.org/0009-0004-0768-7877
Laboratory of Mechanics, Engineering and Innovation, National High School of Electricity and Mechanics, Hassan II University of Casablanca, Morocco

Moussa Mekkaoui
http://orcid.org/0009-0001-5775-970X
Laboratory of Mechanics, Engineering and Innovation, National High School of Electricity and Mechanics, Hassan II University of Casablanca, Morocco

Salah Nissabouri
http://orcid.org/0000-0001-7702-3030
Laboratory of Modelling and Simulation of Intelligent Industrial Systems, ENSET of Mohammedia, Hassan II University of Casablanca, Morocco

Hassan Rhimini
http://orcid.org/0000-0002-5784-9844
Laboratory of Mechanics, Engineering and Innovation, National High School of Electricity and Mechanics, Hassan II University of Casablanca, Morocco

ABSTRACT

This study conducts a comprehensive bibliometric analysis of machine learning (ML) applications in Structural Health Monitoring (SHM) from 2003 to early 2024.

DOI: 10.4018/979-8-3693-6627-1.ch007

© 2027, IGI Global Scientific Publishing. Copying or distributing in print or electronic forms without written permission of IGI Global Scientific Publishing is prohibited. Use of this chapter to ...erative artificial intelligence (AI) technologies is expressly prohibited. The publisher reserves all rights to license its use for generative AI training and machine learning model development.

Utilizing the OpenAlex database, we analyzed 806 publications using keywords related to ML and SHM. The data was processed using VOSviewer software to examine co-citation patterns, keyword frequency, author/institution collaboration networks, and citation analysis. Our findings reveal an exponential growth in research output, with 80% of publications occurring in the last four years. The United States, China, and the United Kingdom emerge as leading contributors. Keyword analysis indicates a shift towards advanced ML techniques like convolutional neural networks. The study identifies key research clusters, influential publications, and emerging trends in the field. This analysis provides valuable insights for researchers and practitioners, highlighting potential areas for future investigation in ML-based SHM.

I. INTRODUCTION

Structural Health Monitoring (SHM) plays a vital role in ensuring the safety and integrity of complex structures across diverse industries, including oil and gas (Centenaro et al., 2019), aeronautics (Lizé et al., 2018; Zitouni et al., 2024), and civil engineering (Z. Sun & Xia, 2022). By employing a network of sensors and data analysis techniques, SHM facilitates continuous monitoring of a structure's health, enabling the early detection of potential damage and defects. This proactive approach is crucial for preventing catastrophic failures and ensuring public safety (An et al., 2019; Z. Sun & Xia, 2022).

Traditional SHM techniques, such as ultrasonic testing (UT) and eddy current testing (ECT), are widely adopted but face inherent limitations. UT, utilizing sound waves for defect identification, requires skilled personnel and can be time-consuming for large structures (Gharehbaghi et al., 2022). Similarly, ECT, employing electromagnetic fields for surface flaw detection, demands meticulous setup and interpretation (Gharehbaghi et al., 2022) . These factors contribute to the high cost and complexity associated with traditional SHM methods, particularly for widespread implementation in complex structures.

The emergence of artificial intelligence (AI), including machine learning (ML) and deep learning (DL), presents a transformative opportunity for SHM particularly in early damage detection (Salah & Barra Ndiaye, 2024). AI encompasses the broad field of creating intelligent machines, while ML is a subset focusing on algorithms that improve through experience. DL, a further specialization of ML, utilizes neural networks with multiple layers to model complex patterns. These AI-powered approaches offer compelling advantages over traditional methods, including enhanced accuracy and efficiency, cost reduction, real-time monitoring, predictive analytics, and scalability (Luckey et al., 2022).

To fully leverage the potential of AI in SHM, a comprehensive understanding of the existing research landscape is crucial. Bibliometrics, the statistical analysis of published works, offers a powerful tool for mapping this landscape. By analyzing factors such as keywords, citations, and co-authorship networks, bibliometric studies can identify emerging trends, influential researchers, and key collaborations. While bibliometric analyses have been conducted separately in the domains of SHM (Mahbubi et al., 2024) and AI (Nur'aeni & Zalsahra, 2024), a recent comprehensive analysis specifically exploring the intersection of AI/ML/DL and SHM is lacking.

This study aims to conduct a thorough bibliometric analysis of research on ML applications in SHM from 2003 to early 2024. By leveraging the OpenAlex database and employing robust bibliometric techniques, we seek to:

1. Map the evolution of research in this domain over time, identifying key milestones and periods of rapid growth.
2. Uncover the most frequently used ML techniques within the context of SHM applications.
3. Identify the most prominent research institutions and authors leading the field.
4. Delineate the major research themes and subfields within AI-powered SHM.
5. Pinpoint potential knowledge gaps and propose promising avenues for future research.

The findings of this bibliometric analysis will provide valuable insights for researchers, engineers, and practitioners working in the field of SHM. By understanding the current research landscape, this study can guide further exploration of ML applications in SHM, ultimately contributing to the development of more efficient, accurate, and cost-effective SHM strategies.

II. METHOD

To gain a deeper understanding of the current research landscape in machine learning (ML) applications for structural health monitoring (SHM), this study employed bibliometric methods. Bibliometrics, a field utilizing statistical techniques to analyze publication patterns and research impact (Benckendorff & Zehrer, 2013), offers valuable insights by examining citation data, co-authorship trends, and keyword co-occurrence (Kim & Zhu, 2018). Through this analysis, we can identify influential articles, prominent researchers, and emerging areas of inquiry within the field of ML-based SHM.

In this study, we will limit ourselves to quantitative analysis; Figure 1 summarizes the methodology followed.

Figure 1. Flowchart of the approach

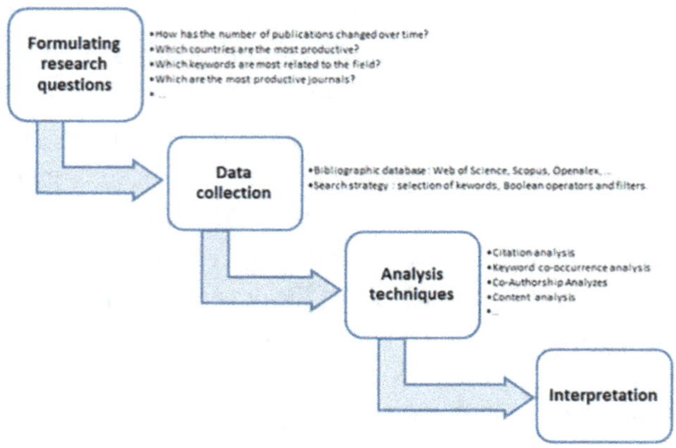

This chapter will follow a four-step approach, outlined in Figure 1, to comprehensively analyze the field of ML-based SHM research. The following sections will detail how each step will be adapted to suit our specific case study.

Research Questions

"Research questions" in a bibliometric analysis are not the same as the research questions we might find in a traditional scientific study. Instead of focusing on specific hypotheses or phenomena, these questions guide the analysis of the data, helping identify patterns, trends, and key insights that can advance the field within the research literature itself.

To validate our objectives and explore the key aspects of ML applications in SHM (ML x SHM), this chapter seeks to answer the following research questions:

- Q 1: What is the current state and historical development of ML x SHM research as reflected in scholarly publications?
- Q 2: Which are the most productive journals in ML x SHM?
- Q 3: Who are the most prolific authors in the field of ML x SHM?
- Q 4: Which countries are the most productive in the field?
- Q 5: What are the potential future directions and emerging trends in ML-based SHM research?
- Q 6: Which are the most popular publications in ML x SHM?

Data Collection

Data mining was carried out between March 25th and April 5th, 2024, using OpenAlex a free and open-source bibliographic database specifically designed for the field of scholarly research it was created as a replacement for the Microsoft Academic Graph (MAG) and is maintained by the OpenResearch Foundation (Culbert et al., 2024).

Figure 2. Venn diagram of the intersection of database records based on exact DOI match, published between 2015 and 2022 from (Culbert et al., 2024)

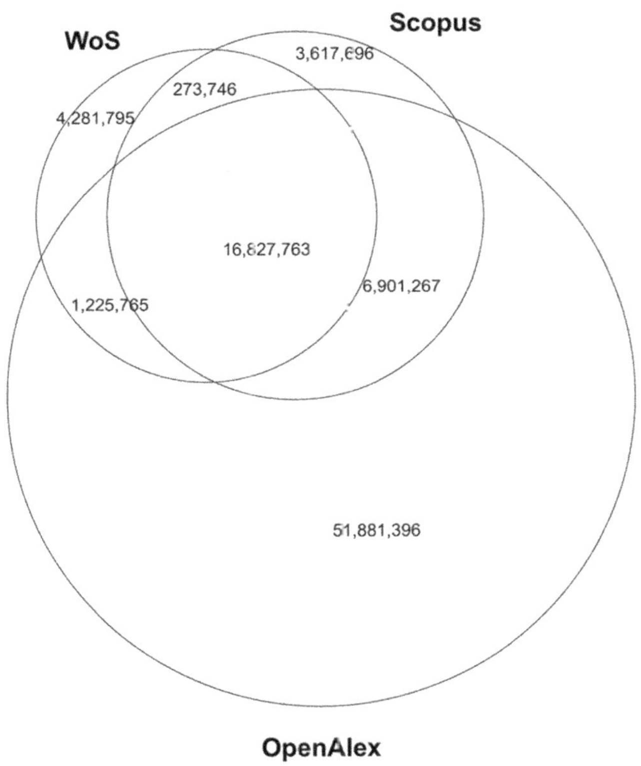

The choice of OpenAlex is justified by the following three criteria:

- Openness and Accessibility: Completely open and free to access, allowing anyone to use and analyze the data without restrictions.
- Coverage: Aims for comprehensive coverage of scholarly works, including many sources not found in Scopus or Web of Science. This is particular-

ly beneficial for researchers interested in interdisciplinary research or niche fields (figure 2).

- Transparency and Data Structure: Provides detailed documentation on its data sources and methodology, promoting transparency and reproducibility of research.

This research focused on reviewing existing literature (peer-reviewed journal articles only, in English) on the application of machine learning in structural health monitoring. The keywords "machine learning" and "structural health monitoring" were searched for in the title and abstract. The publication dates of the reviewed works ranged from 2003 to the end of the first quarter of 2024.

Data Analysis:

To facilitate the interpretation of the results, the collected data is represented by tables or graphically using charts or network maps, created with the Vosviewer software (Kirby, 2023).

Analysis Techniques that will be used in this chapter:

- Publication Trend Analysis: Examine the growth of publications over time to assess the development and research activity in ML-based SHM.
- Citation Analysis: Analyze citation counts and patterns to identify highly cited publications and influential authors.
- Co-citation Network Analysis: Explore the intellectual structure of the field by identifying clusters of highly cited publications that share common themes.
- Keyword Analysis: Identify frequently used keywords and their co-occurrence patterns to uncover major research themes and trends.
- Co-authorship Analysis: Examine collaboration patterns between authors and institutions to understand research networks and international collaborations.

Interpretation:

The results of the bibliometric analysis will be presented as answers to the research questions. The findings will be interpreted in the context of the existing literature and current challenges in SHM to draw meaningful conclusions and identify potential future directions.

III. RESULTS

Q 1: How has the Number of Publications Changed Over Time?

To investigate the current state and historical development of Machine Learning (ML) applications in Structural Health Monitoring (SHM), we conducted a comprehensive bibliometric analysis using data from OpenAlex. The search focused on publications containing both "machine learning" and "structural health monitoring" keywords from 2003 to the first quarter of 2024. Figure 3 presents the results of this analysis, illustrating both the annual publication count and the cumulative number of articles over time. This visualization allows us to track the field's growth, identify key periods of acceleration, and assess the current state of research in this interdisciplinary area.

Figure 3. The annual and cumulative numbers of research articles on ML & SHM in OpenAlex from 2003 until the end of the first quarter of 2024

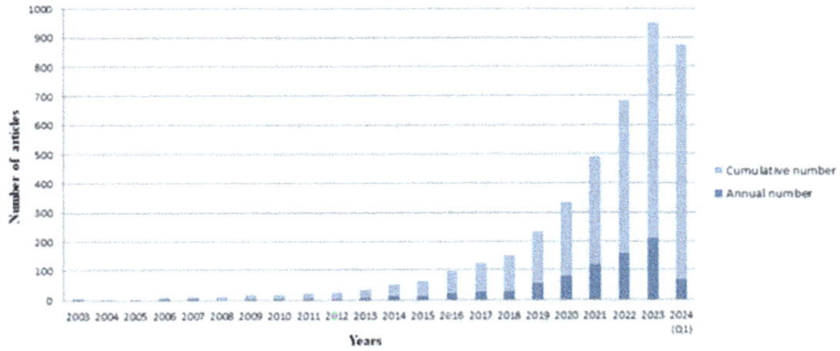

The bibliometric analysis revealed a total of 806 articles published in the last 20 years, with a remarkable 80% of these publications occurring within just the last four years. This recent surge in research output underscores the rapidly growing interest in applying ML techniques to SHM challenges. The cumulative number of articles exhibits an exponential growth trend, indicating not only increased research activity but also suggesting a potential paradigm shift in how SHM is approached. This exponential growth may be attributed to factors such as advancements in ML algorithms, increased availability of SHM data facilitated by robust numerical simulations (Laaz et al., 2025, 2026), and a growing recognition of ML's potential to enhance the accuracy and efficiency of structural monitoring and assessment processes. The trend also reflects the interdisciplinary nature of this research area, potentially drawing increased attention from both the artificial intelligence and SHM communities.

Q 2: Which are the Most Productive Journals?

Before addressing research question Q 2, we asked ourselves: Which are the preferred sources?

To gain a deeper understanding of the publication landscape in the field of Machine Learning applications in Structural Health Monitoring, we analyzed the distribution of sources by type for the 806 papers identified in our bibliometric study. Figure 4 illustrates this distribution, categorizing the publications into journals, conference proceedings, reference articles, and other source types. This analysis provides insights into the preferred publication venues for researchers in this interdisciplinary field.

Figure 4. Distribution of sources by type

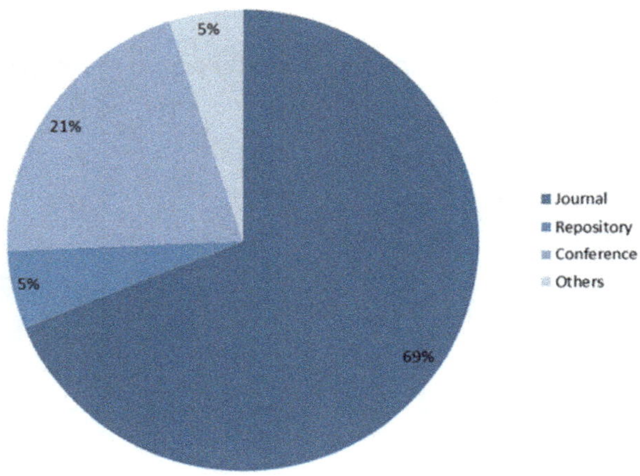

The analysis of source types reveals a clear preference for journal publications in the ML x SHM research domain. Out of the 806 papers examined, 556 (69%) were published in journals, indicating a strong inclination towards peer-reviewed, high-impact venues. Conference proceedings account for 21% of the publications, reflecting the importance of academic conferences in disseminating new ideas and fostering collaborations in this field. Interestingly, 5% (45 papers) were deposited as reference articles, with 36 of these in the arXiv repository hosted principally by Cornell University. This highlights the growing trend of preprint sharing in the scientific community, allowing for rapid dissemination of research findings. The remaining 5% of papers were published in other, less common source types. This distribution underscores the diverse channels through which ML x SHM research is communicated, with a strong emphasis on traditional academic publishing routes while also embracing newer, more accessible platforms for knowledge sharing.

To address research question Q2, "Which are the most productive journals?", we analyzed the publication data and ranked journals according to the number of publications related to Machine Learning (ML) applications in Structural Health Monitoring (SHM). Table 1 presents the top 10 most productive journals in this field, along with key metrics such as total publications, ML x SHM-specific publications, impact factor, and the most cited article for each journal.

Table 1. Top 10 Most Productive Journals in Machine Learning (ML) for Structural Health Monitoring (SHM) Research based on Number of Publications

No.	Journal	Total pub.	Pub. ML x SHM	I.F. (2024)	The most cited article	Time cited	publisher
1	Sensors	53 490	47	3.4	Metal Oxide Gas Sensors: Sensitivity and Influencing Factors (Wang et al., 2010)	2178	Multidisciplinary Digital Publishing Institute
2	Structural health monitoring	2 210	39	5.7	Vibration-based Damage Identification Methods: A Review and Comparative Study (Wei Fan & Pizhong Qiao, 2011)	1562	SAGE Publishing
3	Mechanical systems and signal processing	9 108	32	7.9	A review on machinery diagnostics and prognostics implementing condition-based maintenance (Jardine et al., 2006)	3508	Elsevier BV
4	Applied sciences	49 560	21	2.5	Mini-LED and Micro-LED: Promising Candidates for the Next Generation Display Technology (Wu et al., 2018)	492	Multidisciplinary Digital Publishing Institute
5	Structural control & health monitoring	2 341	13	4.6	Smart sensing technology: opportunities and challenges (Spencer et al., 2004)	483	Wiley-Blackwell
6	Engineering structures	16 340	12	5.6	Detection of structural damage through changes in frequency: a review (Salawu, 1997)	1807	Elsevier BV
7	Smart materials and structures	8 537	11	3.7	A review of power harvesting using piezoelectric materials (2003–2006) (Anton & Sodano, 2007)	2274	IOP Publishing
8	Automation in construction	4 891	8	9.6	Building Information Modeling (BIM) for existing buildings — Literature review and future needs (Volk et al., 2014)	1439	Elsevier BV

continued on following page

Table 1. Continued

No.	Journal	Total pub.	Pub. ML x SHM	I.F. (2024)	The most cited article	Time cited	publisher
9	IEEE sensors journal	19 620	8	4.3	Wearable Sensors for Human Activity Monitoring: A Review (Mukhopadhyay, 2015)	1098	IEEE Sensors Council
10	Ultrasonics	11 270	7	3.8	Synthetic aperture ultrasound imaging (Jensen et al., 2006)	665	Elsevier BV

Table 1 shows that six separate publishers own the top 10 most competitive journals. The Multidisciplinary Digital Publishing Institute owns the journals in the first and fourth positions. SAGE Publishing owns the journal in the second position, and Elsevier BV publishes four journals out of the 10, including one in the third position. "Sensors" with 47 articles is the most productive journal, followed closely by "Structural Health Monitoring" with 39 publications. These two journals, along with "Mechanical Systems and Signal Processing" (32 publications), stand out as the primary outlets for research in this field.

The dominance of multidisciplinary journals, such as "Sensors", "Applied Sciences", and "Smart Materials and Structures", reflects the inherent need for a cross-disciplinary approach, drawing expertise from fields including engineering, computer science, and materials science. This interdisciplinary nature is crucial for developing innovative solutions to enhance structural health monitoring.

It's also interesting to note the emergence of other journals like "Mechanical Systems and Signal Processing" and "Automation in Construction." This indicates the broadening scope of ML x SHM research to encompass areas like machinery diagnostics, prognostics, and building information modeling.

To provide a comparative overview of the journals analyzed in this study, we have included both the Impact Factor (IF) and the total number of publications for each journal. While IF is a widely accepted metric in academic publishing, we acknowledge its limitations. A high IF does not necessarily equate to superior research quality. Therefore, we encourage the reader to consider IF within a broader context. Combining IF with the total number of publications allows for a more nuanced understanding of a journal's influence and research output.

According to the Journal Citation Reports (JCR): Impact Factor June 2024 (Web of Science), four journals had an Impact Factor (IF) of 5 or above. The highest IFs belonged to Automation in Construction (9.6) and Mechanical Systems and Signal Processing (7.9). On the other hand, it is noticeable that the two journals published by Multidisciplinary Digital Publishing Institute had the lowest IFs, despite the large number of articles published. The Specialized journals focused solely on structural health monitoring, such as Structural Health Monitoring and Structural Control & Health Monitoring, boast impressive IFs.

This study provides a snapshot of the most productive journals. A deeper analysis is needed to explore the specific research themes, author networks, and citation patterns to understand the full scope and impact of ML x SHM research.

Q 3: Who are the Most Prolific Authors?

To gain insights into the leading researchers in the field of Machine Learning (ML) applications in Structural Health Monitoring (SHM), we analyzed the publication data to identify the most prolific authors. Table 2 presents a list of the eight most productive researchers who have co-authored at least eleven works in the ML x SHM research area. The Table 2 includes key metrics such as total publications, total citations, h-index, year of first publication, current affiliation, and country. This analysis helps us understand the influential figures shaping the field and their contributions to the advancement of the research.

Table 2. List of the 8 most prolific authors (those who have co-authored at least eleven works published in the ML x SHM research area).

No.	Author	ORCID ID	Total pub.	Total cit.	H-index	Year of First pub.	Current affiliation	country
1	Keith Worden	0000-0002-1035-238X	758	18720	63	1989*	University of Sheffield	UK
2	Nikolaos Dervilis	0000-0002-5712-7323	205	1885	25	2012	University of Sheffield	UK
3	Paul Gardner	0000-0002-1882-9728	67	1432	18	2016	University of Sheffield	UK
4	Robert J. Barthorpe	0000-0002-6645-8482	87	707	13	2008*	University of Sheffield	UK
5	Diego Alexander Tibaduiza Burgos	0000-0002-4498-596X	133	925	15	2007*	Universidad Nacional de Colombia	Colombia
6	Onur Avcı	0000-0003-0221-7126	169	4293	21	2002*	West Virginia University	USA
7	Elói Figueiredo	0000-0002-9168-6903	77	2271	22	1992*	Universidade Lusófona	Portugal
8	Larry Bull	0000-0002-0225-5010	454	3988	34	2009*	University of Cambridge	UK

*First author

The analysis reveals several interesting patterns among the most prolific authors in ML x SHM research. Notably, four of the eight researchers (Keith Worden, Nikolaos Dervilis, Paul Gardner, and Robert J. Barthorpe) are affiliated with the University of Sheffield in the UK, indicating a strong research cluster in this field. Keith Worden stands out with the highest number of total publications (758) and citations (18,720), as well as the highest h-index (63), suggesting his significant influence in the field. The list includes researchers from diverse geographical locations, including the UK, Colombia, USA, and Portugal, highlighting the global nature of ML x SHM research. Interestingly, the years of first publication range from 1989 to 2016, demonstrating that both established researchers and relatively newer entrants are making significant contributions to the field. The high citation counts and h-indices across the board indicate that these researchers are producing impactful work that is widely recognized in the scientific community.

However, it's important to note that this table only considers the top 8 authors. Analyzing a larger sample of authors could reveal additional insights into the broader research landscape, including emerging researchers, diverse institutional contributions, and potential regional trends in ML x SHM research. This study not only identifies the key researchers in ML x SHM but also serves as a starting point for identifying potential collaborations between authors from different countries or institutions.

Furthermore, by analyzing the publications of these authors over time, future research can pinpoint specific research directions that are gaining momentum, as well as areas that require further investigation. This temporal analysis could reveal shifts in research focus, emerging subfields within ML x SHM, and the evolution of methodologies and applications over time. Additionally, examining the network of co-authors and institutions beyond this top tier could uncover important collaborative patterns and identify other significant contributors who may not appear in the top publication counts but play crucial roles in advancing the field.

Q 4: Which Countries are the Most Productive?

To address research question Q 4, “Which countries are the most productive?”, we conducted a comprehensive analysis of international contributions to Machine Learning (ML) applications in Structural Health Monitoring (SHM). Our study applied a threshold of a minimum of 15 documents per country, which revealed the top 15 most successful countries driving the growth of research activity in ML combined with SHM. This approach allows us to focus on the most significant contributors to the field while still capturing a diverse range of international perspectives. Figures 5 and 6 present our findings in two complementary visualizations. Figure 5 offers a bar chart ranking these top 15 countries by the number of publications and international collaborations, providing a clear quantitative comparison. Figure 6 goes further by illustrating the network of connections between countries based on co-authorships, offering insights into the collaborative landscape of ML x SHM research. Together, these visualizations not only identify the most productive countries but also reveal patterns of international cooperation and knowledge exchange in this rapidly evolving field.

Figure 5. Bar chart showing the top 15 countries ranked by number of publications and international collaborations

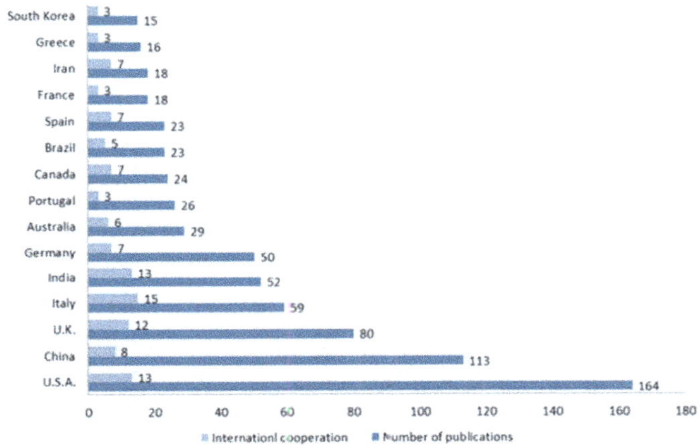

Figure 5 presents a bar chart ranking the top 15 countries by the number of publications and international collaborations in ML-based Structural Health Monitoring (SHM) research. The United States, China, and the United Kingdom emerge as clear leaders in this field, dominating both in terms of total publications and collaborative efforts. This prominence can be attributed to factors such as robust research infrastructure, substantial government funding, and a concentration of expertise within these nations. Following the top three, countries like Italy, Germany, Portugal, and Australia also make significant contributions to the research landscape, albeit to a lesser extent. Interestingly, the chart also highlights the rising influence of countries such as Greece, Iran, and South Korea. While their current output is comparatively smaller, their position on the graph suggests potential for future growth, indicating that ML-based SHM research is becoming increasingly global and diverse, with contributions from a wider range of regions.

Figure 6. The network of connections between countries created based on co-authorships

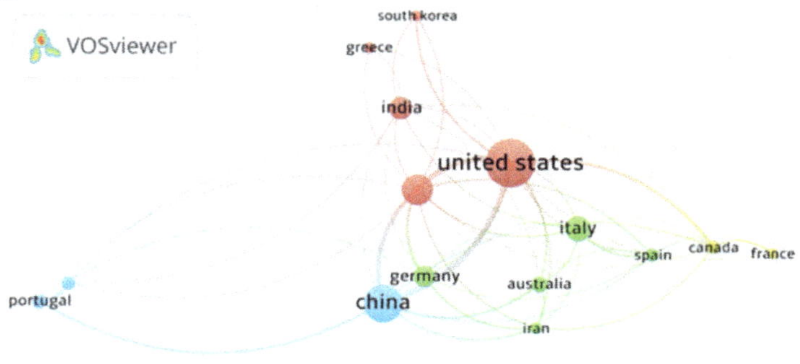

Figure 6 visualizes the network of connections between countries based on co-authorships in ML-based SHM research. This network diagram provides valuable insights into the collaborative landscape of the field. The larger node sizes and central positions of the United States, China, and the United Kingdom reinforce their status as the most productive and influential countries in this research area. The visualization reveals strong collaborative ties among these top three countries, as well as with other major contributors like Germany. This interconnectedness suggests a powerful exchange of knowledge and expertise, likely contributing significantly to the rapid advancements in ML x SHM. Notably, Italy stands out as the most open country to international collaboration, as indicated by both Figures 5 and 6. This suggests that Italian researchers actively seek partnerships beyond their borders, potentially playing a crucial role in bridging different research communities. China and the USA also rank highly in terms of openness to international collaboration, further cementing their positions as global research hubs.

Q 5: Which Keywords are Most Related to the Field?

To address research question Q 5, "Which keywords are most related to the field?", we conducted a comprehensive analysis of keyword co-occurrences in Machine Learning (ML) applications for Structural Health Monitoring (SHM). Our study focused on the period from 2019 to the first quarter of 2024, encompassing 80% of the total publications in this field. This recent timeframe ensures a normalized representation of current research trends. From a total of 1,290 keywords, we selected the 50 keywords with the greatest total link strength, each appearing at least 10 times in the literature. Figure 7 presents a co-occurrence network of these keywords, visually mapping the connections within the ML-based SHM research landscape. This approach allows us to identify key themes, explore how research trends have changed over time, and understand the interconnections between different aspects of the field.

Figure 7. Co-occurrence Network of Keywords in ML-based SHM Research (2019 - 2024 (Q1))

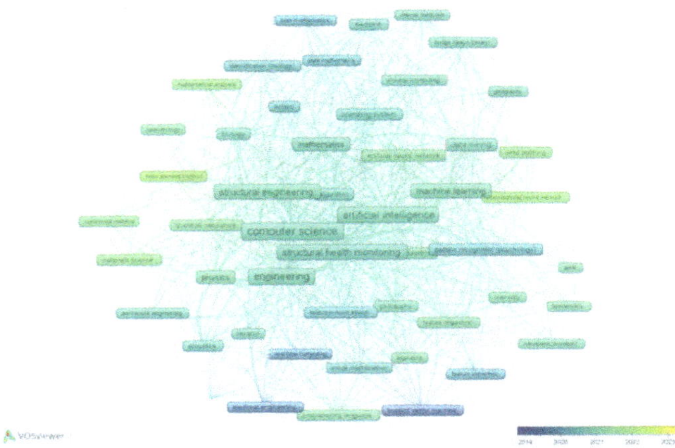

The keyword co-occurrence network in Figure 7 reveals significant insights into the ML-based SHM research landscape. The central position and large node size of “machine learning” and “structural health monitoring” underscore their role as core concepts binding this research field together. Notably, we observe the emergence and interconnection of various disciplines, including engineering fields, computer science, mathematics, material science, and even medicine. This multidisciplinary nature underscores the complexity and broad applicability of ML in SHM. The analysis also highlights the evolution of popular ML algorithms over time. Support Vector Machines (SVMs) were prominent in 2020 and 2021, while Artificial Neural Networks (ANNs) gained traction in 2022. More recently, in 2023, Convolutional Neural Networks (CNNs) have become increasingly prevalent, indicating a shift towards more advanced deep learning techniques. This progression suggests a growing sophistication in the ML approaches applied to SHM problems.

Delving deeper into the network, we can identify key clusters and trends shaping the field. The proximity and strong links between “damage detection,” “vibration analysis,” and “feature extraction” highlight the critical role of these processes in ML-based SHM applications. Keywords such as “optimization,” “pattern recognition,” and “classification” form a cluster, pointing to key methodological approaches in processing and interpreting SHM data. The presence of “internet of things” and “big data” nodes indicates the growing importance of data collection and management, likely driven by the increasing use of sensor networks and real-time monitoring systems. Interestingly, the appearance of "uncertainty quantification" and "reliability" suggests an increasing focus on the robustness and trustworthiness of ML models in SHM applications, a critical consideration for real-world deployments. Furthermore, the presence of specific application areas like "bridge monitoring" and "wind turbines" indicates the diverse range of structures benefiting from ML-based SHM techniques. The analysis addressing Q5 (Which keywords are most related to the field?) not only captures the current state of ML-based SHM research but also points to potential future directions and emerging areas of interest.

Q 6: Which are the Most Popular Publications?

Two-Stage Classification Approach

To identify the most impactful publications in ML-based SHM, a two-stage classification approach was employed. First, publications were ranked based solely on the number of citations, a traditional measure of influence. This initial ranking identified works with widespread adoption and reference within the field. Then, from this list of highly cited publications, the top 10 articles based on co-citation counts were selected.

First Stage of Classification

Table 3 displays the most popular publications based on citation count, It showcases the authors, publication year, number of citations and topic each publication, representing the first stage of our analysis. To limit the scope of the table, only the 10 highest-cited articles are presented.

Table 3. Top 10 most popular publications based on number of citation

No	Article reference	Authors	Year of pub.	Number of citations	Topic
1	1D convolutional neural networks and applications: A survey (Kiranyaz et al., 2021)	Serkan Kıranyaz, Onur Avcı, Osama Abdeljaber, Türker İnce, Moncef Gabbouj, Daniel J. Inman	2021	1 016	Deep Learning in Computer Vision and Image Recognition
2	Guided wave based structural health monitoring: A review (Mitra & Gopalakrishnan, 2016)	Mira Mitra, S. Gopalakrishnan	2016	674	Guided Wave Structural Health Monitoring in Materials
3	1-D CNNs for structural damage detection: Verification on a structural health monitoring benchmark data (Abdeljaber et al., 2018)	Osama Abdeljaber, Onur Avcı, Serkan Kıranyaz, Boualem Boashash, Henry A. Sodano, Daniel J. Inman	2018	338	Automated Pavement Inspection and Maintenance
4	The application of machine learning to structural health monitoring (Worden & Manson, 2007)	Keith Worden, G. Manson	2007	291	Structural Health Monitoring Techniques
5	The State of the Art of Data Science and Engineering in Structural Health Monitoring (Bao et al., 2019)	Yuequan Bao, Zhicheng Chen, Shiyin Wei, Yang Xu, Zhiyi Tang	2019	268	Structural Health Monitoring Techniques
6	Structural damage identification based on autoencoder neural networks and deep learning (Pathirage et al., 2018)	Chathurdara Sri Nadith Pathirage, Jun Li, Ling Li, Hong Hao, Wanquan Liu, Pinghe Ni	2018	250	Structural Health Monitoring Techniques
7	Machine Learning Algorithms in Civil Structural Health Monitoring: A Systematic Review (Flah et al., 2021)	Majdi Flah, Itzel Nunez, Wassim Ben Chaabene, Moncef L. Nehdi	2021	185	Automated Pavement Inspection and Maintenance
8	Machine learning for structural engineering: A state-of-the-art review (Thai, 2022)	Huu-Tai Thai	2022	175	Structural Health Monitoring Techniques

continued on following page

Table 3. Continued

No	Article reference	Authors	Year of pub.	Number of citations	Topic
9	Deep learning in the construction industry: A review of present status and future innovations (Akinosho et al., 2020)	Taofeek Dolapo Akinosho, Lukumon O. Oyedele, Muhammad Bilal, Anuoluwapo O. Ajayi, Manuel Davila Delgado, Olúgbénga O. Akinadé, Ashraf Ahmed	2020	172	Automated Pavement Inspection and Maintenance
10	Machine learning applications for building structural design and performance assessment: State-of-the-art review (H. Sun et al., 2021)	Han Sun, Henry V. Burton, Huan Huang	2021	156	Automated Pavement Inspection and Maintenance

The top 10 most cited publications demonstrate a clear trend towards the adoption of deep learning techniques, particularly convolutional neural networks (CNNs), for SHM applications. This emphasis on advanced algorithms suggests a shift towards more sophisticated models for damage detection and analysis. The publications also showcase the diverse applicability of ML across various SHM challenges, ranging from guided wave analysis and pavement inspection to structural damage identification. The presence of earlier publications, such as (Worden & Manson, 2007) seminal work, underscores the early adoption of ML in SHM research and its lasting influence. Furthermore, the inclusion of recent publications focused on data science and engineering approaches, such as (Bao et al., 2019), reflects the growing integration of these disciplines into SHM research.

This initial citation-based classification, while effective in identifying widely referenced works, did not fully account for the specific focus on ML-based SHM. For instance, (Kiranyaz et al., 2021) presented a broader survey of CNN applications, which, while influential, may not be as directly relevant to the specific challenges of SHM.

The Second Stage of Classification

The second stage of classification focused on co-citation, a measure of intellectual connection between publications. From the initial list of highly cited publications, the top 10 articles with the highest co-citation counts were selected. Co-citation refers to the frequency with which two documents are cited together by other sources (Small, 1973). Unlike simple citation counts, which measure how often a single paper is referenced, co-citation analysis reveals thematic clusters and intellectual connections within a field. This approach can uncover linkages between publications that may not directly cite each other but are frequently mentioned together in the broader scholarly discourse. Generally, co-citation indicates an association, relationship, or similarity between the co-cited documents, providing insight into conceptual connections within the field.

The co-citation network (Figure 8), generated using VOSviewer, provides a visual representation of the intellectual relationships between the most cited publications in ML-based SHM research. The network reveals “Majdi Flah's (2021)” (Flah et al., 2021) publication as a central hub of intellectual connections, linking it to a diverse range of other publications, highlighting its importance and influence within the field. The network can be broken down into distinct clusters, suggesting specific research themes or areas of focus. For instance, the cluster surrounding Flah's work connects to recent publications by Eloi Figueiredo (Figueiredo & Brownjohn, 2022), Alain Gomez-Cabrera (Gomez-Cabrera & Escamilla-Ambrosio, 2022), and Flah himself, potentially representing a specific research area within the field. Other clusters, such as those centered around “Keith Worden (2007)” (Worden & Manson, 2007) and “Mira Mitra (2016)” (Mitra & Gopalakrishnan, 2016), highlight ongoing lines of research and potential research directions.

Figure 8. Co-citation Network of Top Cited Publications in ML-based SHM

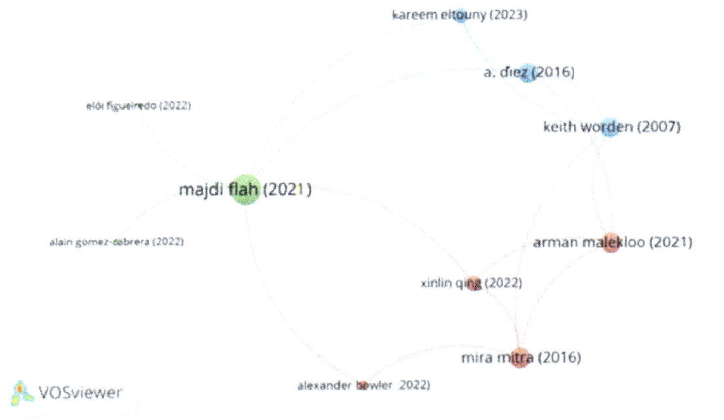

For a deeper understanding of the specific content and impact of each publication, Table 4 is a presentation of the top 10 most popular publications based on co-citation counts; it provides the authors, publication year, number of citations, topic and funding sources associated with each publication.

Table 4. Top 10 most popular publications based on number of co-citation

No	Article reference	Authors	Year of pub.	Number of citations	Topic	Funder
1	Machine Learning Algorithms in Civil Structural Health Monitoring: A Systematic Review (Flah et al., 2021)	Majdi Flah, Itzel Nunez, Wassim Ben Chaabene, Moncef L. Nehdi	2021	185	Automated Pavement Inspection and Maintenance	-
2	Guided wave based structural health monitoring: A review (Mitra & Gopalakrishnan, 2016)	Mira Mitra, S. Gopalakrishnan	2016	674	Guided Wave Structural Health Monitoring in Materials	-
3	Machine learning and structural health monitoring overview with emerging technology and high-dimensional data source highlights (Malekloo et al., 2022)	Arman Malekloo, Ekin Özer, Mohammad AlHamaydeh, Mark Girolami	2021	142	Automated Pavement Inspection and Maintenance	-
4	The application of machine learning to structural health monitoring (Worden & Manson, 2007)	Keith Worden, G. Manson	2007	291	Structural Health Monitoring Techniques	-
5	A clustering approach for structural health monitoring on bridges (Diez et al., 2016)	A. Díez, Nguyen Lu Dang Khoa, Mehrisadat Makki Alamdari, Yang Wang, Fang Chen, Peter Runcie	2016	104	Structural Health Monitoring Techniques	-
6	Machine Learning Based Quantitative Damage Monitoring of Composite Structure (Qing et al., 2022)	Xinlin Qing, Yunlai Liao, Yihan Wang, Binqiang Chen, Fanghong Zhang, Yishou Wang	2022	28	Structural Health Monitoring Techniques	National Natural Science Foundation of China
7	Unsupervised Learning Methods for Data-Driven Vibration-Based Structural Health Monitoring: A Review (Eltouny et al., 2023)	Kareem Eltouny, Mohamed Gomaa, Xiao Liang	2023	20	Structural Health Monitoring Techniques	-
8	A review of ultrasonic sensing and machine learning methods to monitor industrial processes (Bowler et al., 2022)	Alexander Bowler, Michael P. Pound, Nick J. Watson	2022	20	Process Fault Detection and Diagnosis in Industries	Engineering and Physical Sciences Research Council

continued on following page

Table 4. Continued

No	Article reference	Authors	Year of pub.	Number of citations	Topic	Funder
9	Three decades of statistical pattern recognition paradigm for SHM of bridges (Figueiredo & Brownjohn, 2022)	Elói Figueiredo, James Brownjohn	2022	68	Structural Health Monitoring Techniques	-
10	Review of Machine-Learning Techniques Applied to Structural Health Monitoring Systems for Building and Bridge Structures (Gomez-Cabrera & Escamilla-Ambrosio, 2022)	Alain Gomez-Cabrera, Ponciano Jorge Escamilla-Ambrosio	2022	24	Structural Health Monitoring Techniques	Instituto Politécnico Nacional, Consejo Nacional de Ciencia y Tecnología

The second stage of classification reveals several key trends within the field of ML-based SHM research. Notably, the prevalence of systematic reviews and overviews of diverse machine learning techniques and applications reflects the ongoing need for a comprehensive understanding of the state-of-the-art within this rapidly evolving field. A focus on guided wave techniques and pavement inspection highlights the particular relevance and activity in these areas. While influential older publications, such as (Worden & Manson, 2007), underscore the lasting impact of pioneering work, the inclusion of more recent publications exploring unsupervised learning methods (Eltouny et al., 2023) and machine learning applications in industrial processes (Bowler et al., 2022) suggests that these areas are gaining momentum and attracting growing research interest.

Both classifications, based on citation counts and co-citation counts, offer valuable insights into the most influential publications in ML-based SHM research. While both methods are valid, depending on the researcher's objectives, the second classification, focusing on co-citation, yielded more relevant and specific articles. The initial classification, based solely on citation counts, identified some publications with a more generalist focus, such as the work by (Kiranyaz et al., 2021). By incorporating co-citation analysis, the second stage effectively refined the selection, highlighting publications that are not only widely referenced but also contribute to a specific intellectual thread within the field.

It's important to acknowledge that the two-stage classification used in this study is inherently limited by the time it takes for publications to gain citations and co-citations. As a result, recent publications from 2024 such as (Salah et al., 2024) are not included in this list.

IV. SUMMARY OF BIBLIOMETRIC TRENDS

The bibliometric analysis of Machine Learning applications in Structural Health Monitoring (ML x SHM) reveals several significant trends and patterns that merit further discussion.

Firstly, the exponential growth in publications over the past two decades, with 80% of the literature produced in the last four years, underscores the rapidly increasing interest and investment in this field. This surge likely reflects both the maturing of ML technologies and the growing recognition of their potential to enhance SHM practices. The acceleration in research output suggests that we are at a critical juncture where ML is transitioning from an experimental approach to a mainstream tool in SHM.

The dominance of journal publications (69% of the total) over conference proceedings and other formats indicates a preference for more rigorous, peer-reviewed dissemination of research findings. This trend towards journal publications may reflect the field's maturation, with researchers producing more comprehensive and validated studies suitable for in-depth journal articles.

The geographical analysis reveals a clear leadership triad of the United States, China, and the United Kingdom in ML x SHM research. This concentration of productivity likely stems from these countries' robust research infrastructures, substantial funding, and established expertise in both ML and SHM. However, the emerging contributions from countries like Greece, Iran, and South Korea suggest a broadening of the field's geographical base, which could lead to more diverse approaches and applications in the future.

The strong international collaboration networks, particularly evident in Italy's case, highlight the global nature of ML x SHM research. This collaborative approach is crucial for addressing the complex, multidisciplinary challenges in the field and for sharing best practices across different structural contexts and regulatory environments.

The examination of the most prolific authors in the ML x SHM field provides valuable insights into the key contributors shaping this research area. The concentration of four out of the eight most productive researchers at the University of Sheffield underscores the presence of a significant research cluster, potentially indicating a center of excellence in ML x SHM studies. The diverse geographical distribution of the other top authors, spanning the UK, Colombia, USA, and Portugal, reinforces the global nature of this research field. The mix of researchers with first publications dating back to the late 1980s alongside those who entered the field more recently in the 2010s suggests a healthy blend of established expertise and fresh perspectives. This generational diversity likely contributes to the field's dynamic evolution, combining long-standing SHM knowledge with cutting-edge ML techniques. The high citation counts and h-indices of these top authors, particularly

Keith Worden's impressive metrics, indicate that their work is not only prolific but also highly influential, shaping the direction of ML x SHM research.

Expanding this study to a broader set of researchers could reveal additional insights, including emerging talents, diverse institutional contributions, and potential regional or topical specializations within the ML x SHM domain.

The keyword analysis provides valuable insights into the evolving focus of ML-SHM research. The shift from traditional ML algorithms like Support Vector Machines towards more advanced techniques such as Convolutional Neural Networks reflects the field's adoption of state-of-the-art AI methodologies. This progression suggests that researchers are leveraging increasingly sophisticated tools to handle the complex, high-dimensional data typical in SHM applications.

The emergence of keywords related to the Internet of Things, big data, and uncertainty quantification points to new challenges and opportunities in the field. As SHM systems become more interconnected and data-rich, issues of data management, real-time processing, and model reliability come to the forefront. The focus on these areas indicates a growing recognition of the need for robust, scalable ML solutions that can operate reliably in real-world SHM deployments.

The diversity of application areas represented in the keyword analysis, from bridge monitoring to wind turbines, demonstrates the versatility of ML approaches in SHM. This broad applicability suggests that ML-based SHM has the potential to impact a wide range of infrastructure sectors, potentially leading to improved safety, reduced maintenance costs, and extended structural lifespans across various industries.

However, this analysis also reveals potential gaps and areas for future research. For instance, while there is a clear trend towards more advanced ML techniques, there appears to be less focus on interpretability and explainability of ML models in SHM contexts. Given the critical nature of structural health decisions, developing transparent and interpretable ML models could be a crucial area for future work.

Additionally, the relative scarcity of keywords related to regulatory compliance or standardization suggests that the integration of ML-based SHM into existing regulatory frameworks may be an underexplored area. As the field matures, addressing these regulatory and standardization aspects will be essential for widespread adoption of ML in SHM practices.

The analysis of the most popular publications in ML x SHM research, conducted through a two-stage classification approach, provides valuable insights into the field's evolution and current focus areas. The initial citation-based classification highlighted the dominance of deep learning techniques, particularly convolutional neural networks (CNNs), in recent SHM applications. This trend towards more sophisticated algorithms suggests a shift in the field towards leveraging advanced AI capabilities for more accurate and efficient structural health monitoring.

The co-citation analysis, represented in the network visualization, revealed (Flah et al., 2021) publication as a central hub of intellectual connections within the field. This centrality underscores the importance of comprehensive reviews in synthesizing and directing research efforts in this rapidly evolving domain. The distinct clusters observed in the co-citation network point to specific research themes, such as guided wave analysis, automated pavement inspection, and unsupervised learning methods for vibration-based SHM.

Interestingly, while some highly cited publications focus on broad reviews of ML techniques, the co-citation analysis highlighted works more specifically tailored to SHM applications. This suggests that while general ML knowledge is crucial, the field is increasingly valuing research that addresses the unique challenges of applying ML to structural health monitoring.

The presence of both seminal works from the early 2000s and recent publications in the top co-cited list illustrates the field's dynamic nature. It shows how foundational concepts continue to influence current research while new approaches, particularly in unsupervised learning and industrial process monitoring, are gaining traction.

The funding sources noted for some of the top publications, including national science foundations and research councils, highlight the recognition of ML x SHM as a critical area for investment. This governmental and institutional support is likely a key driver in the field's rapid advancement.

However, it's important to note that this analysis has inherent limitations due to the time lag in accumulating citations and co-citations. As a result, the most recent cutting-edge research, particularly from 2024, may not be represented in these rankings. This underscores the need for continual reassessment of the field's direction and the importance of considering both established and emerging research in driving the field forward.

V. DISCUSSION: CRITICAL PERSPECTIVES AND INDUSTRIAL IMPLICATIONS

The bibliometric analysis presented in the preceding sections highlights a field undergoing exponential growth, with 80% of the total literature emerging in the last four years alone. While the quantitative data provides a roadmap of what is being researched—demonstrating a clear shift towards Deep Learning (DL) and an increasing focus on Convolutional Neural Networks (CNNs)—it is equally critical to interpret why these trends are occurring and what they imply for the future of the Structural Health Monitoring (SHM) industry.

The transition of Machine Learning (ML) from academic experimentation to industrial application in civil infrastructure faces unique challenges. Unlike the tech or finance sectors, where ML failure results in financial loss or user inconvenience, the domain of SHM deals with critical safety decisions where failure can lead to catastrophic loss of life. This section discusses the broader implications of our bibliometric findings, focusing on the interpretability vs. accuracy trade-off, the operational shifts in data processing, and the regulatory gaps that currently hinder widespread industrial adoption.

The "Black Box" Dilemma and the Trust Deficit

One of the most significant trends identified in our keyword analysis is the rapid displacement of traditional algorithms, such as Support Vector Machines (SVMs), by Deep Learning architectures like Convolutional Neural Networks (CNNs). While this shift is driven by the superior accuracy of DL in handling high-dimensional data, it introduces a "Black Box" dilemma that poses a substantial barrier to industrial implementation.

Civil engineering is a discipline rooted in physics-based principles and determinism. Engineers and asset managers are accustomed to understanding the physical causality behind a decision—for example, knowing that a bridge is unsafe because a specific crack width exceeds a code-mandated threshold. Deep Learning models, conversely, operate through millions of abstract parameters that are often unintelligible to human operators. As noted in recent literature, high prediction accuracy on a finite dataset does not guarantee performance on unseen data, nor does it prove the rationality of the decision (Hu et al., 2024).

Consequently, an algorithm might correctly predict structural damage with 99% accuracy, but if it cannot explain the feature it used to arrive at that conclusion, a licensed engineer may be legally and ethically unable to act on that information. This creates a "trust deficit" where the opacity of the model blocks its practical application, regardless of its computational efficiency.

To bridge this gap, recent studies emphasize the necessity of Explainable AI (XAI) methods that can visualize the decision-making process of the model. For instance, Hu et al. (2024) demonstrated that by using advanced interpretation techniques like Deep Grad CAM on 1D CNNs, it is possible to verify that the network focuses on physically relevant signal features—such as "bolt reflection waves" and "boundary reflections"—rather than noise. Establishing this alignment between the AI's "attention" and the engineer's physical knowledge of wave propagation is crucial.

However, our bibliometric data reveals a relative scarcity of keywords such as "Explainable AI" compared to the overwhelming presence of generic "Deep Learning" terms. This suggests a potential misalignment between academic research incentives (which prioritize high accuracy scores) and industrial requirements (which prioritize interpretability and causality). For ML-SHM to transition from a research topic to an industry standard, future work must pivot toward XAI, ensuring that algorithms serve as transparent decision-support tools rather than opaque oracles.

The Operational Shift: From Feature Engineering to Feature Learning

The evolution from vibration analysis and traditional ML algorithms (prominent in 2020) to Deep Learning architectures like CNNs (prominent in 2023) signifies a fundamental change in the operational workflow of Structural Health Monitoring (SHM).

In traditional ML-SHM approaches (e.g., SVMs or Random Forests), the workflow relied heavily on "Feature Engineering." This required domain experts to manually process raw sensor data to extract physics-based features such as natural frequencies, mode shapes, or damping ratios. As noted in a comprehensive review by Cha et al. (Cha et al., 2024), this manual process has been a "complex and longstanding challenge" for over three decades, primarily because manually extracted features are often sensitive to environmental noise and temperature variations. The success of these traditional methods depended entirely on the engineer's ability to identify and calculate the correct features beforehand.

The rise of CNNs and Deep Learning, as evidenced by our citation analysis, marks a shift toward "End-to-End Learning" or "Feature Learning." In this paradigm, the algorithm ingests raw data—whether it be 1D vibration signals or 2D images—and automatically learns the optimal hierarchical representations for damage detection. Cha et al. highlight that DL models, such as 1D CNNs applied to acceleration data, can extract robust features that are less susceptible to the noise that plagues manually engineered features.

- **Industrial Advantage:** This reduces the need for manual preprocessing and allows the system to detect subtle, non-linear patterns that human experts might overlook. It potentially lowers the barrier to entry for deploying SHM on complex structures where the physics are not fully modeled.
- **Industrial Disadvantage:** This approach is data-hungry and computationally expensive. It shifts the resource requirement from *structural engineering expertise* to *computational power and data volume*. For the industry, this implies a need to upgrade hardware infrastructure (e.g., GPU-accelerated edge devices) and hire personnel with data science competencies, fundamentally changing the hiring landscape of civil engineering firms.

The Data Scarcity Paradox and the Role of Digital Twins

A critical inference drawn from the literature—and a primary bottleneck for real-world application—is the "Data Scarcity Paradox." Supervised learning algorithms, which dominate the top-cited papers in our analysis, require balanced datasets containing examples of both "healthy" and "damaged" states to learn effectively.

However, in the context of civil and industrial infrastructure, "damage" is a rare event. Bridges, pipelines, and railway tracks operate in a healthy state for the vast majority of their service life. Acquiring real-world training data for a broken rail or a collapsed bridge is often impossible because, in operational settings, damaged sections are immediately repaired or replaced to prevent catastrophe. This creates a severe class imbalance problem where algorithms are trained almost exclusively on healthy data, making them prone to false negatives when novel damage types occur.

This paradox points to the industrial necessity of "Digital Twins" not just as static 3D models, but as dynamic data generators. Recent literature illustrates a shift toward "Hybrid Digital Twins" that combine physics-based modeling with machine learning to synthesize training data for scenarios that have not yet occurred in reality.

For example, the authors in (Ramatlo et al., 2023) addressed this exact challenge in the monitoring of welded railway lines. Recognizing that obtaining real inspection data for various damage evolutions was impossible on operational tracks, they utilized a hybrid Digital Twin. Their approach employed a physics-based Finite Element Method (FEM) to simulate wave propagation and generate "virtual experimental data." To overcome the high computational cost of pure FEM, they trained a Variational Auto-Encoder (VAE)—a deep learning architecture—to rapidly predict complex inspection signals based on the physics-based baseline.

This case study highlights a broader trend: the future of ML in SHM likely relies on Generative AI and Physics-Informed Machine Learning. By training Deep Learning models on synthetic data generated by high-fidelity Digital Twins, engineers can prepare algorithms to recognize failure modes they have never encountered in the field. This synergy between physics-based simulation and data-driven ML solves the data scarcity issue while ensuring the model remains grounded in the physical reality of the structure.

Edge Computing vs. Cloud Processing

While the keyword network (Figure 7) is visually dominated by algorithmic terminology, the underlying bibliometric dataset reveals the emerging presence of connectivity-related terms such as "Internet of Things" (IoT), "Wireless Sensor Networks," and "Big Data." In an industrial context, these terms represent a significant logistical and economic conflict between the increasing complexity of ML models and the constraints of data transmission.

A recent comprehensive review of IoT in SHM by Rahita et al. (Rahita et al., 2024) outlines the standard architecture currently dominating the field: sensors collect data, nodes transmit it via gateways, and deep analysis occurs on centralized cloud platforms. They note that cloud systems are favored for their storage capacity and accessibility, allowing complex algorithms to run on historical data from remote locations. However, they also identify that "gathering, sending, and keeping massive amounts of data" is a primary difficulty for these systems.

This creates a dichotomy for industrial application. Structural health monitoring systems, particularly those using vibration analysis, generate massive volumes of high-frequency data (often sampling at 100Hz or higher across dozens of channels). Transmitting this raw time-series data to a cloud server for processing is often economically unviable due to bandwidth costs and latency issues, particularly for critical infrastructure located in remote areas without stable high-speed internet.

The bibliometric trends identify a potential solution that is beginning to emerge alongside cloud architectures: Edge Computing. As noted in the keyword clustering analysis by (Rahita et al., 2024), "edge computing" and "intelligent systems" have formed a distinct research cluster alongside traditional cloud computing. This suggests a paradigm shift toward Edge AI, where lighter, optimized ML models are deployed directly on the sensor microcontrollers. This allows the system to process data locally and transmit only the actionable insights (e.g., "Crack Detected" or a health index) rather than the raw signal, significantly reducing bandwidth requirements.

However, this presents a technological gap. Our analysis in Section IV shows a trend toward increasingly complex Deep Learning models (like heavy CNNs) which require substantial computational power. There is currently a disconnect between the complex models being developed in academia (which run on powerful workstations) and the low-power, energy-constrained hardware available for field deployment. Future research must therefore focus on "model compression" and "lightweight neural networks" to make the high-accuracy academic models deployable on battery-powered industrial IoT devices.

The Regulatory Gap

Finally, a qualitative assessment of the bibliometric landscape reveals a glaring absence of keywords related to "Standardization," "Regulation," or "Certification." This suggests that while the academic community is aggressively pursuing algorithmic novelty, the pathway to legal and industrial deployment remains under-researched.

Currently, SHM technologies are largely deployed as pilot projects or research demonstrators. For ML-SHM to become a standard industrial practice—mandated by insurance companies or government agencies—it must fit within existing regulatory frameworks. However, as noted by El Mir et al. (El Mir et al., 2023) in their analysis of airworthiness requirements, current certification standards (such as ARP4754 or DO-178 in aviation) were designed for deterministic software, not stochastic Machine Learning models.

El Mir et al. highlight a critical "certification deadlock" facing the industry. On one hand, Supervised Learning models are preferred by regulators because their outputs are predictable; however, certifying them requires "representative labelled training data" for every possible failure mode—data that is operationally impossible to generate for critical infrastructure (the Data Scarcity Paradox discussed in Section 5.3). On the other hand, Unsupervised Learning allows for damage detection without failure data, but it often lacks the "interpretability and explainability" stipulated by safety agencies like EASA (European Union Aviation Safety Agency).

This regulatory gap creates a significant barrier to entry. Until specific standards—such as the emerging SAE G-34 or AS6983 discussed by El Mir et al.—are fully developed and harmonized across industries, ML-SHM will remain an advisory tool rather than a critical safety system. The technology is maturing faster than the governance. Consequently, the next phase of research must move beyond merely improving algorithmic accuracy and focus on "Learning Assurance," developing frameworks that allow ML models to be validated, verified, and legally certified for use in safety-critical environments.

Conclusion of Discussion

In summary, the bibliometric data paints a picture of a vibrant, rapidly expanding research field that has successfully adopted state-of-the-art AI techniques. However, the transition to industrial application requires a shift in focus. The field must move from "Algorithm-Centric" research (chasing higher accuracy on benchmark datasets) to "Application-Centric" research (addressing explainability, data scarcity, edge deployment, and certification). Only by bridging these gaps can the intersection of ML and SHM evolve from a promising academic pursuit into a robust industrial standard that ensures the safety and longevity of global infrastructure.

VI. CONCLUSION

This comprehensive bibliometric analysis of machine learning (ML) applications in structural health monitoring (SHM) has provided valuable insights into the current state and trajectory of this rapidly evolving field. Our study, covering publications from 2003 to early 2024, reveals several key findings and trends:

1. Exponential growth: The field has experienced dramatic growth, with 80% of publications occurring in the last four years, indicating increasing interest and investment in ML-based SHM.
2. Global leadership: The United States, China, and the United Kingdom emerge as clear leaders in research output, while strong international collaborations, particularly from Italy, highlight the global nature of the field.
3. Evolving methodologies: There is a clear shift from traditional ML algorithms towards more advanced techniques such as convolutional neural networks, reflecting the adoption of state-of-the-art AI methodologies in SHM applications.
4. Diverse applications: The keyword analysis reveals a wide range of application areas, from bridge monitoring to wind turbines, demonstrating the versatility of ML approaches in SHM.
5. Emerging trends: The focus on topics such as the Internet of Things, big data, and uncertainty quantification points to new challenges and opportunities in the field, particularly in data management and model reliability.
6. Influential publications: Our two-stage classification approach identified key publications that have shaped the field, with recent works on comprehensive reviews and specific ML applications in SHM gaining significant attention.

While these findings demonstrate the field's rapid advancement, they also highlight areas for future research. These include the need for more interpretable and explainable ML models in SHM contexts, addressing regulatory and standardization aspects, and continuing to explore novel ML techniques for improving SHM practices.

As ML-based SHM continues to mature, it holds the potential to significantly enhance the safety, efficiency, and longevity of critical infrastructure across various sectors. The trends identified in this study not only reflect the current state of the art but also point to exciting future directions for research and application in this crucial domain.

Importantly, the quantitative bibliometric analysis presented in this chapter lays a solid foundation for future research. These results can serve as a springboard for a more in-depth qualitative review of the works related to ML applications in SHM. Such a qualitative analysis could delve deeper into the methodologies, challenges, and innovations presented in key publications, providing a more nuanced understanding of the field's evolution and future directions.

Moving forward, researchers should focus on addressing the identified gaps, such as model interpretability and regulatory integration, while continuing to leverage advancements in ML to tackle the complex challenges in structural health monitoring. By combining the quantitative insights from this bibliometric analysis with future qualitative reviews, the field can further solidify its role in ensuring the reliability and safety of our built environment in an increasingly data-driven world.

REFERENCES

Abdeljaber, O., Avci, O., Kiranyaz, M. S., Boashash, B., Sodano, H., & Inman, D. J. (2018). 1-D CNNs for structural damage detection: Verification on a structural health monitoring benchmark data. *Neurocomputing*, *275*, 1308–1317. DOI: 10.1016/j.neucom.2017.09.069

Akinosho, T. D., Oyedele, L. O., Bilal, M., Ajayi, A. O., Delgado, M. D., Akinade, O. O., & Ahmed, A. A. (2020). Deep learning in the construction industry: A review of present status and future innovations. *Journal of Building Engineering*, *32*, 101827. DOI: 10.1016/j.jobe.2020.101827

An, Y., Chatzi, E., Sim, S., Laflamme, S., Blachowski, B., & Ou, J. (2019). Recent progress and future trends on damage identification methods for bridge structures. *Structural Control and Health Monitoring*, *26*(10). Advance online publication. DOI: 10.1002/stc.2416

Anton, S. R., & Sodano, H. A. (2007). A review of power harvesting using piezoelectric materials (2003–2006). *Smart Materials and Structures*, *16*(3), R1–R21. DOI: 10.1088/0964-1726/16/3/R01

Bao, Y., Chen, Z., Wei, S., Xu, Y., Tang, Z., & Li, H. (2019). The State of the Art of Data Science and Engineering in Structural Health Monitoring. *Engineering (Beijing)*, *5*(2), 234–242. DOI: 10.1016/j.eng.2018.11.027

Benckendorff, P., & Zehrer, A. (2013). A NETWORK ANALYSIS OF TOURISM RESEARCH. *Annals of Tourism Research*, *43*, 121–149. DOI: 10.1016/j.annals.2013.04.005

Bowler, A. L., Pound, M. P., & Watson, N. J. (2022). A review of ultrasonic sensing and machine learning methods to monitor industrial processes. *Ultrasonics*, *124*, 106776. DOI: 10.1016/j.ultras.2022.106776 PMID: 35653984

Centenaro, E., Lany, M., Meyer, T., & Gasparin, E. (2019). *Structural Health Monitoring Solutions for Offshore Platforms*. OMC-2019-1031.

Cha, Y.-J., Ali, R., Lewis, J., & Büyük ztürk, O. (2024). Deep learning-based structural health monitoring. *Automation in Construction*, *161*, 105328. DOI: 10.1016/j.autcon.2024.105328

Culbert, J., Hobert, A., Jahn, N., Haupka, N., Schmidt, M., Donner, P., & Mayr, P. (2024). *Reference Coverage Analysis of OpenAlex compared to Web of Science and Scopus* (No. arXiv:2401.16359). arXiv. https://arxiv.org/abs/2401.16359

Diez, A., Khoa, N. L. D., Makki Alamdari, M., Wang, Y., Chen, F., & Runcie, P. (2016). A clustering approach for structural health monitoring on bridges. *Journal of Civil Structural Health Monitoring*, *6*(3), 429–445. DOI: 10.1007/s13349-016-0160-0

El Mir, H., King, S., Skote, M., & Perinpanayagam, S. (2023). *Machine Learning Requirements for the Airworthiness of Structural Health Monitoring Systems in Aircraft.* 1–13. https://www.icaf.aero/icaf2023/proceedings/documents/128.pdf

Eltouny, K., Gomaa, M., & Liang, X. (2023). Unsupervised Learning Methods for Data-Driven Vibration-Based Structural Health Monitoring: A Review. *Sensors (Basel)*, *23*(6), 3290. DOI: 10.3390/s23063290 PMID: 36992001

Fan, W., & Qiao, P. (2011). Vibration-based Damage Identification Methods: A Review and Comparative Study. *Structural Health Monitoring*, *10*(1), 83–111. DOI: 10.1177/1475921710365419

Figueiredo, E., & Brownjohn, J. (2022). Three decades of statistical pattern recognition paradigm for SHM of bridges. *Structural Health Monitoring*, *21*(6), 3018–3054. DOI: 10.1177/14759217221075241

Flah, M., Nunez, I., Ben Chaabene, W., & Nehdi, M. L. (2021). Machine Learning Algorithms in Civil Structural Health Monitoring: A Systematic Review. *Archives of Computational Methods in Engineering*, *28*(4), 2621–2643. DOI: 10.1007/s11831-020-09471-9

Gharehbaghi, V. R., Noroozinejad Farsangi, E., Noori, M., Yang, T. Y., Li, S., Nguyen, A., Málaga-Chuquitaype, C., Gardoni, P., & Mirjalili, S. (2022). A Critical Review on Structural Health Monitoring: Definitions, Methods, and Perspectives. *Archives of Computational Methods in Engineering*, *29*(4), 2209–2235. DOI: 10.1007/s11831-021-09665-9

Gomez-Cabrera, A., & Escamilla-Ambrosio, P. J. (2022). Review of Machine-Learning Techniques Applied to Structural Health Monitoring Systems for Building and Bridge Structures. *Applied Sciences (Basel, Switzerland)*, *12*(21), 10754. DOI: 10.3390/app122110754

Hu, M., Yue, N., & Groves, R. M. (2024). Damage Classification of a Bolted Connection using Guided Waves and Explainable Artificial Intelligence. *Procedia Structural Integrity*, *52*, 224–233. DOI: 10.1016/j.prostr.2023.12.023

Jardine, A. K. S., Lin, D., & Banjevic, D. (2006). A review on machinery diagnostics and prognostics implementing condition-based maintenance. *Mechanical Systems and Signal Processing*, *20*(7), 1483–1510. DOI: 10.1016/j.ymssp.2005.09.012

Jensen, J. A., Nikolov, S. I., Gammelmark, K. L., & Pedersen, M. H. (2006). Synthetic aperture ultrasound imaging. *Ultrasonics*, *44*, e5–e15. DOI: 10.1016/j.ultras.2006.07.017 PMID: 16959281

Kim, M. C., & Zhu, Y. (2018). Scientometrics of Scientometrics: Mapping Historical Footprint and Emerging Technologies in Scientometrics. In Jibu, M., & Osabe, Y. (Eds.), *Scientometrics*. InTech., DOI: 10.5772/intechopen.77951

Kiranyaz, S., Avci, O., Abdeljaber, O., Ince, T., Gabbouj, M., & Inman, D. J. (2021). 1D convolutional neural networks and applications: A survey. *Mechanical Systems and Signal Processing*, *151*, 107398. DOI: 10.1016/j.ymssp.2020.107398

Kirby, A. (2023). Exploratory Bibliometrics: Using VOSviewer as a Preliminary Research Tool. *Publications/MDPI*, *11*(1), 10. DOI: 10.3390/publications11010010

Laaz, H., Mekkaoui, M., Nissabouri, S., & Rhimini, H. (2026). Investigation of Lamb Wave Excitation and Modal Behavior in Thin Plates for Non-Destructive Testing. In Barkaoui, A., Mabrouki, T., Degiovanni, A., Ben Zineb, T., & El Otmani, R. (Eds.), *Advances in Mechanics, Materials, and Energy* (pp. 117–136). Springer Nature Switzerland., DOI: 10.1007/978-3-032-12980-2_11

Laaz, H., Rhimini, H., Nissabouri, S., & Mekkaoui, M. (2025). Signal Processing Analysis of Lamb Wave Propagation in Stainless Steel Plates with Thickness Defects: Implications for AI-Based Non-Destructive Testing. *2025 5th International Conference on Innovative Research in Applied Science, Engineering and Technology (IRASET)*, 1–10. DOI: 10.1109/IRASET64571.2025.11008279

Lizé, E., Rébillat, M., Mechbal, N., & Bolzmacher, C. (2018). Optimal dual-PZT sizing and network design for baseline-free SHM of complex anisotropic composite structures. *Smart Materials and Structures*, *27*(11), 115018. DOI: 10.1088/1361-665X/aad534

Luckey, D., Fritz, H., Legatiuk, D., Peralta Abadía, J. J., Walther, C., & Smarsly, K. (2022). Explainable Artificial Intelligence to Advance Structural Health Monitoring. In Cury, A., Ribeiro, D., Ubertini, F., & Todd, M. D. (Eds.), *Structural Health Monitoring Based on Data Science Techniques* (Vol. 21, pp. 331–346). Springer International Publishing., DOI: 10.1007/978-3-030-81716-9_16

Mahbubi, K., Zaki, A., & Nugroho, G. (2024). Bibliometric and Scientometric Trends in Structural Health Monitoring Using Fiber-Optic Sensors: A Comprehensive Review. *Journal of Civil and Hydraulic Engineering*, *2*(1), 51–64. DOI: 10.56578/jche020104

Malekloo, A., Ozer, E., AlHamaydeh, M., & Girolami, M. (2022). Machine learning and structural health monitoring overview with emerging technology and high-dimensional data source highlights. *Structural Health Monitoring*, *21*(4), 1906–1955. DOI: 10.1177/14759217211036880

Mitra, M., & Gopalakrishnan, S. (2016). Guided wave based structural health monitoring: A review. *Smart Materials and Structures*, *25*(5), 053001. DOI: 10.1088/0964-1726/25/5/053001

Mukhopadhyay, S. C. (2015). Wearable Sensors for Human Activity Monitoring: A Review. *IEEE Sensors Journal*, *15*(3), 1321–1330. DOI: 10.1109/JSEN.2014.2370945

Nur'aeni, R., & Zalsahra, R. (2024). Bibliometric Analysis of Artificial Intelligence. *West Science Interdisciplinary Studies*, *2*(1), 119–128. DOI: 10.58812/wsis.v2i01.563

Pathirage, C. S. N., Li, J., Li, L., Hao, H., Liu, W., & Ni, P. (2018). Structural damage identification based on autoencoder neural networks and deep learning. *Engineering Structures*, *172*, 13–28. DOI: 10.1016/j.engstruct.2018.05.109

Qing, X., Liao, Y., Wang, Y., Chen, B., Zhang, F., & Wang, Y. (2022). Machine Learning Based Quantitative Damage Monitoring of Composite Structure. *International Journal of Smart and Nano Materials*, *13*(2), 167–202. DOI: 10.1080/19475411.2022.2054878

Rahita, A. C., Zaki, A., Nugroho, G., & Yadi, S. (2024). Internet of Things (IoT) in Structural Health Monitoring: A Decade of Research Trends. *Instrumentation Mesure Métrologie*, *23*(2), 123–139. DOI: 10.18280/i2m.230205

Ramatlo, D. A., Wilke, D. N., & Loveday, P. W. (2023). Digital Twin Hybrid Modeling for Enhancing Guided Wave Ultrasound Inspection Signals in Welded Rails. *Mathematical & Computational Applications*, *28*(2), 58. DOI: 10.3390/mca28020058

Salah, N., & Barra Ndiaye, E. (2024). Towards an optimization of automatic defect detection by artificial neural network using Lamb waves. *IAES International Journal of Artificial Intelligence (IJ-AI)*, *13*(2), 1459. DOI: 10.11591/ijai.v13.i2.pp1459-1468

Salah, N., Hamza, E., Mustapha, J., & Mansouri, K. (2024). Towards an automatic classification of welding defect by convolutional neural network and robot classifier. *Indonesian Journal of Electrical Engineering and Computer Science*, *33*(3), 1768. DOI: 10.11591/ijeecs.v33.i3.pp1768-1774

Salawu, O. S. (1997). Detection of structural damage through changes in frequency: A review. *Engineering Structures*, *19*(9), 718–723. DOI: 10.1016/S0141-0296(96)00149-6

Small, H. (1973). Co-citation in the scientific literature: A new measure of the relationship between two documents. *Journal of the American Society for Information Science, 24*(4), 265–269. DOI: 10.1002/asi.4630240406

Spencer, B. F., Ruiz-Sandoval, M. E., & Kurata, N. (2004). Smart sensing technology: Opportunities and challenges. *Structural Control and Health Monitoring, 11*(4), 349–368. DOI: 10.1002/stc.48

Sun, H., Burton, H. V., & Huang, H. (2021). Machine learning applications for building structural design and performance assessment: State-of-the-art review. *Journal of Building Engineering, 33*, 101816. DOI: 10.1016/j.jobe.2020.101816

Sun, Z., & Xia, Y. (2022). Study on Structural Health Monitoring and Damage Detection in the Field of Civil Engineering. *Highlights in Science. Engineering and Technology, 10*, 92–98. DOI: 10.54097/hset.v10i.1231

Thai, H.-T. (2022). Machine learning for structural engineering: A state-of-the-art review. *Structures, 38*, 448–491. DOI: 10.1016/j.istruc.2022.02.003

Volk, R., Stengel, J., & Schultmann, F. (2014). Building Information Modeling (BIM) for existing buildings—Literature review and future needs. *Automation in Construction, 38*, 109–127. DOI: 10.1016/j.autcon.2013.10.023

Wang, C., Yin, L., Zhang, L., Xiang, D., & Gao, R. (2010). Metal Oxide Gas Sensors: Sensitivity and Influencing Factors. *Sensors (Basel), 10*(3), 2088–2106. DOI: 10.3390/s100302088 PMID: 22294916

Worden, K., & Manson, G. (2007). The application of machine learning to structural health monitoring. *Philosophical Transactions of the Royal Society A: Mathematical, Physical and Engineering Sciences, 365*(1851), 515–537. DOI: 10.1098/rsta.2006.1938

Wu, T., Sher, C.-W., Lin, Y., Lee, C.-F., Liang, S., Lu, Y., & Chen, H. (1557). S.-W., Guo, W., Kuo, H.-C., & Chen, Z. (2018). Mini-LED and Micro-LED: Promising Candidates for the Next Generation Display Technology. *Applied Sciences (Basel, Switzerland), 8*(9), 1557. Advance online publication. DOI: 10.3390/app8091557

Zitouni, I., Rhimini, H., & Chouaf, A. (2024). Modeling the Propagation of Ultrasonic Guided Waves in Four-Layer Laminated Composites. In Azari, Z., El Had, K., Ait Ali, M. E., El Mahi, A., Chaari, F., & Haddar, M. (Eds.), *Advances in Applied Mechanics* (pp. 25–33). Springer Nature Switzerland., DOI: 10.1007/978-3-031-49727-8_3

Chapter 8
Piezoelectric Energy Harvesting From Linear Vibrations of Conventional and Nanocomposite Materials

Mustapha Fouaidi
Mechanical and Energy (ME) Group, LPNAMME Laboratory, Department of Physics, Faculty of Sciences, Chouaib Doukkali University, El Jadida, Morocco

Ouassim Mehdi Bilah
Laboratory of Modelling and Simulation of Intelligent Industrial Systems, ENSET of Mohammedia, Hassan II University of Casablanca, Casablanca, Morocco

Laidi Zahiri
http://orcid.org/0000-0003-2685-6088
Laboratory of Modelling and Simulation of Intelligent Industrial Systems, ENSET of Mohammedia, Hassan II University of Casablanca, Casablanca, Morocco

ABSTRACT

Piezoelectric energy harvesting has emerged as a promising technology for transforming mechanical energy from vibrations into electrical power. This paper examines the performance of piezoelectric energy harvesters using both traditional and nanocomposite materials exposed to linear vibrations. The study investigates how material characteristics, including stiffness, damping, and piezoelectric coefficients, affect energy harvesting efficiency. Numerical analyses are employed to compare the energy conversion capabilities of conventional materials like metals and ceram-

DOI: 10.4018/979-8-3693-6627-1.ch008

© 2027, IGI Global Scientific Publishing. Copying or distributing in print or electronic forms without written permission of IGI Global Scientific Publishing is prohibited. Use of this chapter to [...]erative artificial intelligence (AI) technologies is expressly prohibited. The publisher reserves all rights to license its use for generative AI training and machine learning model development.

ics with those of nanocomposites reinforced with nano-scale fillers. The findings highlight the potential advantages of nanocomposite materials in enhancing energy harvesting efficiency by utilizing their distinctive mechanical and electrical properties at the nano-level. This research contributes to advancing the comprehension of piezoelectric energy harvesting mechanisms and offers insights for optimizing the design of effective energy harvesting systems for diverse applications.

1) INTRODUCTION

Energy harvesting is the method of collecting ambient energy from the environment and converting it into a usable electrical form. Among the different sources for power generation, mechanical vibrations are advantageous due to their common occurrence in various engineering structures. Numerous studies have investigated the use of piezoelectric materials for converting ambient environmental vibrations into energy (Duan et al., 2005 ; Wang & Wu, 2012 ; Xie et al., 2013). The use of graphene oxide powder GOP as reinforcement in polymer composites is increasing due to their exceptional physical and mechanical properties (Fouaidi et al., 2021 ; 2023 ; Sobhani & Safaei, 2023).

Recently, with the advent of functionally graded (FG) materials, graphene oxide particles (GOP) have been used as reinforcement in a functionally graded manner, leading to the development of FG-GOP-reinforced materials. The development of GOP-reinforced nanocomposites aims at their utilization in real structural applications, emphasizing the importance of research on FG-GOP-reinforced harvesters. Apart from the use of CNT (Heshmati & Amini, 2020 ; Mahesh & Mangalasseri, 2024 ; Niazi et al., 2022), to the best of the authors' knowledge, no research has been conducted on the feasibility of using functionally graded piezoelectric GOP-reinforced harvesters under harmonic vibration to generate electrical power. Consequently, this study focuses on analyzing GOP-reinforced harvesters in a unimorph configuration under harmonic base excitation, and exploring the feasibility of using these novel piezoelectric energy harvesters to capture vibration energy. In the theoretical formulation, the generalized Hamilton's principle is applied based on Euler–Bernoulli beam theory to derive the governing equations. The study also explores the uniform distribution of GOP within the polymer matrix of the substructure layer. The effective material properties of the FG-GOP substructure layer are determined using the modified Halpin-Tsai model. The finite element method is used to simulate piezoelectric harvesters in a unimorph configuration. Analyses are conducted in the frequency domain based on the finite element equations. The study then examines how the weight fraction of GOP affects the energy harvesting performance of the FG-GOP-reinforced harvester under harmonic base excitation.

2) THEORETICAL FORMULATION

This section is dedicated to the mechanical modeling of a unimorph piezoelectric beam using the Euler-Bernoulli beam model.

2-1) Problem Definition

Fig. 1 shows the schematic drawing of a unimorph piezoelectric cantilever beam, of length L_1, width b and total thikness $h = h_p + h_{np}$.

Figure 1. Piezoelectric unimorph beam under base excitation with attached variable load resistance

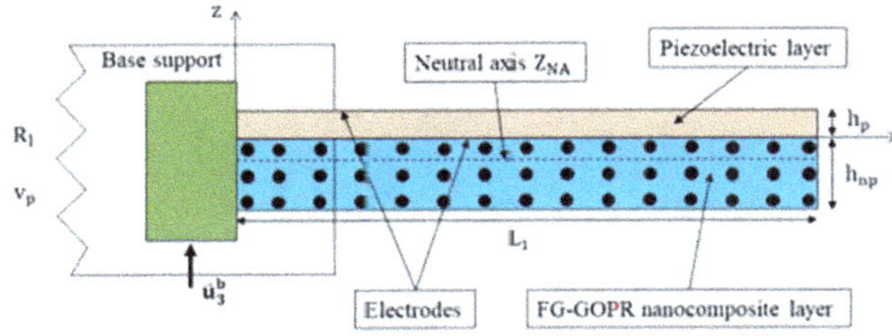

The origin of the reference coordinate system is situated on the beam's neutral axis. The beam is subjected to base excitation due to the motion of the host structure and variable load resistance, as shown in Fig. 1. The unimorph piezoelectric beam is composed of two layers: the first layer, located at the top of the beam, is made from a piezoelectric material of thickness h_p; the second non-piezoelectric layer, located at the top of the beam and with thickness h_{np} is made from a nanocomposite material reinforced with graphene oxide powder fillers uniformly distributed within the polymer matrix material. The two layers are supposed bonded to each other. It is assumed that the electrodes attached to the piezoelectric layer are perfectly conductive, fully covering both the top and bottom surfaces.

2-2) Displacement field

Utilizing the Euler–Bernoulli beam model, the total displacement vector at a specific point (x, y, z) within the beam can be written as:

$$\mathbf{U}(x,z,t) = \mathbf{Z}_d(z)\mathbf{u}(x,t) \tag{1}$$

here, **U** stands for the three-dimensional displacement vector that depends on the coordinates (x, y, z), $\mathbf{Z_d}$ signifies the transformation matrix which changes with the thickness coordinate z, and **u** refers to the generalized displacement vector expressed as:

$$\mathbf{U}(x, z, t) = [\mathbf{U}_1 \quad \mathbf{U}_3]^T \; \mathbf{Z_d} = \begin{bmatrix} 0 & -z \\ 1 & 0 \end{bmatrix}$$

;

$$\mathbf{u} = [u_3 \quad \theta]^T \tag{2}$$

with u_3 represents the transverse displacement of the unimorph piezoelectric beam axis and $\theta(x, t) = \frac{\partial u_3(x, t)}{\partial x}$ is the slope of the beam.

2-3) Strain Field

From the displacement field (Eq. (1)), the non-zero strain component can be determined as follows:

$$S_1^{(i)}(x, z, t) = -z\frac{\partial^2 u_3(x, t)}{\partial x^2} \tag{3}$$

With $i = np$ for non-piezoelectric layer and $i = p$ for piezoelectric layer.

2-4) Constitutive Equations

In the present work, the unimorph piezoelectric cantilevered beam is composed of a piezoelectric layer and a non-piezoelectric layer. This subsection is dedicated to providing the constitutive equations for the two layers.

2-4-1) Constitutive Equations for the Non-Piezoelectric Layer (Bottom Layer)

The constitutive equations of the substrate layer can be expressed as follows:

$$T_1^{(np)}(x, z, t) = Y_{eff}^{(np)}(z) S_1^{(np)}(x, z, t) \tag{4}$$

Where $Y_{eff}^{(np)}$ represents the effective Young's modulus of the FG-GOP reinforced composite beams, and it can be calculated using a modified Halpin–Tsai model as follows:

$$Y_{eff}^{(np)}(z) = \alpha_L Y_L^{(np)}(z) + \alpha_T Y_T^{(np)}(z) \tag{5}$$

with $\alpha_L = 0.49$ and $\alpha_T = 0.51$ indicate the GOP reinforcement efficiency in the longitudinal and transverse directions, respectively. The Young's modulus $Y_L^{(np)}$ and $Y_T^{(np)}$ correspond to the longitudinal and transverse Young's modulus of the composite material, respectively, and are calculated as follows:

$$Y_\gamma^{(np)}(z) = \frac{1 + \xi_\gamma \eta_\gamma V_{GOP}(z)}{1 - \eta_\gamma V_{GOP}(z)} Y_{PM} \tag{6}$$

with $\gamma = (L, T)$ and the coefficients η_γ are determined as follows:

$$\eta_\gamma = \frac{(Y_{GOP}/Y_{PM}) - 1}{((Y_{GOP}/Y_{PM})) + \xi_\gamma} \tag{7}$$

where Y_{GOP} and Y_{PM} represent the Young's modulus of the GOP and polymer matrix, respectively. The geometry factors ξ_γ, with $\gamma = (L, T)$ can be computed using the following formula:

$$\xi_\gamma = 2\left(\frac{d_{GOP}}{h_{GOP}}\right) \tag{8}$$

in which d_{GOP} and h_{GOP} refers to the diameter and thickness of GOP, respectively. The total volume fraction of GOPs can be calculated as:

$$V_{GOP}(z) = \frac{W_{GOP}}{W_{GOP} + (\rho_{GOP}/\rho_{PM})(1 - W_{GOP})} \tag{9}$$

in which ρ_{GOP} and ρ_{PM} denote the density of graphene oxide powder and matrix, respectively, and W_{GOP} is the GOP's weight fraction. The volume fraction of the PM can be determined as:

$$V_{PM}(z) = 1 - V_{GOP}(z) \tag{10}$$

The effective mass density of the FG-GOP reinforced PM is given as:

$$\rho_{eff}^{(np)}(z) = \rho_{GP} V_{GOP}(z) + \rho_{PM} V_{PM}(z) \tag{11}$$

2-4-2) Constitutive equations for the piezoelectric layer (Top layer)

For the piezoelectric layer (i = p) made from Lead Zirconate Titanate (PZT), the constitutive equation, based on linear piezoelectricity, is given as follows:

$$T_1^{(p)}(x,z,t) = Y_{eff}^{(p)} S_1^p(x,z,t) - e_{31} E_3(z,t) \tag{12}$$

with, the parameters $Y_{eff}^{(p)}$, e_{31} and E_3 denote the piezoelectric elastic stiffness at a constant electric field, the piezoelectric coefficient, and electric field, respectively. The nonzero electric field component, as a function of the electrical potential with a negative gradient operator, can be represented as follows:

$$E_3(z,t) = -\Omega(z) v_p(t) \tag{13}$$

The electric displacement component is defined as

$$D_3(x,z,t) = e_{31} S_1^{(p)}(x,z,t) + \varepsilon_{33}^S E_3(z,t) \tag{14}$$

with ε_{33}^S is the permittivity at constant strain.

2-5) Governing Equation

The generalized Hamilton's principle can be written as:

$$\int_{t_1}^{t_2} \left[\delta(K_E - P_E + E_E) + \delta E_w\right] dt = 0 \tag{15}$$

Where K_E, P_E, E_E and E_W design the kinetic energy, the potential energy, the electrical energy, and the external work due to input base excitation and electrical charge output, respectively given as:

$$K_E = \frac{1}{2} \int_{V^{(p)}} \dot{\mathbf{U}}^T \rho^{(p)} \dot{\mathbf{U}} \, dV^{(p)} + \frac{1}{2} \int_{V^{(np)}} \dot{\mathbf{U}}^T \rho^{(np)} \dot{\mathbf{U}} \, dV^{(np)}$$

$$P_E = \frac{1}{2} \int_{V^{(p)}} \mathbf{S}^{(p)T} \mathbf{T}^{(p)} dV^{(p)} + \frac{1}{2} \int_{V^{(np)}} \mathbf{S}^{(np)T} \mathbf{T}^{(np)} dV^{(np)}$$

$$E_E = \frac{1}{2} \int_{V^{(p)}} \mathbf{E}^T \mathbf{D} \, dV^{(p)}$$

$$E_W = -U_3 F + v_p q_p \tag{16}$$

3) FINITE ELEMENT DESCRETIZATION

In this work two-noded C^1 continuous beam elements of length L with two degrees of freedom u_3^i and θ^i per node are used for the discretization of kinematics variables. The transverse displacement u_3 is interpolated using C^1 shape functions as:

$$\tilde{u}_3(x,t) = \Phi(x)u^{(e)}(t) \tag{17}$$

where $\mathbf{u}^{(e)}$ is the nodal displacement vector for the eth element given by:

$$\mathbf{u}^{(e)} = \left[u_3^1, \theta^1, u_3^2, \theta^2\right]^T \tag{18}$$

with Φ is row vector containing C^1 shape functions, and defined as:

$$\Phi = \left[\Phi_1, \Phi_2, \Phi_3, \Phi_4\right] \tag{19}$$

where the element shape functions $\Phi_i (i = 1,\ldots,4)$ are given as:

$$\begin{cases} \Phi_1(x) = 1 - \frac{x^2}{L^2} + \frac{2x^3}{L^3} \\ \Phi_2(x) = x - \frac{2x^2}{L} + \frac{x^3}{L^2} \\ \Phi_3(x) = \frac{3x^2}{L^2} - \frac{2x^3}{L^3} \\ \Phi_4(x) = \frac{x^3}{L^2} - \frac{x^2}{L} \end{cases} \tag{20}$$

The discretized strain vector can be written as:

$$\tilde{S}_1^{(i)}(x,z,t) = -z\Psi(x)u^{(e)}(t) \tag{21}$$

where Ψ represnts the row vector of second derivative of the shape functions given as:

$$\Psi(x) = \frac{d^2\Phi(x)}{dx^2} = \left[\Psi_1(x),\ \Psi_2(x),\ \Psi_3(x),\ \Psi_4(x)\right] \tag{22}$$

where the functions $\Psi_i (i = 1,\ldots,4)$ are written as:

$$
\begin{cases}
\Psi_1(x) & = & -\frac{2}{L^2} + \frac{12x}{L^3}, \\
\Psi_2(x) & = & -\frac{4}{L} + \frac{6x}{L^2} \\
\Psi_3(x) & = & \frac{6}{L^2} - \frac{12x}{L^3} \\
\Psi_4(x) & = & \frac{6x}{L^2} - \frac{2}{L}
\end{cases}
\tag{23}
$$

The global matrix forms of electromechanical dynamic equations can be written as:

$$
\begin{cases}
M\ddot{u}(t) + C\dot{u}(t) + Ku(t) + P_\theta v_p(t) & = & F(t) \\
P_\theta^T\dot{u}(t) + P_D\dot{v}_p(t) & = & i_p(t)
\end{cases}
\tag{24}
$$

where **M** is global mass matrix, **K** is the global stiffness matrix and **C** is the damping matrix. Other parameters $\mathbf{P}_\theta$, $\mathbf{P}_D$ and **F** indicate global electromechanical coupling matrices, global capacitance matrices, global mechanical forces, respectively. Moreover, variable $\mathbf{i}_p$ is global current output, **u** is global nodal displacements vector, and $\mathbf{v}_p$ is global voltage output. These matrices and vectors are given as:

$$
\left(\mathbf{K}, \mathbf{M}, \mathbf{C}, \mathbf{P}_\theta, \mathbf{P}_\theta^T, \mathbf{P}_D, \mathbf{F}\right) = \sum_{e=1}^{N_E}\left(\mathbf{K}^{(e)}, \mathbf{M}^{(e)}, \mathbf{C}^{(e)}, \mathbf{P}_\theta^{(e)}, \mathbf{P}_\theta^{(e)T}, \mathbf{P}_D^{(e)}, F^{(e)}\right)
\tag{25}
$$

where local matrices and vectors are given as:

$$\int_{V^{(p)}} z^2 \Psi^T(x) Y_{eff}^{(p)} \Psi(x) dV^{(p)} + \int_{V^{(np)}} z^2 \Psi^T(x) Y_{eff}^{(np)} \Psi(x) dV^{(np)}$$

$$\int_{V^{(p)}} \rho^{(p)} \Phi^T(x)\Phi(x) dV^{(p)} + \int_{V^{(np)}} \rho^{(np)} \Phi^T(x)\Phi(x) dV^{(np)}$$

$$\alpha \mathbf{M}^{(e)} + \beta \mathbf{K}^{(e)}$$

$$-\int_{V^{(p)}} z \Psi^T(x) e_{31} \Omega(z) dV^{(p)}$$

$$-\int_{V^{(p)}} z \Omega^T(z) e_{31} \Psi(x) dV^{(p)}$$

$$-\int_{V^{(p)}} \Omega^T(z) \varepsilon_{33}^s \Omega(z) dV^{(p)}$$

$$\int_{V^{(p)}} \rho^{(p)} \Phi^T(x) dV^{(p)} \ddot{u}_b + \int_{V^{(p)}} \rho^{(np)} \Phi^T(x) dV^{(p)} \ddot{u}_b$$

4) FREQUENCY-DOMAIN ANALYSIS

The piezoelectric response under base excitation can be assumed to be harmonic. Therefore, solutions of the mechanical displacement and voltage responses can be formulated as:

$$\begin{cases} \mathbf{u} = ae^{i\omega t} \\ \mathbf{u}_3^b = u_3^b e^{i\omega t} \\ \mathbf{v}_p = be^{i\omega t} \end{cases} \tag{27}$$

Here, variable a is the amplitude of the base translation and b is the amplitude of the harmonic voltage and ω is the driving frequency. The multi-mode voltage FRFs can be formulated after simplifying the derivations to give:

$$\frac{v_p}{-\omega^2 u_3^b e^{i\omega t}} = \begin{bmatrix} i\omega C_p - \frac{1}{R_l} - i\omega\,\Theta^T \\ \times\,[\mathbf{K} - \omega^2\mathbf{M} + i\omega\mathbf{C}]^{-1}\,\Theta \end{bmatrix}^{-1} \times j\omega\,\Theta^T[\mathbf{K} - \omega^2\mathbf{M} + i\omega\mathbf{C}]^{-1}\mathbf{Q} \tag{28}$$

The multi-mode FRFs of the electric current output related to the input base transverse excitation can also be formulated by dividing the voltage response with the load resistance to yield:

$$\frac{i_p}{-\omega^2 u_3^b e^{i\omega t}}$$

$$= \frac{1}{R_l}\left\{\begin{bmatrix} i\omega C_p - \frac{1}{R_l} - i\omega\,\Theta^T \\ \times\,[\mathbf{K} - \omega^2\mathbf{M} + i\omega\mathbf{C}]^{-1}\,\Theta \end{bmatrix}^{-1} \times i\omega\,\Theta^T[\mathbf{K} - \omega^2\mathbf{M} + i\omega\mathbf{C}]^{-1}\mathbf{Q}\right\} \tag{29}$$

The multi-mode power output FRFs can also be formulated as:

$$\frac{P_p}{\left(-\omega^2 u_3^b e^{i\omega t}\right)^2}$$

$$= \frac{1}{R_l}\left\{\begin{bmatrix} i\omega C_p - \frac{1}{R_l} - i\omega\,\Theta^T \\ \times\,[\mathbf{K} - \omega^2\mathbf{M} + i\omega\mathbf{C}]^{-1}\,\Theta \end{bmatrix}^{-1} \times i\omega\,\Theta^T[\mathbf{K} - \omega^2\mathbf{M} + i\omega\mathbf{C}]^{-1}\mathbf{Q}\right\}^2 \tag{30}$$

where $\Theta = \mathbf{P}_\theta \vartheta^T$is a transformed electromechanical coupling vector and $\vartheta = [1 \;\ldots\; 1]$.

REFERENCES

Duan, W. H., Quek, S. T., & Wang, Q. (2005). Free vibration analysis of piezoelectric coupled thin and thick annular plate. *Journal of Sound and Vibration, 281*(1-2), 119–139. DOI: 10.1016/j.jsv.2004.01.009

Fouaidi, M., Belfallah, K., Jamal, M., & Belouaggadia, N. (2023). Transient analysis of functionally graded graphene oxide powders-reinforced porous composite beams resting on elastic foundations using the reproducing kernel particle meshless method. *Engineering Analysis with Boundary Elements, 146*, 460–472. DOI: 10.1016/j.enganabound.2022.10.029

Fouaidi, M., Jamal, M., Zaite, A., & Belouaggadia, N. (2021). Bending analysis of functionally graded graphene oxide powder-reinforced composite beams using a meshfree method. *Aerospace Science and Technology, 110*, 106479. DOI: 10.1016/j.ast.2020.106479

Heshmati, M., & Amini, Y. (2020). An electromechanical finite element model for new CNTs-reinforced harvesters subjected to harmonic and random base excitations. *Iranian Journal of Science and Technology. Transaction of Mechanical Engineering, 44*(1), 163–181. DOI: 10.1007/s40997-018-0254-x

Mahesh, V., & Mangalasseri, A. S. (2024). Agglomeration effects of CNTs on the energy harvesting performance of multifield interactive magneto-electro-elastic/nanocomposite unimorph smart beam. *Mechanics Based Design of Structures and Machines, 52*(3), 1324–1350. DOI: 10.1080/15397734.2022.2144886

Niazi, K., Parsi, M. K., & Mohammadi, M. (2022). Nonlinear Dynamic Analysis of Hybrid Piezoelectric-Magnetostrictive Energy-Harvesting Systems. *Journal of Sensors, 2022*(1), 8921779. DOI: 10.1155/2022/8921779

Sobhani, E., & Safaei, B. (2023). Dynamic analysis of functionally graded embedded graphene oxide powder nanocomposite conical shells resting on flexible foundations. *Engineering Analysis with Boundary Elements, 152*, 757–773. DOI: 10.1016/j.enganabound.2023.05.006

Wang, Q., & Wu, N. (2012). Optimal design of a piezoelectric coupled beam for power harvesting. *Smart Materials and Structures, 21*(8), 085013. DOI: 10.1088/0964-1726/21/8/085013

Xie, X. D., Wu, N., Yuen, K. V., & Wang, Q. (2013). Energy harvesting from high-rise buildings by a piezoelectric coupled cantilever with a proof mass. *International Journal of Engineering Science, 72*, 98–106. DOI: 10.1016/j.ijengsci.2013.07.004

Chapter 9
Predicting Mohs Hardness Using Machine Learning

Kalp Prajapati
http://orcid.org/0009-0004-2564-3886
VIT Bhopal University, India

Ansh Patel
VIT Bhopal University, India

Milan Chauhan
http://orcid.org/0009-0001-7296-0946
VIT Bhopal University, India

D. Lakshmi
http://orcid.org/0000-0003-4018-1208
VIT Bhopal University, India

ABSTRACT

This chapter offers a thorough approach to predicting Mohs hardness using machine learning techniques. Various models, including Random Forest, Linear Regression, and Gradient Boosting, were assessed to determine the most effective method for this task. Mohs hardness is a crucial property in identifying and characterizing minerals and solid materials. Accurate prediction is vital for both industrial and research purposes. The study explores how machine learning can be used to predict Mohs hardness based on different physicochemical properties of materials, highlighting the trade-offs between model complexity and prediction accuracy. It emphasizes the importance of finding a balance for practical applications and provides insights from

DOI: 10.4018/979-8-3693-6627-1.ch009

© 2027, IGI Global Scientific Publishing. Copying or distributing in print or electronic forms without written permission of IGI Global Scientific Publishing is prohibited. Use of this chapter to [...]erative artificial intelligence (AI) technologies is expressly prohibited. The publisher reserves all rights to license its use for generative AI training and machine learning model development.

feature importance analysis to better understand the factors affecting Mohs hardness. The results show that machine learning can greatly enhance the accuracy of Mohs hardness predictions, offering a valuable tool for material scientists and engineers.

1.0 INTRODUCTION

The Mohs hardness scale is an essential tool in material science, providing a consistent way of assessing the durability of materials and minerals against scratches (Gerberich et al., 2015). However, traditional methods of determining Mohs hardness often involve labour-intensive and time-consuming procedures. In recent years, the emergence of machine learning has fundamentally changed predictive analytics in many areas. The article examines the intersection of these two domains, and how Machine Learning techniques can be applied for accurate Mohs hardness prediction. To facilitate the characterization of materials, researchers want to use large data sets and advanced algorithms which will provide new opportunities for science, engineering, and industrial applications. As we use machine learning to unlock the mysteries of Mohs hardness, join us on this journey.

1.1 Understanding Mohs' Hardness

Mohs hardness is a fundamental concept in material science and provides an objective assessment of the durability of materials against scratch. This scale, which goes from 1 to 10, describes the relative scratchability of minerals. While the Mohs hardness scale provides a qualitative measure, the mechanical properties such as microhardness, toughness, and 2modulus offer a more quantitative assessment of mineral hardness (Broz, Cook, & Whitney, 2006).To evaluate the material's resistance to scratch, it is found that the Mohs Hardness provides an effective and practicable means. Knowing the Mohs hardness of the material can be used by researchers and engineers to determine its suitability for different applications.

Historically, the assessment of Mohs hardness was carried out by labour-intensive techniques such as manually cutting mineral samples with known hardness or using specialized testing instruments. However, there are some drawbacks to these techniques. Depending on the state of the testing apparatus and methodology used by the tester, these tests can be difficult to carry out in a qualitative way or prone to variation.

The Mohs hardness prediction is essential for industrial applications, product development, and material selection. Choose building materials for construction projects or to develop a cutting instrument. Machine learning for Mohs hardness prediction is a systematic and efficient way of evaluating material properties and performance characteristics. This predictive capability can be used by researchers and engineers to improve the design of products, select the most suitable materials or guarantee quality control procedures in a wide range of sectors such as electronics, and aerospace.

In all of this, precise Mohs hardness prediction based on machine learning is not just accelerating the discovery but also promoting innovation and progress in a wide range of industries. We can open new avenues for research in materials science and engineering, shaping the future of technology and innovation by using machine learning's ability to predict.

1.2 How to Calculate Mohs Hardness?

Mohs hardness is a qualitative ordinal scale that characterizes the scratch resistance of various minerals through the ability of harder materials to scratch softer ones.The Mohs hardness scale, traditionally used to rank minerals based on scratch resistance, can be interpreted through physical principles as described by Tabor (1954). It is not calculated using a specific formula but rather determined through direct comparison against a standard set of minerals. The hardness scale is shown in the Table 1:

Table 1. The Hardness Scale

Mohs Hardness	Reference Mineral	Chemical formula	Absolute Hardness
1	Talc	$Mg_3Si_4O_{10}(OH)_2$	1
2	Gypsum	$CaSO_4{\cdot}2H_2O$	2
3	Calcite	$CaCO_3$	14
4	Fluorite	CaF_2	21
5	Apatite	$Ca_5(PO_4)_3(OH^-,Cl^-,F^-)$	48
6	Orthoclase feldspar	$KAlSi_3O_8$	72
7	Quartz	SiO_2	100
8	Topaz	$Al_2SiO_4(OH^-,F^-)_2$	200
9	Corundum	Al_2O_3	400
10	Diamond	C	1500

1.3 Determining Mohs Hardness

1. Scratch Test: Scratch the mineral with a reference material from the Mohs hardness scale.

a. If the mineral is scratched by the reference material, it is softer than that reference material.
b. If the mineral scratches the reference material, it is harder than that reference material.

2. Comparison: Continue comparing against harder and softer reference materials to narrow down the mineral's hardness.
   a. For instance, if a mineral can scratch fluorite (hardness of 4) but not apatite (hardness of 5), its Mohs hardness is between 4 and 5.

Example of Determining Mohs Hardness:

1. Start with the softest reference material (talc).
2. Attempt to scratch the unknown mineral with talc.
   a. If talc scratches the unknown mineral, the unknown mineral's hardness is less than 1.
   b. If talc does not scratch the unknown mineral, move to gypsum.
3. Continue with the next hardest reference materials until you find the hardest reference material that can scratch the unknown mineral or the softest reference material that the unknown mineral can scratch.
4. Figure 1 shows the interrelationship between the Mohs scale and the absolute hardness.

Figure 1. Absolute Hardness Vs Mohs Scale

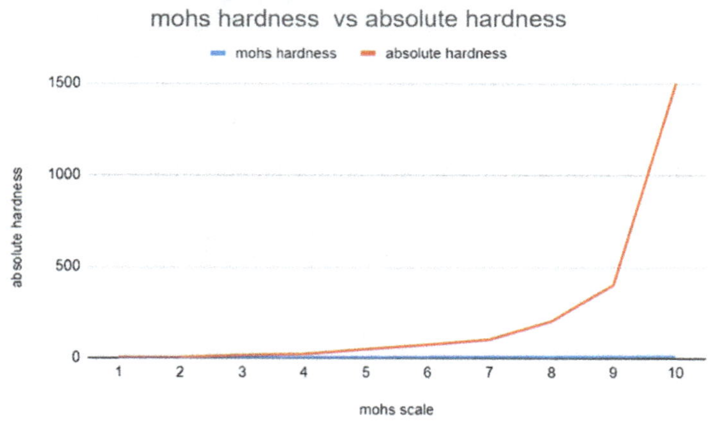

1.4 Practical Use

- Geologists use the Mohs hardness scale in the field to identify minerals by scratching them with common objects like a fingernail (hardness 2.5), a copper coin (hardness 3), a glass plate (hardness 5.5), or a steel file (hardness 6.5).

In conclusion, while Mohs hardness is not calculated via a formula, it is an empirically determined value through physical testing.Accurate measurement of hardness is crucial in materials science, particularly for traditional ceramics (Kim & Kim, 2002).

2.0 MACHINE LEARNING FUNDAMENTALS

Machine learning involves a variety of algorithms and techniques that enable computers to learn from the data, and make predictions or decisions that are not explicitly programmed. Machine learning tools are available to analyse complex relationships among different factors that affect the hardness and make accurate predictions, to forecast Mohs hardness.Machine learning has been increasingly employed to predict Mohs hardness by analyzing compositional features, as highlighted by Garnett (2019). An overview of the basic concepts is given below:

2.1 Overview of Machine Learning Algorithms and Techniques

a. It is possible to classify machine learning algorithms into supervised, unsupervised, and reinforcement learning. In supervised learning, each example is matched to a target label such as Mohs hardness in this case and the algorithm learns from those labels. Supervised learning algorithms include regression to predict continuous values and classification to predict discrete categories. Unsupervised learning consists of learning patterns from unlabelled data, while reinforcement learning learns to make decisions through interaction with the environment.
b. Techniques such as decision trees, support vector machines, neural networks, and ensemble methods (like Random Forests) are commonly used in supervised learning tasks. The strength and weakness of each algorithm depend on the nature of the problem, the size and quality of the data, or the computational resources at your disposal.

2.2 Explanation of Supervised Learning

c. The most appropriate approach for predicting hardness is supervised learning, which involves learning to map input features, such as mineral composition, and crystal structure, to the target hardness, based on labelled training data.
d. Three regression models are selected for predicting Mohs hardness:
   i. Random Forest Regressor
   ii. Linear regression
   iii. LightGBM
e. In this context, regression techniques such as linear regression and Random Forest regression are used to predict continuous values (Mohs hardness) based on input features.Grömping (2009) provides a detailed comparison of variable importance assessment in linear regression and random forests, offering valuable insights into their respective strengths and limitations.Our methodology employs linear regression techniques as detailed by Weisberg (2005) to model the relationship between compositional features and Mohs hardness.
f. Supervised learning aims to learn a model that aggregates well the unseen data and makes accurate predictions for new samples.

2.3 Introduction to Feature Selection and Feature Engineering:

a. Identifying the most relevant features contributing to the prediction of Mohs hardness is part of the feature selection process. Domain knowledge, statistical tests, or automated feature selection algorithms could be used to do this.
b. Feature engineering is a fundamental aspect of building effective machine learning models, as detailed in the comprehensive work by Dong and Liu (2018). To represent the underlying patterns in data, feature engineering is a process of developing new features or updating current ones. Combining or scaling features, including the inclusion of categorical variables and obtaining useful information from raw data. Li, Ma, and Xin (2017) provide insights into feature engineering for chemisorption models, emphasising its importance in optimising catalyst design through machine learning.

2.3 Discussion on Training Data Collection, Preprocessing, and Validation Techniques

c. It is crucial to collect high-quality training data for the development of accurate forecasts. Training data from CSV files is loaded into the provided code and explored to identify its characteristics.

d. Data preprocessing plays a crucial role in machine learning, influencing the performance and interpretability of models (Zelaya, 2019).The model should be able to learn from pure and properly formatted data through preprocessing techniques, like the elimination of values, scale features with StandardScaler, and splitting data into training and test sets via train_test_split.

   i. Standardisation: using the standardscaler, features shall be standardised. For each feature with a mean of 0 and a standard deviation of 1, the technique scales it. It ensures that all features contribute equally to model training and prevents features with larger scales from dominating the learning process.

   ii. Train-Test Split: The dataset is split into training and testing sets using the train_test_split function. This technique helps to evaluate the performance of models based on unobservable data and assess their generalisation capacity.

e. For the assessment of model performance and generalisation ability, validation techniques such as cross-validation or holdout validation are essential. The code uses metrics such as mean squared error, mean absolute error, and Rsquared on the test set to evaluate model performance.

   i. After training, the models make predictions on the test data (x_test) using the predict() method.

   ii. Evaluation metrics such as Mean Squared Error (mean_squared_error), Mean Absolute Error (mean_absolute_error), and R-squared Score (R^2_score) are calculated to assess model performance. These metrics quantify the accuracy and goodness-of-fit of the models.

Machine learning allows for the development of predictive models for Mohs hardness and other complex problems in different areas by applying these basic concepts and techniques. Building reliable and accurate predictive models involves a systematic approach to data collection, preprocessing, model training, evaluation, and validation.

2.4 Data Collection and Preprocessing

The foundation of our machine learning efforts to predict Mohs hardness lies in a dataset that we use and the way it's processed for meaningful insights. In this section, the complexities of data collection and preprocessing, the essential steps in the path to accurate predictive modelling, are explored.

2.5 About the Dataset

Preprocessing steps are crucial for enhancing the accuracy of machine learning algorithms, as demonstrated in sentiment analysis studies (Alam & Yao, 2019). Our analysis is based on a dataset used for Mohs hardness prediction. It consists of a collection of samples, each with specific characteristics that are important for the Mohs hardness assessment. These characteristics include a variety of factors, including elemental composition, crystal structure, and physical properties, all of which contribute to the material's scratch resistance. Using this dataset, we'll be able to discover basic patterns and relationships governing Mohs hardness in diverse materials.

2.6 Description of Features

The features that are taken into account for Mohs hardness prediction are central to our predictive modelling effort. These characteristics constitute a wealth of information about the materials subject to examination, which provides insights into their intrinsic properties. The composition of the elementals elucidates the chemical constituents present in a material and influences its structural character and hardness. The crystal structure determines the arrangement of atoms and ions in a material lattice, which affects its physical properties. A wide range of physical characteristics, such as density, porosity, and thermal conductivity, contribute to the material's scratch resistance. We are trying to achieve the multidirectional nature of Mohs hardness and improve prediction accuracy by integrating these different features into our forecast models.

2.6 Model Selection and Training

Selecting and training models is a critical step in the process of machine learning, as it determines the architecture and characteristics of these models that will best capture data patterns for accurate prediction. The model selection process involves choosing the most appropriate machine learning algorithm and fine-tuning its hyperparameters to build an accurate predictive model for Mohs hardness prediction. The model selection process, including hyperparameter tuning and cross-validation, is described in detail below:

1. Algorithm Selection

a. Start by considering many Machine Learning algorithms that can be used to perform regression tasks, e.g. Linear Regression, Decision Trees, Random Forests, Support Vector Machines and Neural Networks.
b. Assess the pros and cons of each algorithm by taking into account aspects such as interpretation, scalability, handling of nonlinear relationships, correlation, and computer efficiency.
c. Choose an algorithm that best matches the characteristics of the data set and requirements for a given problem.
d. A wide variety of algorithms and techniques to predict Mohs hardness are available from machine learning, each with its specific strengths and weaknesses. Some common models are given in this overview:
    i. Linear Regression

- Linear regression is a simple yet powerful algorithm that assumes a linear relationship between the input features and the target variable (Mohs hardness).
- It's interpretable and computationally efficient, making it a good starting point for modelling tasks where the relationship between features and target is approximately linear.
- However, it may not capture complex non-linear relationships present in the data.

ii. Random Forests

- Random Forest is a method of learning which consists in building various decision trees while training, and combining their forecasts to produce more reliable and accurate models.
- By aggregating predictions from multiple trees, Random Forest reduces overfitting and improves generalisation performance.
- It is capable of handling complex relationships and interactions between features, making it suitable for predicting Mohs hardness where the underlying factors are highly interconnected.

iii. LightGBM

- LightGBM (Light Gradient Boosting Machine) is a gradient boosting framework that builds models in a sequential manner, where each new model corrects the errors made by the previous ones. It is designed to be highly efficient and scalable, making it particularly suitable for large datasets and complex problems.

e. The interpretability of the model, the size and complexity of the dataset, the computational resources available, and other criteria determine which machine learning model is best suited for a given set of circumstances. All of these models have special benefits to offer. Successful Mohs hardness prediction requires experimenting with several models and choosing the one that best strikes a compromise between interpretability and predictive performance.

2. Hyperparameter Tuning
    a. Hyperparameters are variables that control how the algorithm learns; they are not directly determined by the input. The quantity of trees in a random forest, the neural network's learning rate, and the kind of kernel in an SVM are a few examples.
    b. To determine the ideal settings that optimise the model's performance on unknown data, undertake hyperparameter tuning.
    c. Bayesian optimization, random search, and grid search are methods for hyperparameter tuning. While random search selects hyperparameters at random from predetermined distributions, grid search entails thoroughly scanning a predefined hyperparameter grid.
    d. Bayesian optimization uses probabilistic models to more effectively direct the search procedure.
    e. Use a validation set or cross-validation to assess each hyperparameter configuration's performance.
3. Cross-Validation
    a. One method for evaluating the model's generalisation performance and reducing the chance of overfitting is cross-validation.
    b. Divide the dataset into many folds (usually k folds) and utilise the remaining data for training, with each fold acting as a validation set.
    c. Utilising k-1 folds as training data, assess the model's performance using the held-out fold. Carry out this procedure k times, utilising a distinct fold as the validation set each time.
    d. To get a fair assessment of the model's performance, compute the average performance metrics over all folds.
    e. K-fold cross-validation, leave-one-out cross-validation (LOOCV), and stratified k-fold cross-validation (for classification tasks) are examples of common cross-validation approaches.
4. Model Evaluation
    a. Evaluate the chosen model's performance using evaluation measures including mean squared error, mean absolute error, and R-squared score after hyperparameter adjustment and cross-validation.
    b. To choose the final model that performs best on the validation data, compare the performance of several methods and hyperparameter settings.

c. When selecting the ultimate choice, it is critical to take into account both prediction performance and model interpretability.

5. Final Model Selection and Deployment:
   a. Retrain the model with the chosen hyperparameters on the complete training dataset after determining which one performs the best.
   b. Make sure the final model performs well in terms of generalising to new data by testing it on the test dataset.
   c. Deploy the model in a production setting to make predictions on fresh data if you are happy with its performance.

3.0 RESULTS AND EVALUATION

Now that we have painstakingly trained and adjusted our machine learning models, it is time to present the results of our predictive project and assess their effectiveness. This section includes the experimental results, an examination of the accuracy metrics, a comparison of the effectiveness of various machine learning models, an analysis of the most important features for Mohs hardness prediction, and a discussion of the predictive models' shortcomings and potential future developments.

3.1 Presentation of Experimental Results

Key accuracy measures were used to assess the effectiveness of our predictive models, Linear Regression, Random Forest Regressor, and LightGBM Classifier. We found that the Random Forest Regressor had an R-squared (R^2) score of 0.4325, a mean absolute error (MAE) of 0.9203, and a mean squared error (MSE) of 1.5423. An MSE of 2.1054, an MAE of 1.1720, and an R^2 score of 0.2253 were also obtained using the Linear Regression model. These measures offer important insights into our models' predictive power: higher R^2 scores denote better data fit, and lower MSE and MAE indicate better performance. Figure 2 shows the correlation heatmap.

- LightGBM Classifier Experiment Results

The LightGBM classifier was evaluated using Repeated Stratified K-Fold cross-validation with 10 splits. The model's performance was measured using the accuracy score for each fold. The overall out-of-fold (OOF) accuracy was approximately 0.5366. These results highlight that while the LightGBM model offers moderate performance, there is room for further optimization and improvement.

Figure 2. Correlation Heatmap

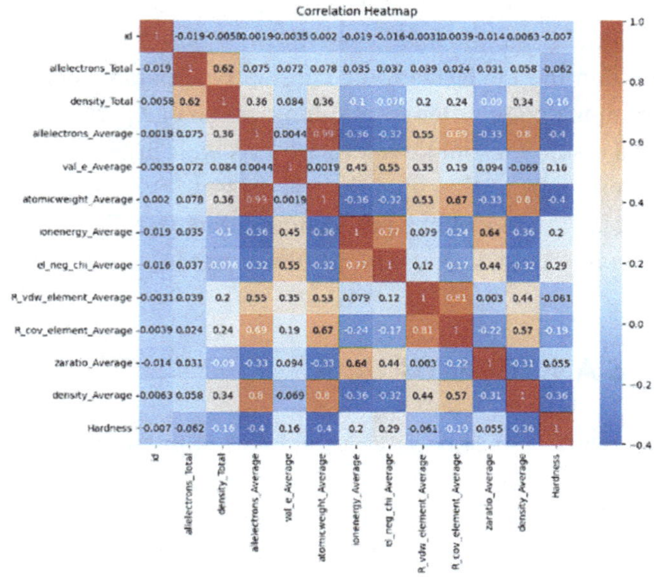

Based on the provided correlation heatmap, we can analyze the feature importance for predicting Mohs hardness using various machine learning models. The heatmap displays the correlation coefficients between features and the target variable (Hardness). Higher absolute values of these coefficients indicate stronger relationships with the target variable.

3.2 Feature Importance Analysis Based on Correlation Heatmap

- Correlation with Hardness

  - val_e_Average (0.29): This feature has the highest positive correlation with Mohs hardness, indicating that the average valence electrons are a significant predictor.
  - density_Average (0.36): This feature also shows a strong positive correlation with Mohs hardness, suggesting that the average density is another important predictor.
  - ionenergy_Average (0.45): This feature has the strongest positive correlation with Mohs hardness, making it a crucial feature for predicting hardness.
  - atomic weight_Average (-0.36): This feature shows a strong negative correlation with Mohs hardness, indicating that average atomic weight inversely affects the hardness.

- el_neg_chi_Average (0.32): This feature also shows a strong positive correlation with hardness.
- R_cov_element_Average (0.1): This feature shows a moderate positive correlation with hardness.
- R_vdw_element_Average (0.3): This feature also shows a moderate positive correlation with hardness.

3.3 Feature Importance Analysis for Each Model

1. Linear Regression: Linear Regression models often benefit from features with high correlation coefficients. Therefore, the model will likely assign significant weights to ionenergy_Average, density_Average, val_e_Average, atomicweight_Average, and el_neg_chi_Average.
2. Random Forest: Random Forest models can capture non-linear relationships and interactions between features. Features like ionenergy_Average, density_Average, val_e_Average, atomicweight_Average, and el_neg_chi_Average are expected to have higher importance scores due to their strong correlations with the target variable. Random Forest algorithms can also identify the importance of less correlated features if they provide useful information in combination with others.
3. LightGBM Classifier:

The LightGBM model, similar to the Random Forest, can capture non-linear relationships and interactions between features. The feature importance analysis from LightGBM indicated the following significant features:

- ionenergy_Average: The strongest predictor, similar to other models.
- density_Average: Important for predicting Mohs hardness.
- val_e_Average: High positive correlation with hardness.
- atomic weight_Average: Shows an inverse relationship with hardness.
- el_neg_chi_Average: Positively correlated with hardness.

Additionally, feature importance analysis was conducted to discern the most influential factors driving Mohs hardness prediction. For the Random Forest model, [discuss insights from feature importance analysis]. Similarly, for the Linear Regression model, [discuss insights from feature coefficients]. The feature importance analysis highlights the significance of various features in predicting Mohs hardness. Features such as ionenergy_Average, density_Average, val_e_Average, atomicweight_Average, and el_neg_chi_Average consistently show strong correlations with the target variable and are expected to have higher importance scores across different models. Understanding these important features aids in refining predictive models and offers valuable insights for materials science research. The model validation metrics are shown in the Table 2.

Table 2. Machine Model Validation Metrics

Model	R-squared (R^2)	Mean Absolute Error (MAE)	Mean Squared Error (MSE)
Random Forest Regressor	0.4325	0.9203	1.5423
Linear Regression	0.2253	1.17201	2.1054

3.4 Discussion on Limitations and Future Directions

Notwithstanding the encouraging outcomes, it's critical to recognise the shortcomings of our prediction models and pinpoint opportunities for further development. A constraint pertains to the dependence on conventional characteristics like crystal structure and elemental composition, which might not encompass all facets of material hardness. Adding sophisticated features or domain-specific expertise could improve the forecast accuracy and performance of the model.

Furthermore, even though our models were effective in predicting Mohs hardness, they might not work as well with new materials or situations that weren't included in the training set. To meet this challenge, we must continuously improve and enhance our predictive frameworks, either by using transfer learning strategies or creating adaptive models that can pick up new information.

Moreover, interpretability is still a major issue, especially for sophisticated machine learning models like Random Forests. Improving the interpretability of the model may make it easier to find relevant insights and build confidence in the forecasted results, which will make it easier for real-world applications to adopt the model.

Future research projects might concentrate on utilising cutting-edge technology like ensemble methods or deep learning to further improve the robustness and accuracy of predictions. We can improve our knowledge of Mohs hardness prediction and its applications in material science and engineering by adopting a multidisciplinary approach and encouraging collaboration across fields. This will pave the way for revolutionary developments in the subject.

4.0 APPLICATIONS AND IMPLICATIONS

Based on machine learning Mohs hardness prediction has great potential in several fields, including industrial applications, material science, and engineering. This section examines the various uses and consequences of precise hardness prediction and talks about how it might transform quality control procedures, product design optimisation, material selection, and material discovery and development.

4.1 Exploration of Potential Applications

There are a lot of opportunities in material science, engineering, and business when machine learning is used to forecast Mohs hardness. Accurate hardness prediction can help characterise novel materials in material science research, helping scientists better grasp their mechanical characteristics and applicability for particular uses. It can help engineers create stronger, longer-lasting components, which will enhance the functionality and lifespan of their products. Accurate hardness prediction can help industries like electronics, automotive, and aerospace ensure the performance and dependability of their materials and products.

4.2 Facilitating Material Selection and Product Design Optimization

Precise hardness estimation is essential for optimising product design and choosing materials. Engineers and designers can determine which materials are best suited for a given application by using machine learning algorithms to estimate Mohs hardness. This allows them to maximise variables like weight, cost, and manufacturability while selecting materials that fulfil the required performance requirements. Furthermore, precise hardness prediction makes it possible to spot any flaws or places of failure in product designs, allowing for the proactive implementation of strategies to improve durability and dependability.

4.3 Enhancing Quality Control Processes

Accurate hardness prediction can expedite inspection and testing processes in manufacturing and quality control processes, saving money and time compared to manual testing methods. Manufacturers may automate hardness testing and guarantee consistency and accuracy in their processes by incorporating machine learning models into their quality control systems. This boosts consumer happiness and brand reputation in addition to increasing efficiency and improving product quality and reliability.

4.4 Implications for Accelerating Material Discovery and Development

Material research and development procedures can be greatly accelerated by machine learning. Researchers can more effectively find novel materials with suitable hardness qualities by utilising vast datasets and sophisticated algorithms. Complex interactions between a material's composition, structure, and qualities can be analysed by machine learning algorithms, which can find new materials with previously unheard-of performance and properties. This has significant ramifications for a number of sectors, including advanced manufacturing, healthcare, and renewable energy, where the development of innovative materials is essential to innovation and advancement.

To sum up, Mohs hardness prediction using machine learning has broad applications and implications in material science, engineering, and business. By utilising machine learning, we can improve product performance and dependability, provide new avenues for innovation, and quicken the rate of material discovery and development—all of which will have a significant impact on how society and technology develop in the future.

5.0 CONCLUSION

We have examined the great importance of this predictive tool in material science and industry throughout our exploration of Mohs hardness prediction using machine learning. Reliability, performance, and innovation of materials and products are greatly dependent on accurate hardness prediction, which can help with material selection, product design optimisation, and quality control process simplification.

5.1 Recap of Significance

A key component of material science and industry, Mohs hardness prediction provides information on the mechanical characteristics and scratch resistance of a wide range of materials. Researchers, engineers, and manufacturers can improve product performance, longevity, and customer happiness by using precise hardness predictions to guide their material selection selections. This predictive capacity is the key to progress in a number of industries, including aerospace, automotive, electronics, and healthcare.

5.2 Summary of Key Findings

Machine learning-based methods have shown to be effective tools for predicting Mohs hardness during our investigation. Recent advancements in machine learning, particularly deep learning, have been applied to mineral identification and hardness prediction, as demonstrated by Zeng et al. (2021).We've gained important insights into the characteristics and behaviour of materials by utilising sophisticated algorithms and sizable datasets, which have allowed us to create predictive models that are remarkably accurate and effective. Every method, from linear regression models to Random Forest Regressors, has added to our knowledge of hardness prediction and provided a different viewpoint.

5.3 Call to Action

It is critical to acknowledge that our pursuit of predictive greatness is far from ended as we come to the end of our trip. To advance predictive models for Mohs hardness and beyond, further research and cooperation are needed. We may push the limits of predictive modelling and open up new vistas in material science, engineering, and other fields by embracing emerging technology and forming interdisciplinary partnerships. Let's set out on a joint voyage of exploration, creativity, and cooperation to create a future where predictive models enable us to fully use the potential of materials and bring in a new period of prosperity and scientific growth.

REFERENCES

Alam, S., & Yao, N. (2019). The impact of preprocessing steps on the accuracy of machine learning algorithms in sentiment analysis. *Computational & Mathematical Organization Theory, 25*(3), 319–335. DOI: 10.1007/s10588-018-9266-8

Broz, M. E., Cook, R. F., & Whitney, D. L. (2006). Microhardness, toughness, and modulus of Mohs scale minerals. *The American Mineralogist, 91*(1), 135–142. DOI: 10.2138/am.2006.1844

Dong, G., & Liu, H. (Eds.). (2018). *Feature engineering for machine learning and data analytics*. CRC press.

Garnett, J. C. (2019). Prediction of Mohs hardness with machine learning methods using compositional features. In *Machine Learning in Chemistry: Data-Driven Algorithms, Learning Systems, and Predictions* (pp. 23-48). American Chemical Society. DOI: 10.1021/bk-2019-1326.ch002

Gerberich, W. W., Ballarini, R., Hintsala, E. D., Mishra, M., Molinari, J. F., & Szlufarska, I. (2015). Toward demystifying the Mohs hardness scale. *Journal of the American Ceramic Society, 98*(9), 2681–2688. DOI: 10.1111/jace.13753

Grömping, U. (2009). Variable importance assessment in regression: Linear regression versus random forest. *The American Statistician, 63*(4), 308–319. DOI: 10.1198/tast.2009.08199

Kim, H., & Kim, T. (2002). Measurement of hardness on traditional ceramics. *Journal of the European Ceramic Society, 22*(9-10), 1437–1445. DOI: 10.1016/S0955-2219(01)00457-5

Li, Z., Ma, X., & Xin, H. (2017). Feature engineering of machine-learning chemisorption models for catalyst design. *Catalysis Today, 280*, 232–238. DOI: 10.1016/j.cattod.2016.04.013

Tabor, D. (1954). Mohs's hardness scale-a physical interpretation. *Proceedings of the Physical Society. Section B, 67*(3), 249–257. DOI: 10.1088/0370-1301/67/3/310

Weisberg, S. (2005). *Applied linear regression* (Vol. 528). John Wiley & Sons. DOI: 10.1002/0471704091

Zelaya, C. V. G. (2019, April). Towards explaining the effects of data preprocessing on machine learning. In *2019 IEEE 35th international conference on data engineering (ICDE)* (pp. 2086-2090). IEEE.

Zeng, X., Xiao, Y., Ji, X., & Wang, G. (2021). Mineral identification based on deep learning that combines image and mohs hardness. *Minerals (Basel)*, *11*(5), 506. DOI: 10.3390/min11050506

Chapter 10

Policy and Strategic Perspectives on Green Hydrogen and Renewable Energy in Emerging Countries: A Case Study of Morocco

Mariem Bibih

Laboratory of Electric Engineering and Intelligent Systems (EEIS), ENSET of Mohammedia, Hassan II University of Casablanca, Casablanca, Morocco

Karim Choukri

Laboratory of Electric Engineering and Intelligent Systems (EEIS), ENSET of Mohammedia, Hassan II University of Casablanca, Casablanca, Morocco

Houssam Eddine Chakir

Laboratory of Electric Engineering and Intelligent Systems (EEIS), ENSET of Mohammedia, Hassan II University of Casablanca, Casablanca, Morocco

Mehdi Et-Taoussi

Laboratory of Electric Engineering and Intelligent Systems (EEIS), ENSET of Mohammedia, Hassan II University of Casablanca, Casablanca, Morocco

ABSTRACT

This chapter provides a policy and strategic overview of green hydrogen and renewable energy deployment in emerging economies, using Morocco as a focused case study. It synthesizes the national legal, institutional, and market context shap-

DOI: 10.4018/979-8-3693-6627-1.ch010

© 2027, IGI Global Scientific Publishing. Copying or distributing in print or electronic forms without written permission of IGI Global Scientific Publishing is prohibited. Use of this chapter to nerative artificial intelligence (AI) technologies is expressly prohibited. The publisher reserves all rights to license its use for generative AI training and machine learning model development.

ing investment conditions, and links these frameworks to implementation realities across infrastructure, finance, and governance. The chapter reviews Morocco's renewable energy trajectory, the evolving electricity market architecture, and the Green Hydrogen Morocco Offer, including the role of public private partnerships, as enabling platforms for integrated projects and export-oriented value chains. It highlights the opportunities and constraints created by Morocco's evolving regulatory and market reforms, the role of the regulator, and the need for clear grid reception capacity and permitting procedures. It then examines cross cutting deployment dimensions and concludes with actionable, transferable lessons for other emerging countries pursuing green hydrogen scale up.

I. INTRODUCTION

Emerging countries are increasingly recognizing the importance of transitioning to renewable energy to address their energy security concerns, reduce environmental impacts, and promote sustainable economic development. These countries often face unique challenges, such as high dependency on energy imports, substantial energy bills, and the volatility of international energy prices. As they strive to build resilient and sustainable energy systems, the integration of green hydrogen and other renewable energy sources becomes a pivotal aspect of their energy policies and strategies.

Morocco, a country with limited conventional energy resources, has embarked on an ambitious journey towards renewable energy to tackle these challenges. With a rapidly growing population and increasing energy demand, Morocco's transition to renewable sources and its recent foray into green hydrogen present a compelling case study for understanding the policy and strategic dimensions of sustainable energy development in emerging economies. This chapter aims to provide a comprehensive analysis of Morocco's renewable energy policies and strategies, with a particular emphasis on the integration of green hydrogen, to extract lessons and insights applicable to other countries undergoing similar transitions.

The Green Hydrogen Morocco Offer (GHMO) represents a comprehensive, transparent, and practical approach aimed at providing investors with a clear roadmap for sustainable energy investments. With a multifaceted framework spanning six key components, it encompasses the scope of application, land mobilization, infrastructure development, incentive measures, investor selection process, and sector governance. This holistic approach not only facilitates investment but also ensures alignment with national development priorities and sustainability objectives.

The GHMO targets integrated projects encompassing renewable energy generation, electrolysis, and the production of green hydrogen and its derivatives, including ammonia, methanol, and synthetic fuels. The offer aims to attract both national and international investors keen on contributing to the domestic market, export opportunities, or a combination of both, reflecting the global interest in Morocco's green hydrogen potential.

Against this backdrop, this chapter aims to dissect Morocco's renewable energy policies and strategies. Firstly, it entails a detailed examination of the evolution of Morocco's renewable energy policies, spanning regulatory frameworks, incentive mechanisms, institutional arrangements, and strategic initiatives. By analyzing the policy landscape, the chapter aims to identify key drivers, objectives, and challenges shaping Morocco's renewable energy transition.

Secondly, the chapter assesses the strategic considerations underpinning Morocco's integration of green hydrogen within its energy mix. This includes an analysis of policy incentives, market mechanisms, research and development initiatives, infrastructure development, and international collaborations aimed at promoting green hydrogen as a clean energy carrier.

Furthermore, the chapter identifies and analyzes the key challenges hindering the effective implementation of renewable energy and green hydrogen policies in Morocco. These challenges may include regulatory barriers, financing constraints, technological limitations, institutional capacity gaps, and socio-economic considerations.

Additionally, the chapter evaluates the role of public-private partnerships (PPPs) and strategic collaborations in advancing renewable energy projects in Morocco. By examining successful PPP models, innovative financing mechanisms, and stakeholder engagement strategies, the chapter aims to draw lessons and insights for fostering collaboration between the public and private sectors in other emerging economies.

Lastly, the chapter involves extracting policy lessons, best practices, and recommendations from Morocco's experience to inform policy formulation and implementation strategies for other emerging economies undergoing energy transitions. By synthesizing key findings and insights, the chapter aims to provide actionable recommendations for policymakers, industry stakeholders, and development practitioners.

II. RENEWABLE ENERGY POLICIES

1. Overview of Morocco's Renewable Energy Policies and Strategic Initiatives

Morocco faces significant energy challenges, including high dependency on energy imports (around 90% as of 2023), a substantial energy bill, and the volatility of international energy prices impacting the domestic market. The primary objectives of the Moroccan energy policy are to ensure a reliable energy supply to the economy and population while promoting environmental preservation. This policy is structured around five strategic axes: security of supply, energy accessibility, mobilization of domestic resources, promotion of energy efficiency, and regional energy integration (Szima et al., 2021).

a) Legal and Institutional Framework

The National Energy Strategy, launched in 2009, set Morocco on a transformative path to increase the contribution of renewable energies to 52% of the national electricity mix and achieve a 20% reduction in energy consumption by 2030 through enhanced energy efficiency (Abdel-Karim et al., 2018a).

Figure 1. Breakdown of electricity production by energy source, end of 2021[1]

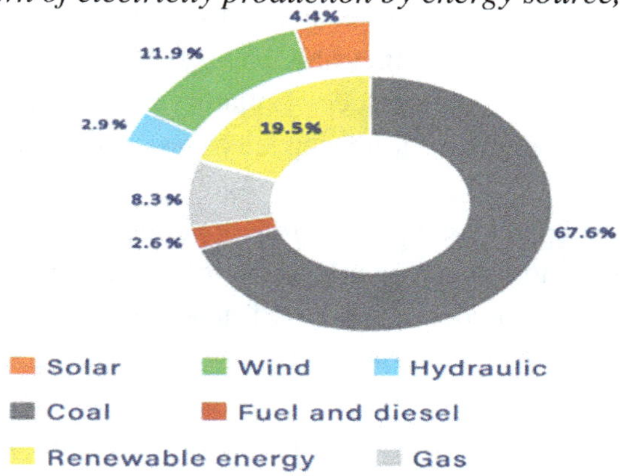

Today, Morocco boasts an operational installed capacity exceeding 4.1 GW, representing 39% of its energy mix. This includes 1,553 MW from wind power, 842 MW from solar power, and 1,771 MW from hydroelectric power. These achievements reflect a substantial investment of approximately \$5.3 billion (excluding pre-2009 hydroelectric power plants) and position Morocco as a progressive leader in the global renewable energy sector (Abdel-Karim et al., 2018b).

b) Strategic Renewable Energy Programs

Morocco is strategically positioning itself to become a leader in the green hydrogen sector, capitalizing on its significant renewable energy resources and ambitious climate goals. The country's roadmap outlines a clear vision to harness its abundant solar and wind energy to develop a robust green hydrogen industry. Morocco aims to establish itself as a major player in the global green hydrogen market, with production targets set to reach 2 million tons annually by 2030 and scale up to 5 million tons by 2050 (Bahetta et al., 2021). The goal is to not only meet domestic energy needs but also to become a key exporter of green hydrogen, particularly targeting European markets and other regions with high demand (Meliani, El Barkany, et al., 2021).

Figure 2. Evolution of Renewable Energy Sources Installed Capacity, MW (2000-2021)[2]

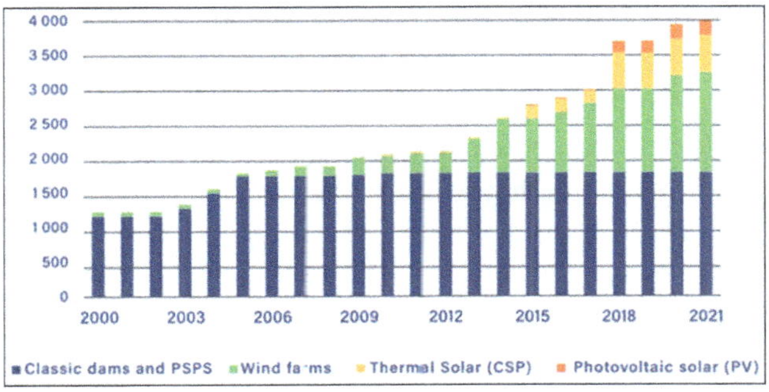

To achieve these objectives, Morocco is focusing on several strategic measures. This includes significant investments in infrastructure to support the production, storage, and distribution of green hydrogen (Meliani, El, et al., 2021). The road-map highlights the importance of constructing hydrogen production facilities and developing transport networks. Additionally, Morocco is prioritizing research and innovation to enhance the efficiency and reduce the costs of hydrogen production. Strategic partnerships with international companies and governments are also a cornerstone of this plan, aiming to foster collaboration, share expertise, and access global markets. Investment and financing play a crucial role in Morocco's green hydrogen strategy (Zakaria et al., 2024). The government is allocating public funds to support green hydrogen projects, providing subsidies and incentives to attract private sector participation. Private investment is being encouraged through favorable policies and tax incentives, alongside public-private partnerships. International support is also sought, with Morocco looking to secure funding and technical assistance from global organizations and financial institutions (Makan et al., 2022).

Performance indicators are essential to track the progress of Morocco's green hydrogen roadmap. Key metrics include the growth in hydrogen production capacity, the economic impact through job creation and investment inflows, and the environmental benefits, particularly the reduction in carbon emissions. These indicators will help assess whether the country is meeting its production milestones and aligning with its climate goals under the Paris Agreement (Zarkik & Ouhnini, 2022).

The implementation of Morocco's green hydrogen roadmap is planned in phases. In the short term, from 2023 to 2025, the focus will be on initial investments, pilot projects, and building foundational infrastructure (Kasseh et al., 2023a). The medium-term phase, from 2026 to 2030, will see the expansion of production capacity and scaling up of projects, with an emphasis on strengthening international collaborations. In the long term, from 2031 to 2050, Morocco aims to achieve its production targets, reinforce its global market position, and continue advancing technological innovations (Adun et al., 2022).

Overall, Morocco's roadmap for green hydrogen represents a comprehensive strategy to leverage its renewable energy potential, develop a competitive green hydrogen industry, and contribute significantly to global sustainability goals. The country is committed to becoming a leading player in the green hydrogen sector through strategic investments, innovation, and international cooperation(Rahmani et al., 2022).

2. Recent Developments

Since the announcement of the renewable energy programs, significant progress has been made. By 2023 (figure 3), Morocco had already achieved around 1,800 MW of installed solar capacity and 1,900 MW of wind capacity, putting it on track to meet and possibly exceed its 2024 targets. Additionally, Morocco's ambitious plans for green hydrogen production have started to take shape, with pilot projects being initiated to explore the potential of this emerging technology (Belmahdi et al., 2022).

Figure 3. Evolution of Renewable Energy Source Distribution from 2009 to 2030

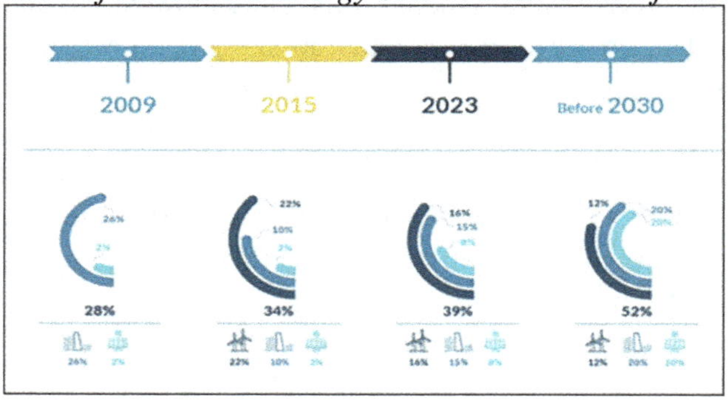

Furthermore, Morocco is actively working towards its 2030 goal of achieving 52% renewable energy in its energy mix. This involves continuous expansion of solar, wind, and hydro projects, supported by robust legal and regulatory frameworks. The recent tenders and projects, such as the Noor PV II and Noor Midelt projects, underscore the country's commitment to enhancing its renewable energy capacity (Okpanachi et al., 2022).

3. Strategic Assessment

Morocco's energy strategy is a comprehensive and multi-faceted approach aimed at reducing dependency on imported energy, enhancing energy security, and promoting sustainable development. By diversifying energy sources and investing in renewables, Morocco mitigates the risks associated with global energy market volatility, thereby enhancing energy security and supporting economic stability (Kouzzi et al., 2022). The focus on renewable energy and energy efficiency brings significant economic and social benefits, including job creation, stimulation of technological innovation, and attraction of foreign investment. Additionally, improving energy efficiency helps reduce costs for consumers and businesses. Morocco's commitment to renewable energy also significantly contributes to reducing greenhouse gas emissions and combating climate change ("Morocco's Quest for Stronger and Inclusive Growth," 2023). This transition to renewable energy sources aligns with global sustainability goals and enhances Morocco's reputation as a leader in green energy. Furthermore, by fostering regional energy integration, Morocco leverages shared resources and infrastructure, which enhances energy security and creates opportunities for cross-border collaboration and trade.

4. Future Policy Priorities

To maximize the potential of renewable energy, Morocco must focus on several key policy priorities. Strengthening institutional capacity to efficiently manage and implement renewable energy projects is essential, along with creating jobs through local manufacturing and R&D activities (Hafner et al., 2023). Accessing external funding through international climate funds, development banks, and public-private partnerships is crucial to support large-scale renewable energy projects. Enhancing grid infrastructure and negotiating market access to facilitate clean energy exports to Europe will leverage Morocco's competitive advantage in renewable energy production. Additionally, Morocco should actively participate in global climate negotiations and access climate finance mechanisms to support its renewable energy and climate adaptation goals.

III. GENERAL OVERVIEW OF GREEN HYDROGEN INTEGRATION IN MOROCCO

1. Hydrogen Use in Traditional Applications

In 2022, global hydrogen demand reached 95 million tonnes (Mt), showing a 3% increase from the previous year. The majority of this demand remains concentrated in traditional sectors like refining and chemical production, with limited penetration into newer applications (Ammari et al., 2022).

Hydrogen use in refining surpassed 41 Mt in 2022, driven largely by increases in North America and the Middle East. However, about 80% of hydrogen used in refining is produced onsite from fossil fuels, leading to significant CO2 emissions. In 2022 alone, refining hydrogen production emitted between 240 and 380 Mt of CO2 (Adeli et al., 2024). The sector also shows limited integration of low-emission hydrogen, with only 250 kilotonnes (kt) sourced from four facilities using fossil fuels with carbon capture, utilization, and storage (CCUS) (Dunlap & Laratte, 2022).

2. Hydrogen Production Technologies

The primary methods for hydrogen production include steam methane reforming (SMR) and electrolysis. SMR, particularly without CCUS, remains the largest emitter of CO2 in hydrogen production. Electrolysis, which relies on renewable energy, presents a path toward low-emission hydrogen but faces challenges related to high costs and scalability (Slimani et al., 2021).

a) Understanding Green Hydrogen Production: A Scientific Perspective

Green hydrogen is produced via the electrolysis of water using renewable energy sources such as solar, wind, or hydroelectric power, making it a sustainable and carbon-free alternative. Its production requires renewable electricity to ensure a zero-carbon footprint, contributing significantly to reducing global greenhouse gas emissions (Ait Errouhi et al., 2022). Environmentally, green hydrogen supports sustainability and energy security by minimizing fossil fuel dependence. It possesses high energy density by weight, advantageous for energy storage and transport, although its low volumetric energy density necessitates efficient storage methods like compressed gas, liquid hydrogen, or metal hydrides, each presenting unique challenges (Chennaif et al., 2022).

b) Comparing Green Hydrogen with Other Production Methods

Hydrogen can be produced through several methods, each with distinct environmental implications:

Table 1. Comparison of Hydrogen Types Based on Production Methods and Environmental Impact

Type of Hydrogen	Production Method	Environmental Impact
Gray Hydrogen	Produced via steam methane reforming (SMR), where methane reacts with steam to produce hydrogen and CO_2.	High carbon footprint; relies on fossil fuels with substantial CO_2 emissions.
Blue Hydrogen	Produced using SMR with carbon capture and storage (CCS) to capture and store CO_2 emissions.	Lower carbon footprint than gray hydrogen, but still depends on fossil fuels; effectiveness of CCS varies
Turquoise Hydrogen	Produced via methane pyrolysis, splitting methane into hydrogen and solid carbon, avoiding direct CO_2 emissions.	Potential for low emissions; sustainability depends on solid carbon management and energy source used.
Brown/Black Hydrogen	Produced from coal gasification, involving the reaction of coal with oxygen and steam to produce hydrogen and CO_2.	Highly carbon-intensive; significant CO_2 emissions due to coal-based production.

Table I compares hydrogen production methods, highlighting their techniques and environmental impacts. Gray hydrogen, produced from steam methane reforming without carbon capture, has high emissions. Blue hydrogen reduces emissions by incorporating carbon capture but still uses fossil fuels. Turquoise hydrogen, created through methane pyrolysis, offers potential for lower emissions if energy and carbon management are optimized. Brown/black hydrogen, made from coal gasification, is the most carbon-intensive. This overview clarifies the sustainability of each method (Kousksou et al., 2015).

3. Sustainable Pathways for Green Hydrogen: Utilizing Renewable Energy Sources for Electrolysis

a) Integration of Renewable Energy Sources

Green hydrogen production is a critical element in the transition to a low-carbon economy. Sustainable pathways for green hydrogen focus on leveraging renewable energy sources to power the electrolysis process, which splits water into hydrogen and oxygen (Idrissi Kaitouni et al., 2023). This section explores various strategies for integrating renewable energy into green hydrogen production, addressing technological advancements, economic considerations, and environmental impacts.

Photovoltaic (PV) systems are increasingly utilized to provide electricity for hydrogen electrolysis. Solar energy is abundant and widely available, making it an ideal source for generating green hydrogen. PV systems can be installed on-site at electrolysis facilities or integrated with existing solar farms to ensure a steady supply of renewable electricity. Combining solar PV with other renewable sources, such as wind or hydropower, can enhance the reliability of energy supply for electrolysis (Elmansour et al., 2022). Hybrid systems help mitigate the intermittency of solar power and ensure continuous hydrogen production.

Onshore wind farms offer significant potential for generating renewable electricity for hydrogen production. The scalability and maturity of onshore wind technology make it a viable option for integration with electrolysis facilities. Offshore wind farms, with their higher and more consistent wind speeds, provide a robust source of renewable energy. Offshore wind can be particularly advantageous in regions with limited land availability or where onshore wind is less effective (Ouchani et al., 2022).

Hydropower plants, including run-of-river and reservoir systems, can supply a stable and continuous source of electricity for green hydrogen production. The predictability and low variability of hydropower make it a reliable partner for electrolysis. Additionally, pumped hydro storage can balance the intermittency of other renewable sources by storing excess energy during periods of high generation and releasing it during periods of low generation, thereby supporting consistent hydrogen production (Dadashi et al., 2022).

b) Electrolysis Technologies

Various electrolyzer technologies for green hydrogen production are summarized in the accompanying Table II. It details the operational principles, advantages, and efficiency of Proton Exchange Membrane (PEM), Alkaline, Solid Oxide Electrolyzers (SOE), and Anion Exchange Membrane (AEM) electrolyzers(Derouez & Ifa, 2024). The table offers a concise comparison, highlighting how each technology integrates with renewable energy sources and their overall performance in hydrogen production.

Table 2. Comparison of Electrolyzer Types: Technology Overview and Advantages

Electrolyzer Type	Technology Overview	Advantages
Proton Exchange Membrane (PEM)	Uses a proton exchange membrane to separate hydrogen and oxygen. Operates at high current densities and is responsive to power supply fluctuations.	High efficiency, compact design, operates at high pressures, reducing the need for additional compression.
Alkaline	Utilizes an alkaline electrolyte (KOH or NaOH) for electrolysis. Well-established technology with reliable performance.	Cost-effective, proven reliability, suitable for large-scale production, integrates well with renewable energy.
Solid Oxide Electrolyzers (SOE)	Uses a solid oxide ceramic electrolyte and operates at high temperatures (700-1000°C). Achieves high efficiencies, especially with high-temperature heat sources.	Highly efficient, utilizes heat from renewable sources like concentrated solar power, higher overall system efficiencies.
Anion Exchange Membrane (AEM)	Employs an anion exchange membrane for the electrolysis process. Newer technology with promising benefits.	Combines high efficiency and lower cost materials, potential for cost reductions and improved performance.

c) Technological and Economic Considerations

Technological advances, such as improvements in electrolyzer efficiency and durability, are critical for optimizing green hydrogen production. Innovations in materials and designs enhance performance and reduce operational costs. Integrating hydrogen production with energy storage systems, like batteries or thermal storage, can help smooth out fluctuations in renewable energy supply and improve overall system efficiency (Almulla et al., 2022).

Reducing the costs of renewable energy and electrolyzer technologies is essential for making green hydrogen economically viable. Economies of scale, technological advancements, and supportive policies can drive down costs and increase the competitiveness of green hydrogen. Government incentives, subsidies, and investment in research and development play a crucial role in supporting the growth of green hydrogen production. Policies that encourage renewable energy integration and hydrogen infrastructure development are critical for fostering a sustainable hydrogen economy (Ouhammou et al., 2023).

Utilizing renewable energy sources for electrolysis ensures that the hydrogen production process remains zero-emission, significantly reducing the overall carbon footprint and supporting global climate goals. Renewable energy sources, such as solar and wind, have minimal environmental impacts compared to fossil fuels, making them key to sustainable resource management (Haidi & Cheddadi, 2022). Conducting lifecycle analyses of green hydrogen production systems helps to assess the overall environmental impact, including energy use, emissions, and resource consumption. These evaluations are essential for optimizing production pathways and ensuring sustainability.

d) Emerging Applications and Future Prospects

Table III summarizes the key areas where hydrogen is expected to have a significant impact in the future. It highlights the potential applications and developments across various sectors, emphasizing hydrogen's role in reducing emissions, enhancing energy efficiency, and supporting sustainable growth (Bakkari et al., 2024).

Table 3. Key Areas and Developments in Hydrogen Applications

Category	Details
Heavy Industry	- Steel production with hydrogen, reducing emissions by up to 95%
	- Pilot projects in Europe and Asia
Transport	- Long-haul trucking, maritime shipping, aviation
	- Hydrogen fuel cells as alternatives to batteries
Power Generation	- Hydrogen turbines and fuel cells for zero-emission power generation
Infrastructure & Trade	- Repurposing natural gas pipelines for hydrogen transport
	- New hydrogen pipelines, underground storage, international trade

4. Investment and Innovation

a) Cost Challenges

The cost of hydrogen production remains a significant challenge, especially for electrolysis. Capital expenditure for electrolysers has risen due to inflation and disruptions in the supply chain. Government support through subsidies and tax incentives is crucial to bridge the cost gap between low-emission hydrogen and fossil fuel-based alternatives (Boharb et al., 2022).

b) Technological Innovation

Advances in electrolyser efficiency, hydrogen storage, and fuel cell technologies are pivotal for reducing costs and improving performance. Research and development efforts focus on scaling up production capacity, enhancing durability, and minimizing energy consumption (El Hathat et al., 2023).

5. Policy and Regulatory Landscape

a) Production Incentives

Governments worldwide are implementing incentives to stimulate hydrogen production, such as the US Hydrogen Production Tax Credit and the EU's Important Projects of Common European Interest (IPCEI). These measures aim to lower production costs and attract private investment (Bensouda & Benali, 2023).

b) Demand Stimulation

Policies to promote the use of low-emission hydrogen include mandates for existing applications and incentives for new applications. Public procurement policies and technology-neutral regulations in sectors like steel, shipping, and aviation are expected to drive adoption (Fragkos, 2023).

c) Certification and Standards

Certification schemes for hydrogen's environmental attributes are critical for ensuring transparency and building market confidence. Harmonizing certification standards and facilitating mutual recognition across countries can enhance international trade and regulatory consistency (Ayad et al., 2023).

IV. STRATEGIES, LAWS, AND ROADMAP FOR GREEN HYDROGEN IN MOROCCO AND EMERGING COUNTRIES

1. Global Green Hydrogen Laws and Regulations

Some key laws and regulations regarding green hydrogen in different countries demonstrate a global commitment to integrating green hydrogen into energy systems to achieve decarbonization goals and enhance energy security. Each country tailors its policies to its specific energy landscape and economic priorities, fostering innovation and investment in hydrogen technologies (Kharrich et al., 2019).

In the European Union, detailed rules for the production of renewable hydrogen, known as renewable fuels of non-biological origin (RFNBOs), have been established under the Renewable Energy Directive. These regulations define the conditions under which hydrogen can be considered renewable, focusing on the use of renewable electricity and achieving significant greenhouse gas emissions savings (Lebrouhi et al., 2024). The EU targets 10 million tonnes of domestic renewable hydrogen production and an additional 10 million tonnes of imports by 2030 (Ersoy et al., 2021).

France regulates green hydrogen projects through Ordinance No. 2021-167, which categorizes hydrogen into renewable, low-carbon, and carbonaceous types (Meliani et al., 2022). The Energy Code includes specific provisions for the production, storage, distribution, and transport of hydrogen, alongside public funding mechanisms to support hydrogen production via water electrolysis. The government offers investment and operational subsidies for hydrogen projects and expedited access to land resources (Alaoui, 2019).

Germany promotes green hydrogen under its Renewable Energy Sources Act (EEG), providing financial incentives such as market premiums for renewable energy. The country has ambitious targets to increase renewable energy generation and integrate green hydrogen into its energy mix. The legal framework supports the sale and purchase of renewable electricity and includes specific measures for grid access and connection for hydrogen production facilities (Bouyghrissi et al., 2022).

In the United States, green hydrogen is supported through various federal initiatives and state-level policies. The bipartisan Infrastructure Investment and Jobs Act and the Inflation Reduction Act include provisions to fund hydrogen projects, develop hydrogen hubs, and offer tax incentives for clean hydrogen production. The Department of Energy is actively working on regulatory frameworks to facilitate hydrogen infrastructure development and integration into the energy system (de Arce et al., 2012).

Namibia is developing a regulatory framework to support green hydrogen production, which is expected to drive economic growth and sustainability. The government is focusing on creating a conducive environment for investments in hydrogen projects, including regulations for the production, transportation, and trade of hydrogen(Mohammadshahi et al., 2022).

Japan's Basic Hydrogen Strategy includes targets for hydrogen production, utilization, and fuel cell deployment, with government support for research and development, infrastructure development, and international cooperation. Japan aims to achieve a hydrogen society by 2050, with extensive use of hydrogen in various sectors (Gunawan & Monaghan, 2022).

Australia's National Hydrogen Strategy outlines the development of hydrogen production, infrastructure, and export capabilities. The government supports research, industry development, and international partnerships, aiming to position Australia as a major global player in hydrogen production and export (El-Emam et al., 2022).

These regulations and incentives collectively reflect a global commitment to green hydrogen, tailored to national energy landscapes and economic priorities to foster innovation and investment in hydrogen technologies(Glanville et al., 2022).

2. Global Strategies for Green Hydrogen: Comparative Analysis

To provide a comprehensive understanding, the following Table IV summarizes the visions, objectives, strategic measures, investments, and performance indicators of several leading countries in the hydrogen sector. These global insights will help contextualize Morocco's efforts and highlight the common goals and innovative approaches taken by various nations to integrate hydrogen into their energy systems and achieve decarbonization targets.

Table 4. Comparative Table of Hydrogen Strategies

Country	Vision	Objectives	Strategic Measures	Investments	Performance Indicators
Japan(Stamatakis et al., 2022)	Hydrogen society by 2050	Extensive use of hydrogen across sectors	Support R&D, infrastructure, and international cooperation	Not specified	Progress in hydrogen adoption, infrastructure development
Australia(Romeo et al., 2022a)	Major player in hydrogen production and export	Develop production, infrastructure, and export capabilities	Support research, industrial development, and international partnerships	Not specified	Position in global hydrogen market
South Korea(Romeo et al., 2022b)	Zero-carbon country with hydrogen economy by 2050	Secure 1 million tons of clean hydrogen annually by 2030	Green hydrogen from wind and electrolysis, blue and yellow hydrogen	Not specified	Green hydrogen production cost, offshore wind capacity
European Union(Nordio et al., 2021)	40 GW electrolyzer capacity by 2030	6 GW electrolyzer capacity by 2024, emphasize renewable hydrogen	Government support, quotas, carbon difference contracts	220-340 billion euros	Electrolyzer capacity, hydrogen production cost
United States(Valverde et al., 2016)	Clean hydrogen production of 10 million tons/ year by 2030	Achieve 50 million tons/year by 2050	Regional infrastructures, hydrogen hub networks	Supported by Bipartisan Infrastructure Law and Inflation Reduction Act	Annual hydrogen production, cost reductions
India(Mohseni et al., 2020)	Global center for green hydrogen production, use, and export	Reduce hydrogen production cost to below 100 INR/kg	Reduce investment and operational costs, develop PEM electrolyzers	19,744 crore rupees	Hydrogen production cost, technological advancements
Canada(Quarton & Samsatli, 2020)	Global leader in hydrogen by 2050	Produce 5 million tons of clean hydrogen/year by 2030	Infrastructure development, cost reduction, public-private partnerships	Not specified	Clean hydrogen production, GHG emissions reduction

continued on following page

Table 4. Continued

Country	Vision	Objectives	Strategic Measures	Investments	Performance Indicators
Chile (Okundamiya, 2021)	Global leader in green hydrogen production and export	5 GW electrolyzer capacity by 2025, cost-effective production by 2030	Pilot projects, financial support, international collaboration	Not specified	Electrolyzer capacity, production cost, export ranking
Turkey (Dispenza et al., 2019)	Competitive hydrogen production costs by 2053	Reduce production costs to $2.4/kg by 2035, 2 GW electrolyzer capacity by 2030	Pilot projects, R&D, international collaboration	Not specified	Hydrogen production cost, electrolyzer capacity

3. Green Hydrogen Road Map of Morocco

Morocco is strategically positioning itself to become a leader in the green hydrogen sector, capitalizing on its significant renewable energy resources and ambitious climate goals. The country's roadmap outlines a clear vision to harness its abundant solar and wind energy to develop a robust green hydrogen industry (Haines et al., 2023). Morocco aims to establish itself as a major player in the global green hydrogen market, with production targets set to reach 2 million tons annually by 2030 and scale up to 5 million tons by 2050. The goal is to not only meet domestic energy needs but also to become a key exporter of green hydrogen, particularly targeting European markets and other regions with high demand.

To achieve these objectives, Morocco is focusing on several strategic measures, including significant investments in infrastructure to support the production, storage, and distribution of green hydrogen. The roadmap highlights the importance of constructing hydrogen production facilities and developing transport networks (El Majaty et al., 2023). Additionally, Morocco is prioritizing research and innovation to enhance the efficiency and reduce the costs of hydrogen production. Strategic partnerships with international companies and governments are also a cornerstone of this plan, aiming to foster collaboration, share expertise, and access global markets.

Investment and financing play a crucial role in Morocco's green hydrogen strategy. The government is allocating public funds to support green hydrogen projects, providing subsidies and incentives to attract private sector participation. Private investment is being encouraged through favorable policies and tax incentives, alongside public-private partnerships. International support is also sought, with Morocco looking to secure funding and technical assistance from global organizations and financial institutions (Kasseh et al., 2023b).

Performance indicators are essential to track the progress of Morocco's green hydrogen roadmap. Key metrics include the growth in hydrogen production capacity, the economic impact through job creation and investment inflows, and the environmental benefits, particularly the reduction in carbon emissions. These indicators will help assess whether the country is meeting its production milestones and aligning with its climate goals under the Paris Agreement (El Hafdaoui et al., 2024).

The implementation of Morocco's green hydrogen roadmap is planned in phases. In the short term, from 2023 to 2025, the focus will be on initial investments, pilot projects, and building foundational infrastructure (Nakach et al., 2023). The medium-term phase, from 2026 to 2030, will see the expansion of production capacity and scaling up of projects, with an emphasis on strengthening international collaborations. In the long term, from 2031 to 2050, Morocco aims to achieve its production targets, reinforce its global market position, and continue advancing technological innovations (Wang et al., 2018a).

Overall, Morocco's roadmap for green hydrogen represents a comprehensive strategy to leverage its renewable energy potential, develop a competitive green hydrogen industry, and contribute significantly to global sustainability goals . The country is committed to becoming a leading player in the green hydrogen sector through strategic investments, innovation, and international cooperation.

4. Comparative Analysis of Green Hydrogen Strategies: Morocco vs. Other Leading Countries

Morocco is strategically positioning itself as a leader in the green hydrogen sector by capitalizing on its abundant renewable energy resources and ambitious climate goals. The country aims to produce 2 million tons of green hydrogen annually by 2030, scaling up to 5 million tons by 2050. This ambitious target places Morocco among the global frontrunners in green hydrogen production. Morocco's strategy includes significant investments in infrastructure, research, and international partnerships, and emphasizes job creation, investment inflows, and carbon emission reductions as key performance indicators (Stöckl et al., 2021).

In comparison, Japan envisions becoming a hydrogen society by 2050 with extensive use of hydrogen across sectors, supported by R&D, infrastructure development, and international cooperation. However, Japan has not specified a production target for 2030. Similarly, Australia aims to become a major player in hydrogen production and export, focusing on research, industrial development, and international partnerships, but lacks specific short-term production targets (Wang et al., 2018b).

South Korea aims to secure 1 million tons of clean hydrogen annually by 2030, leveraging green hydrogen from wind and electrolysis, and blue and yellow hydrogen. The European Union has set a target of 40 GW electrolyzer capacity by 2030, with

6 GW by 2024, supported by government initiatives, quotas, and carbon difference contracts. The United States targets 10 million tons of clean hydrogen production per year by 2030, scaling up to 50 million tons by 2050, driven by regional infrastructures and hydrogen hub networks (Rabiee et al., 2021).

India seeks to become a global center for green hydrogen production, aiming to reduce production costs below 100 INR/kg, with a target of 5 million tons annually by 2030. Canada also aims to produce 5 million tons of clean hydrogen annually by 2030, focusing on infrastructure development, cost reduction, and public-private partnerships. Chile and Turkey have set ambitious targets for electrolyzer capacity, aiming for 5 GW and 2 GW by 2030, respectively, with a strong focus on pilot projects, financial support, and international collaboration (Armijo & Philibert, 2020).

Overall, while the specific targets and strategic measures vary, all these countries share a common goal of integrating green hydrogen into their energy systems to achieve decarbonization and enhance energy security. Morocco's comprehensive strategy, with its clear production targets and focus on infrastructure and international partnerships, positions it well within the competitive landscape of the global green hydrogen market (Ge et al., 2020).

V. POLICY IMPLEMENTATION CHALLENGES

1. Electricity Market: Administered Market

a) Global Configuration of Electricity Market

In a regulated or administered electricity market, the prices for electricity are controlled by the government or a designated regulatory authority. This authority could be a government agency or an independent regulatory body, tasked with ensuring that electricity prices are fair and equitable for all consumers. Suppliers operating within this market are required to adhere to the pricing guidelines and rules established by the regulatory entity, which are designed to maintain stability and fairness in the market (Cao et al., 2024).

The primary objective of a regulated electricity market is to protect consumers by ensuring they have access to electricity at reasonable and predictable prices (Jensen et al., 2020). This protection is crucial, as electricity is an essential service for households, businesses, and industries. By regulating prices, the government or regulatory authority prevents electricity suppliers from charging excessively high rates that could exploit consumers, especially vulnerable populations who may not have the means to absorb significant price increases (Koirala et al., 2021). Additionally, regulated markets often impose requirements on suppliers to maintain consistent pricing for all consumer categories, ensuring that individuals, small businesses, and large industries all pay similar rates for electricity (Deininge & Gren, 2022).

Figure 4. Legislative reforms bearing on the opening of the electricity market[3]

This regulated market model stands in stark contrast to unregulated or deregulated electricity markets, where prices are determined by the forces of supply and demand. In an unregulated market, electricity suppliers have the freedom to set prices based on market conditions and customer demand (Idllalène, 2021). This can lead to a more dynamic pricing environment, where consumers might benefit from lower prices during periods of low demand. However, it also introduces the risk of price volatility, where prices can spike during peak demand times or due to supply constraints.

Unregulated markets can offer the advantage of allowing consumers to negotiate personalized prices based on their consumption patterns and needs. Large industrial consumers, for instance, might secure lower rates through bulk purchasing agreements (Young, 2015). On the other hand, individual consumers or small businesses that lack bargaining power might find themselves paying higher prices. This lack of uniform pricing can create disparities and lead to situations where certain consumer groups are disadvantaged.

Moreover, unregulated markets can foster competition among suppliers, potentially driving innovation and efficiency. Suppliers might develop new services, offer better customer support, or invest in renewable energy sources to attract customers (Weimann et al., 2021). While this competition can benefit consumers, it also requires a level of vigilance and market oversight to ensure that competitive practices do not devolve into unfair or exploitative behavior (Wodon, 2023). The main technological and economic characteristics of the electricity sector activities mentioned above are presented in Table V.

Table 5. Electricity sector activities

Activity	Description	Features
Production	Transformation of primary energy into electrical energy Economics of this activity is characterized by a trade-off between capital costs and variable costs (fuels)	Potentially competitive activity Opportunities for competition in the market or for the market
Transportation	Transport of electricity from production to transformation centers. "Capital intensive" activity with high proportion of fixed costs.	Natural monopoly of national scope
Distribution	Transport from processing centers to end consumers "Capital intensive" activity with high proportion of fixed costs.	Natural monopoly of regional scope Possibility of comparison competition
Marketing	Purchase and sale of electricity and customer service "Labour-intensive" activity.	Potentially competitive activity

Both transportation and distribution of electricity are capital-intensive activities with a significant proportion of fixed costs. These activities are considered natural monopolies, typically at a national or regional level, due to the substantial infrastructure investments required and the economic characteristics unique to these sectors (Belaid et al., 2021). The high share of fixed costs in transportation and distribution impacts the calculation of their breakeven income, as we will explore further.

When it comes to electricity production, in addition to these inherent characteristics, the availability and relative costs of primary energy sources (such as hydroelectric, nuclear, hydrocarbons, and wind) play a crucial role (Bigerna, 2024). These factors influence the cost structure of electricity production and determine the level of competition within the market. The availability of these resources and their associated costs directly affect the feasibility and competitiveness of different energy sources (Norouzi, 2023).

Over the past three decades, the Moroccan electricity sector has undergone significant transformations, marked by a notable increase in private sector involvement in electricity production and distribution(Yan et al., 2023). This shift began in 1994 with the introduction of independent electricity producers and continued with the granting of distribution concessions starting in 1997. These changes have gradually reshaped the landscape of the Moroccan electricity market (D. Xu & Hussain, 2023).

Despite the increasing presence of private sector players, the administered market remains dominant in terms of overall capacity and energy volume (Adeleke et al., 2023). However, this situation is evolving, particularly in specific energy sectors. For instance, in the wind energy market, the free market has surpassed the regulated market in terms of capacity and energy volume, as evidenced by recent data (refer to the previous table).

In contrast, the solar energy market has faced delays due to the prolonged publication of the solar zoning decree. This decree, which came 12 years after the promulgation of Law 13-09 on renewable energies, has hindered the development of a free market for solar energy. Nevertheless, several solar projects have recently been authorized and are currently underway, indicating potential growth in this sector (Mena-Cervantes et al., 2023).

b) Institutional Reforms

i. The Public Utility in Energy Sector

Law 40.09 was enacted to integrate the operations of the National Office of Electricity (ONE) and the National Office of Potable Water (ONEP). This integration aims to harmonize national strategies across these two critical sectors, which share areas of synergy (Hejazi et al., 2023). The law represents a significant milestone and is essential for the reorganization of production, transportation, distribution, and marketing of electricity and drinking water. This reorganization is necessary to ensure the continuity of public services for these vital resources and to maintain a stable and efficient supply to the country, adhering to optimized standards of quality and cost (Utomo et al., 2021).

The Electricity Branch (BE) of the National Office of Electricity and Water (ONEE), the traditional operator in this sector, remains a key player. Its mission, assigned by the Moroccan government, includes the production, transportation, and distribution of electricity, particularly outside of distribution concession areas and in rural regions (Hossin et al., 2023).

Initially, the sector was organized around a single-buyer model, where ONEE-BE purchased all electricity generated within Morocco through Power Purchase Agreements (PPA) and imported electricity from Spain and Algeria. Since 1994, ONEE has been permitted to enter contracts with private electricity producers (Al-Najjar et al., 2022). The landscape shifted further in 2010 with the introduction of Law 13-09 on renewable energies, which liberalized the renewable energy market. This law allowed any investor to establish their own electricity production facilities and sell the electricity directly to consumers (Aghahosseini et al., 2020).

Additionally, a recent agreement between ONEE and the Moroccan Agency for Sustainable Energy (Masen) has been established (Shahrtash, 2023). This agreement facilitates the transfer of all ONEE's renewable energy production activities to Masen, aligning with Law No. 38-16, which amends and supplements Article 2 of Dahir No. 1-63-226 dated August 5, 1963. This law transformed the National Office of Electricity into the National Office of Electricity and Drinking Water (ONEE). According to Article 2 of this law, all real estate properties used for existing or under-construction renewable energy installations, as well as all project-related documentation and real estate assets, will be transferred to Masen progressively and no later than five years from the law's publication date (Timilsina & Deluque Curiel, 2023).

ii. Moroccan Agency for Sustainable Development (MASEN)

The Moroccan Agency for Solar Energy (MASEN) was established under Law 57-09. It is a state-majority-owned limited company, created with the purpose of developing integrated solar energy projects for electricity production, aiming to achieve a minimum total capacity of 6,000 MW by 2030. As stipulated in Article 3 of Law 57-09, all electricity generated from MASEN's projects must be purchased in full by the National Office of Electricity (ONE) or any other designated public or private entity. This acquisition process is defined by agreements between the State, MASEN, and the purchasing organization, with provisions allowing for the export of the produced energy (Lu et al., 2024).

Law 37-16, which amends and supplements Law 57-09, enhances MASEN's framework by transforming it into a public limited company with a board of directors. This amendment broadens MASEN's responsibilities to include the construction of renewable energy plants. MASEN's extended mandate now covers the entire lifecycle of renewable energy projects, including study, design, financing, construction, operation, and maintenance (Topalović et al., 2024).

Under this law, the Moroccan Solar Energy Agency is called the "Moroccan Agency for Sustainable Energy", and its missions are focused in particular on the following:

- Identification of the electricity production capacities from renewable sources to be achieved, on the basis of the planning agreed with the manager of the national electricity system and the carrying out of the studies necessary for the qualification of sites and the design of renewable energy projects, their achievements, operations and maintenance(Lee & Im, 2024);
- Development of integrated electricity production installations using renewable energies, and contribution to research and mobilization of the financing necessary for their creation and operation;
- Conducting promotion, awareness-raising, advisory and service activities related to renewable energies and design, implementation, operation and maintenance activities of renewable energy projects;
- Creation of the infrastructures necessary for the development of ENR Projects, allowing them to be connected to the road, electricity and water supply networks, subject to the attributions devolved in this matter to any other public or private body;
- Proposal to the administration the terms of development of competitive industrial sectors linked to renewable energy projects, as well as the terms of local industrial integration (Choukri et al., 2017a).

c) Electricity Policy Reforms

i. Reform of the Electricity Transmission Segment

Moreover, the accounting separation between ONEE's transportation activities and its other operations, as mandated by Law 48-15 on the regulation of the electricity sector, has been initiated by the National Regulatory Authority for Electricity (ANRE). This measure aims to ensure fairness among all stakeholders, allowing ONEE to treat the activities of independent developers and its own production activities equally (Ismagilov et al., 2019).

To facilitate this process, a joint working group comprising representatives from ANRE and ONEE has been established. This committee is tasked with overseeing the implementation of the accounting separation project, which is a crucial step towards ensuring a fair and competitive market. The project is fundamental for the gradual opening of the electricity market, while also maintaining high standards of quality and security in the supply of electrical energy (Amorim et al., 2024).

ii. New Reform of the Electricity Distribution Segment

The government council has recently approved Bill No. 83-21, which focuses on the establishment of Regional Multiservice Companies (SRM) to manage electricity distribution, drinking water supply, and liquid sanitation services (Onyinyechukwu Chidolue et al., 2024). This initiative aims to enhance the effectiveness and efficiency of public investments and ensure that each region is equipped with an SRM. This effort is part of the broader reform initiated in July 2021, which was formalized through a memorandum of agreement signed by the Ministry of the Interior, the Ministry of Economy and Finance, the Ministry of Energy, Mines, Water, and the Environment, along with the National Office of Electricity and Drinking Water (ONEE) (Olusegun Gbenga Odunaiya et al., 2024).

Figure 5. Evolution of the distribution market share (2012-2021)[4]

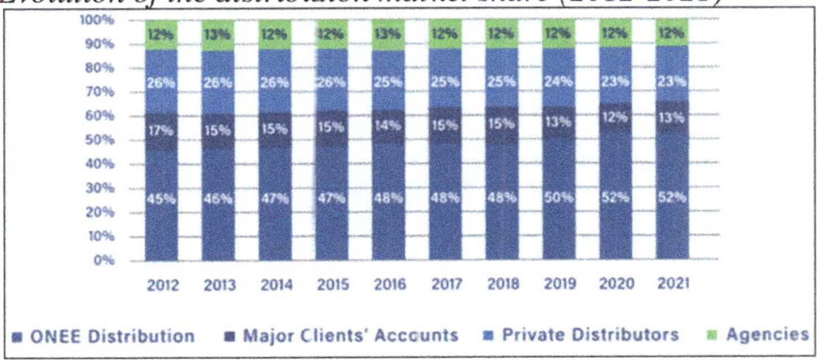

The primary goals of this reform are threefold: first, to address the challenges posed by the involvement of multiple actors in the management of certain municipal public services, including municipalities, ONEE, authorities, delegates, and associations; second, to enhance the quality of infrastructure and services provided to users by optimizing investments and resources; and third, to ensure equitable coverage of these services across the national territory, with a particular focus on rural areas (Asiaban et al., 2021).

This reform is aligned with the advanced regionalization process enshrined in Article 1 of the Constitution and follows the reform of public establishments and enterprises outlined in Framework Law No. 50-21. Promulgated by Dahir No. 1-21-89 on July 26, 2021, this law seeks to resize the public sector and rationalize public expenditure. The reform's foundation includes enacting legislation to create the SRMs, transferring assets and resources currently allocated to Public Distribution Services within each region to these SRMs, gradually dissolving municipal authorities, and reorganizing ONEE as it disengages from Public Distribution Services to facilitate their takeover by SRMs (Badran, 2017).

In summary, the creation of SRMs is a strategic move to streamline the management of vital public services, improve service delivery, and ensure that investments are used efficiently across the regions. This reform signifies a substantial shift towards regional autonomy and better resource management, aimed at delivering high-quality services to all citizens, particularly in underserved rural areas(Mahlooji et al., 2020).

2. Open Market

In the past, Morocco's expansion of renewable energy was primarily driven and centralized by the state. The market was dominated by two major state or semi-state institutions: the National Office of Electricity and Drinking Water (ONEE), which also manages the Moroccan electricity grid, and the Moroccan Agency for Sustainable Energy (MASEN), established to promote the development and operation of solar and wind power plants. Consequently, Morocco's renewable energy production has predominantly originated from large-scale solar power stations and wind farms (H. Xu & Shi, 2022).

Table 6. Framework of Wind Energy Production[5]

Framework of Wind Energy Production	Installed capacity in MW	Production in GWh	Rate of injection into the transmission network in %
Production governed by law 13-09	663	2643,8	6,1
Production by ONEE	255	606.76	1,5
Production by IPP	511	1773,73	4,2
Auto-production	37	-	-

Despite these developments, the energy market in Morocco remains highly restricted. Moroccan companies, municipalities, and private households are permitted to generate electricity for their own consumption but are not allowed to sell any excess power back to the public grid. This limitation poses a significant barrier to the acceleration of the country's energy transition (Choukri et al., 2018). Therefore, the call for liberalizing the energy market in Morocco is more pressing than ever (Hajar et al., 2023).

Efforts to open the electricity market (Table VI) to additional players have been ongoing for several years, with the legal framework still evolving. Notably, Law 13-09 was developed to facilitate this transition and has since been amended by Laws 58-15 and 40-19. These legislative changes aim to foster a more inclusive and competitive energy market. Additionally, frameworks for self-production of electricity are being established to further support these initiatives (Rafik et al., 2019).

Law 13-09 establishes a legal framework for the construction and operation of installations for the production of electrical energy from a renewable energy source. Its provisions and implementing texts allow private investors to develop renewable energy projects for the production of electricity by concluding contracts directly with customers connected to the Very High Voltage, High Voltage and Medium Voltage networks. electricity sales (Park et al., 2023).

i. Operating Regimes

The law establishes three regimes depending on the capacity of the installation:

- Free operation for an installed capacity less than 20 kW (if electric) and 8 MW (if thermal);
- Declaration regime between 20 kW and 2 MW (if electric) and greater than or equal to 8 MW (if thermal);
- Authorization regime for 2 MW and more .

In this last category, and in the case of solar or wind projects, the installation must be carried out in specific areas proposed by the organization responsible for the development of renewable energies, the local authorities concerned and the manager of the national electricity transport network. The wind zoning has already been published since 2011 while the solar zoning has just been published in October 2022 and opens the photovoltaic solar market to private investors.

ii. Marketing and Export of the Electrical Energy Produced

Law 13-09 provides a framework for the national marketing of electrical energy produced through agreements between the state (or its designated representative) and the energy operator. These agreements outline the validity period and commercial terms for supplying the produced energy. Under this law, operators can sell their electricity to specific consumers connected to the network through a contract, which primarily requires the consumer to commit to exclusively using the supplied electricity for their own needs (Zhao et al., 2022).

Additionally, the law permits the export of electricity, contingent upon technical approval from the national electricity grid manager. If necessary, the operator can establish and utilize a direct line for export purposes, provided they meet the concession requirements and discuss the feasibility and necessary measures with the grid manager. The national electricity grid manager oversees and regulates access to the national electricity network, including medium, high, and extra-high voltage networks, interconnections, direct transport lines, and all renewable energy export operations (Odenweller et al., 2022).

Figure 6. Flowchart of the procedure for granting provisional and final authorizations.

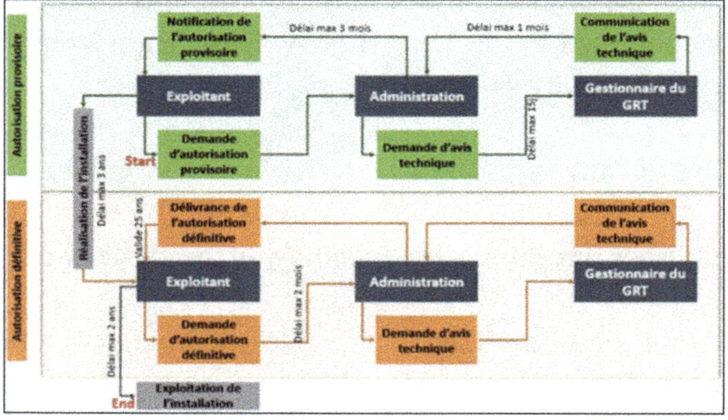

Exported electricity is subject to an annual operating fee payable to the state, calculated based on the exported energy volume and according to regulations yet to be published. This fee can be paid in cash, in kind, or a combination of both, as requested by the state (Tleubergenova et al., 2023).

iii. Control Measures

In order to ensure compliance with the clauses of this law by operators, the law describes a set of infrastructure control and observation measures. To do this, it allows various authorized agents to have free access to the construction or operating work of any installation. These agents can request any document necessary for the verification and, if necessary, the start-up of the installation. All findings are issued in the form of a report(Cloete & Hirth, 2020).

iv. Access to Medium Voltage

Essential and innovative guarantees are granted to investors in renewable energy projects on the medium voltage electricity network, inspired by the highest international standards(Cárdenas et al., 2021).

A transparent, non-discriminatory and clear connection procedure is organized. The connection request is preceded by an orientation study aimed at "orienting" the investor, informing them about the positioning of the project in the trajectory and envelopes of the distribution zone concerned or offering them solutions. alternative routes. For production installations greater than 2 MW, the orientation study is followed by a detailed study, which results in a technical and financial proposal (PTF) binding for the distribution network operator (Morton et al., 2023).

As such, an obligation for consultation is provided for between the distribution network managers and the transmission network manager and also the costs of the aforementioned studies are the responsibility of the developer, hence the need to clarify, legally, the The order of magnitude or the method of calculating these costs to guarantee transparency and fair treatment either between developers or also between distribution network managers (Ghaemi et al., 2023).

Then, a "capacity reservation" mechanism is organized to guarantee the investor a capacity to accommodate the electricity produced by their installation at the time of its commissioning. A "provisional" capacity reservation allows, from the start of the project, the investor to ensure the feasibility of the project (Ruhnau & Schiele, 2023). The provisional reservation becomes definitive following the completion of certain formalities.

In addition, non-discriminatory "priority" rules for electricity production installations from renewable energy sources are taken, firstly, in the processing of connection files (priority compared to conventional installations, the rule of "first come, first served" between ENR installations), and then, in the operational rules for electricity injection (priority compared to conventional electricity, between ENR installations, the rule of pro rata) (Zhou et al., 2018).

v. Industrial Customers

Taking into consideration the prerogatives of the framework law for renewable energies 13-09, only consumers connected to THT/HT lines were able to really benefit from the purchase of electricity from renewable sources. Industrialists focused on purchasing green electricity were mainly attracted by the price of wind kWh which is 10 to 15% cheaper than conventional kWh. The following paragraph gives an overview of the main industrial customers:

- Lafarge [(6,7)]: Lafarge was able to satisfy 85% of its electricity needs from wind projects (self-production '32 MW' and under contracts with IPPs)
- OCP: the OCP group intends to satisfy 95% of its electricity needs from cogeneration and wind power[8].
- Sonasid[9]: The company has entered into a purchase contract for 20% of the needs of its Jorf Lasfar rolling mill based on future wind farms developed by NAREVA, from 2018. This contract would allow the company to improve its performance environmental, but also economic[10].
- Maghreb Steel[11]: the group announced that it could start its supply of wind-generated electricity from Enel Green Power, up to 90% of its needs in 2018.
- Cements from Morocco[12]: The group is considering diversifying its electrical energy supplies by purchasing clean energy from wind sources for its own production needs. At least 80% of its electrical energy mix should be satisfied from renewable energy sources.
- Managem: The Managem group[13]satisfies 47% of its electricity consumption from renewable energy sources. Four sites are powered by wind farms.

Despite the publication of its decree relating to medium voltage (MV) in 2015, no project has been authorized due to the absence of publication of the decree establishing the trajectory of the renewable capacities reserved for the latter. In November 2021, the joint decree of the Ministers of Energy Transition and the Interior establishing the development trajectory of renewable energies for the next ten years (2022-2031), including the injection envelopes for electricity produced from of renewable sources on the medium voltage network, has been published. This publication, as well as the new directions in favor of the decarbonization of the industry, should stimulate the medium voltage market (Stanley et al., 2023).

3. Self-Production

Until 1994, the National Electricity Office (ONE) held a monopoly on the production of electrical energy. Dahir No. 1-63-226 of June 5, 1963 creating the National Electricity Office (ONE) in fact charged this office with the public service mission of production and transport1 of electrical energy on the the entire national territory and has, for this purpose, established for its benefit the exclusivity (that is to say the monopoly) of the "arrangement of the means" of this production "of a power greater than 300 kW " (Firdous et al., 2024).

Self-production has therefore been established in Moroccan regulations since 1963 (Table VII). Although limited to a power of 300 kW, self-production was therefore completely free and did not obey any particular conditions within the limit of this threshold (Kim et al., 2024).

The year 1994 marked the beginning of the policy of opening up the electrical energy production sector through two instruments:

- The concession by call for tenders for the production of electricity within the framework of agreements with the obligation to sell it exclusively to the ONE.
- The increase in the self-production threshold to a power of 10 MW (by decree-law -294-503 of September 23, 1994), then to 50 MW (by law 16-08).

Table 7. History of self-production

1956	**Order regulating the autonomous production of electrical energy** • **The connection diagram must be validated by the distributor** • **A specific subscription contract with the Distributor**
1963	**Dahir of creation of the ONE**: ONE was responsible for Production, Transport and Distribution (where there is no concession) • ONE has the exclusivity for the development of the means of production with P > 10 KW
2008	Imbalance between supply and demand (private participation in meeting demand) • Increasing the self-production threshold from 10 KW to 50 MW • Network access is not permitted
Self-production is permitted without authorization from the administration	
2015	Law 54-14 on self-production (ONEE text) • Self-production with P < 50 MW without access to the network • Self-production with P > 300 MW with access to the network • Authorization by the administration is imposed by this text, which requires an authorization procedure
Self-production is permitted with authorization from the administration	
2023	Comprehensive reform of self-production in a single regulatory framework (law 82.21) Modernization of self-production with a system of procedures inspired by law 13-09 relating to renewable energies

3.1. Law 16-08

Law 16-08 (as maintained in force by Law 40-09 relating to ONEE) constituted the only legal framework of reference applicable to the self-production of electrical energy (before Law 54-14) whatever or the nature of the energy sources used, whether "conventional" (coal, fuel oil, gas) or renewable (solar, wind, hydraulic.) (Mewafy et al., 2024). This regime was based on the following elements:

- The right of self-production by "public or private persons" within the limit of 50 MW;
- Obtaining authorization from ONE(E);
- The allocation of production to the "exclusive use" of the self-producer;
- The transfer of excess electrical energy not consumed exclusively to ONE;
- The compatibility of the project with "the electrical energy supply plans of the area concerned";
- The conclusion (case by case) of an agreement to connect the installation to the national electricity network.
- Likewise, this production must not disrupt the electrical energy supply plans for the area concerned.

3.2. Law 54-14

The dahir creating the National Electricity Office was modified and also supplemented by law 54-14 which gives the possibility to large national self-producers of electricity (whose cumulative installed power needs exceed 300 MW) to construct electricity production installations exceeding the capacity of 50 MW and to access the electricity transmission network to transport their energy produced from the production site to the consumption sites (Refaat et al., 2024).

3.3. Law 82-21

Law 82-21 aims to regulate the self-production of electrical energy for self-consumption, whatever the nature of the network, the voltage level and the capacity of the installation used, while ensuring the safety and security of the national electricity network, and ensuring compliance with the principles of transparency and non-discrimination between all stakeholders. The law also provides, and for the first time, the right of access to electrical energy storage services, in addition to the right to sell the surplus to electricity network managers, as well as the expansion of the domain of access to the national electricity network to transport electricity from the production site to the consumption site (Li et al., 2017).

4. Regulation of Electricity Market

The National Electricity Regulatory Authority ANRE is considered a legal entity under public law with financial autonomy. It was created by Law No. 48.15 relating to the regulation of the electricity sector published in Official Bulletin No. 6472 of June 9, 2016. The text sets out the principles necessary for regulation, in particular those governing the management of the electricity network. national transport and management of medium voltage electrical networks, the principles of access to these networks already exist within the framework of Law 13-09 relating to renewable energies. This authority is responsible in a first phase for regulating the electricity sector while awaiting the implementation of the Gas Plan and the development of natural gas activity in our country (Hassan et al., 2023).

a) Missions of the Regulator

According to Law No. 48-15, which pertains to the regulation of the electricity sector, the National Electricity Regulatory Authority (ANRE) ensures the proper functioning of the free electricity market and regulates the access of self-producers to the national electricity transport network. This law grants self-producers the right to access the network.

ANRE's main missions encompass several key areas. Firstly, in terms of network access, ANRE is responsible for approving the multi-year investment program in the national electricity transport and interconnections network and monitoring its implementation. It also approves the national electricity transport network code developed by the General Transmission Network (GRT) and the quality indicators that the network must meet. Additionally, ANRE approves codes of good conduct for managing the national electricity transport network and medium voltage electricity distribution networks. It provides opinions on requests for provisional authorization to construct renewable energy installations and on requests for authorization to construct and use direct transmission lines.

Secondly, regarding the accounting separation of the national electricity transport network manager (GRT), ANRE approves the respective scopes of activities related to the production, transport, and distribution of electrical energy, as well as other activities of the National Office of Electricity and Drinking Water. It also approves the rules for attributing asset and liability items, expenses, and income among these scopes. Furthermore, ANRE approves the principles that determine the financial relationships between these accountably separated activities.

Thirdly, in terms of pricing, ANRE is tasked with setting the tariff for the use of the national electricity transport network and the rates for using medium voltage electricity distribution networks. It also provides notices regarding sales rates for electrical energy following requests from the administration authorized to set these rates under current regulations.

Lastly, ANRE has the power to resolve disputes and impose sanctions. It addresses disputes between the manager of the national electricity transport network and its users, as well as disputes between operators of the electricity distribution network and medium voltage electricity distribution network users. ANRE also holds sanctioning power to enforce compliance and resolve issues within the sector.

b) Supply of Electricity to the Private Sector

According to Law No. 48-15, relating to the regulation of the electricity sector and the creation of the National Electricity Regulatory Authority ANRE, the electricity supplier is defined as any natural person or legal entity which produces or purchases electricity with a view to its partial or total resale.

c) New Attribution: Reception Capacity

In 2021, Law 48-15 was amended by Law 40.19 to take into consideration the amendments made to Law 13.09 relating to renewable energies. This involves calculating and updating the absorption capacity of the manager of the National Electricity Transport Network (GRT), its adoption and publication by ANRE according to a frequency set in January of each year (K, 2021).

VI. EVALUATING PUBLIC-PRIVATE PARTNERSHIPS (PPPS) IN ADVANCING RENEWABLE ENERGY PROJECTS

1. Role and Impact of PPPs in Renewable Energy

Public-Private Partnerships (PPPs) have become a crucial mechanism for advancing renewable energy projects globally. In Morocco, PPPs have played a significant role in the development and implementation of sustainable energy initiatives. This section evaluates the impact of PPPs and strategic collaborations in Morocco's renewable energy sector, focusing on successful models, innovative financing mechanisms, and effective stakeholder engagement strategies. By examining these elements, valuable lessons and insights can be drawn to foster collaboration between the public and private sectors in other emerging economies (Cherkasova, 2023a).

2. The Role of PPPs in Renewable Energy Development

PPPs in Morocco have been instrumental in bridging the gap between public sector goals and private sector capabilities. These partnerships leverage the strengths of both sectors: the public sector's regulatory and policy-making power and the private sector's efficiency, innovation, and capital resources (Cherkasova, 2023b).

2.1. Infrastructure Development

PPPs have facilitated the construction of large-scale renewable energy infrastructures, such as solar and wind power plants. These projects require substantial investments and technical expertise, which private partners often provide (IRENA, 2021).

2.2. Technology Transfer

Through PPPs, Morocco has gained access to advanced technologies and expertise in renewable energy, accelerating the country's transition to sustainable energy sources (Jalbout et al., 2022).

2.3. Capacity Building

PPPs contribute to local capacity building by involving Moroccan firms and the workforce in renewable energy projects, enhancing local knowledge and skills (Szemat-Vielma et al., 2023).

3. Successful PPP Models in Morocco

Several PPP models in Morocco have demonstrated significant success in the renewable energy sector (Boullouz et al., 2022).

3.1. Noor Solar Complex

One of the world's largest solar power complexes, the Noor Solar Complex, is a prime example of a successful PPP. This project was developed through a partnership between the Moroccan Agency for Sustainable Energy (MASEN) and various international private investors. The collaboration involved innovative financing mechanisms, such as green bonds and international loans, which helped mitigate financial risks (Lopez et al., 2023).

3.2. Tarfaya Wind Farm

The Tarfaya Wind Farm, a joint venture between Nareva Holding (a Moroccan private company) and GDF SUEZ (a French multinational), highlights the effective collaboration between public and private entities. This project benefitted from long-term power purchase agreements (PPAs), ensuring financial stability and investor confidence (Neumann et al., 2023).

4. Innovative Financing Mechanisms

The success of PPPs in Morocco's renewable energy sector is partly due to the adoption of innovative financing mechanisms (Slaoui et al., 2023).

4.1. Green Bonds

Issuing green bonds has provided a sustainable financing option for large-scale renewable energy projects. These bonds attract environmentally conscious investors and offer competitive interest rates (Galimova et al., 2023).

4.2. Blended Finance

Combining public funds with private investments, blended finance structures have reduced risks and increased the bankability of renewable energy projects. Public funds often cover initial high-risk stages, encouraging private sector participation (Raab & Dietrich, 2023).

4.3. International Loans and Grants

International financial institutions, such as the World Bank and the African Development Bank, have provided loans and grants that lower the financial barriers for renewable energy projects. These funds often come with favorable terms that make projects more attractive to private investors (Touili et al., 2022).

5. Stakeholder Engagement Strategies

Effective stakeholder engagement is crucial for the success of PPPs in Morocco. Several strategies have been employed to ensure all stakeholders are aligned and committed (Gado & Hassan, 2023).

5.1. Community Involvement

Engaging local communities in the planning and implementation phases of renewable energy projects has fostered public support and minimized resistance. This includes conducting public consultations and ensuring community benefits (Amrani et al., 2023).

5.2. Transparency and Communication

Maintaining transparent communication channels between public and private partners has built trust and facilitated smoother project execution. Regular updates and open dialogue are key components of this strategy (Rosenstiel et al., 2021).

5.3. Regulatory Support

The Moroccan government has provided strong regulatory support to PPPs, creating a conducive environment for investment. Clear policies, incentives, and a robust legal framework have been critical in attracting private sector participation (Khouya, 2020).

6. Insights from the OECD Symposium on PPPs in Morocco

The OECD symposium on PPPs highlighted Morocco's achievements and future prospects. Mr. Abdelaziz Talbi emphasized Morocco's extensive experience with concessions dating back to the early 20th century. Modern PPPs have driven significant projects like the Jorf Lasfar power plant and liberalized sectors such as telecommunications and agriculture. These initiatives have improved risk assessment, service delivery, and attracted both national and international investments (Lindner, 2023).

6.1. Strategic Sector Focus

Future PPP opportunities in Morocco are being explored in strategic sectors including transportation, infrastructure, social services, water and sanitation, and energy. A comprehensive study aims to identify and develop these opportunities, enhancing economic growth and public service delivery (Plank et al., 2023).

6.2. Policy Recommendations

To successfully implement PPPs, a multi-tiered policy approach is recommended:

- An economic strategy that places responsibility for growth on the private sector
- Sector-specific liberalization with fair competition and transparency
- Empowerment of public operators to compete or collaborate with the private sector
- Transfer of commercial expertise to non-commercial sectors like education and health

7. Lessons and Insights for Emerging Economies

The evaluation of PPPs in Morocco's renewable energy sector offers valuable lessons for other emerging economies (Smith, 2020). Establishing a clear and supportive regulatory framework is essential for attracting private investments and ensuring project success (Hamza et al., 2022). Leveraging diverse and innovative financing mechanisms can mitigate risks and enhance the financial viability of renewable energy projects (Bahou, 2023). Building strong partnerships and maintaining transparent communication with all stakeholders, including local communities, is crucial for project acceptance and sustainability(Choukri et al., 2017b). Investing in local capacity building through training and involvement of local firms can ensure long-term benefits and project success(Choukri et al., 2018).

VII. DISCUSSION & CONCLUSION

Renewable energies and energy efficiency now occupy a prominent place in the new national energy policy. This is reflected in a number of measures taken by public authorities in order to achieve the energy saving objectives of 20% and the creation of an electricity production park representing 52% of the total power installed at the site. horizon 2030.

This is in particular a vast regulatory project dedicated to renewable energies opened by the Ministry of Energy Transition and Sustainable Development and in continuous improvement to follow the rapid evolution of the energy sector both at national level and also at an international level.

The regulatory framework (Law 40-19) creates the necessary conditions for increased use of renewable energies and an evolution in the involvement of the private sector for progressive liberalization. It thus contributes to achieving the objectives linked to greenhouse gas emissions and toxic emissions.

By supporting the liberalization of the energy market and the effective regulation of networks, it also contributes to sustainable economic development and promotes social participation by giving households and private companies the opportunity to actively participate in the development of renewable energies. This regulatory framework provides essential regulatory references for the implementation of Law 40.19.

To this end, the aforementioned objectives will not be achieved without the publication of these regulatory references, in particular the references directly linked to the authorization procedure.

Also, the provisions of the law allow an operator connected to the high and very high voltage network (THT-HT) to supply electricity (i) to customers connected to the THT-HT network, (ii) to customers supplied with medium voltage (MV) and (iii) electricity distributors (including utilities).

This is a broader liberalization of the electrical energy supply market for different types of customers (including distributors) connected to the national electricity transmission network (THT-HT) and distribution networks (MT). Such an opening will naturally lead the free electricity market, based on renewable energies, to a structural imbalance between a production supply quite limited by the capacity of the system and a greater demand for electricity encompassing the needs distributors, VHT-HT customers and MV customers. This would then result in a convergence of free market prices towards public electricity prices in the regulated market, which will not allow eligible customers to benefit from the continued and significant reduction in production costs based on renewable energies.

In addition, this configuration of the free market, governed by laws 13-09 and 40-19, is likely to give rise to a significant development of large centralized renewable projects, connected to the THT-HT electrical network, to the detriment of decentralized regional industrial integration promoting the development of small and medium-sized businesses, based on renewable energies, at the level of medium voltage networks.

In a logic of greater openness than that provided for within the framework of law 13-09 (opening of marketing to distributors), the existence of guaranteed purchasing contracts for certain producers who occupy a strong position (nearly 54% of the market) limits the potential Moroccan market shares for new entrants.

Public-Private Partnerships have proven to be a vital component in advancing renewable energy projects in Morocco. By examining successful PPP models, innovative financing mechanisms, and effective stakeholder engagement strategies, this section highlights the potential for PPPs to drive sustainable energy development in other emerging economies. The lessons learned from Morocco's experience can serve as a blueprint for fostering collaboration between the public and private sectors, ultimately contributing to the global transition towards renewable energy

The existence of long-term PPAs between private producers and a single buyer prevents the free play of real competition and limits the flexibility of the electricity system necessary to manage the vagaries of renewable energy production. It would be desirable to study the flexibility of the Moroccan electricity system using indicators developed by the IEA for example.

On the other hand, the regulatory authority is a crucial element to ensure the stability of the electricity market in Morocco. By guaranteeing its credibility, it provides essential security for investors and potential customers, thus promoting confidence in the sector. In addition, the regulatory framework put in place to support the creation of this regulatory authority contributes significantly to the development of renewable energies in the country.

The following table provides a better understanding of the responsibilities of all the players involved in the Moroccan electricity market, and highlights the key role of the regulator in this dynamic.

Indeed, by defining appropriate standards and regulations for private investors in the energy sector, the regulatory authority helps ensure fair and transparent competition, while protecting the interests of consumers.

It is important to emphasize that the procedure for determining and publishing the reception capacity announced by Law 40-19 is ensured by ANRE. However, this procedure did not clarify the scope of the trajectories and envelopes provided for in the Medium Voltage Decree. In principle, the carrying capacity must replace the trajectories system to avoid redundancy and contradiction between the two mechanisms.

In other words, Law 40-19 provides for the determination and publication of the carrying capacity, which is the maximum amount of electricity that a network can support at a given time. However, the Medium Voltage Decree also includes trajectories and envelopes which define the limits of the quantity of electricity that can be injected into the network. These two mechanisms can conflict and create redundancy in the management of the electricity network.

It is therefore essential to clarify the role of reception capacity in relation to the trajectories and envelopes of the Medium Voltage Decree in order to ensure effective and coherent management of the electricity network. By eliminating redundancy and contradiction between the two mechanisms, we can ensure optimal use of the electricity network, guarantee the security of installations and avoid power cuts.

Also, Law 40-19 clearly announces that capping beyond the regulated rate, the network manager must re-list the operator of the energy production installation from renewable sources. Whereas, article 12 of the medium voltage decree stipulates that "The refusal of injection by distribution network managers, for reasons of security and network integrity, does not entail for the latter any obligation to financial compensation ".

Overall, the creation and maintenance of a credible and effective regulatory authority are essential elements to guarantee the stability and growth of the electricity market in Morocco, as well as to promote the transition towards renewable and more sustainable energies.

REFERENCES

Abdel-Karim, N., Preston, E., Moura, J., & Coleman, T. (2018a). Variable energy resource capacity contributions consistent with reserve margin and reliability. *IEEE Power and Energy Society General Meeting, 2018-January*. DOI: 10.1109/PESGM.2017.8274569

Abdel-Karim, N., Preston, E., Moura, J., & Coleman, T. (2018b). Variable energy resource capacity contributions consistent with reserve margin and reliability. *IEEE Power and Energy Society General Meeting, 2018-January*. DOI: 10.1109/PESGM.2017.8274569

Adeleke, A. A., Petrus, N., Ayuba, S., Yahya, A. M., Ikubanni, P. P., Okafor, I. S., Emmanuel, S. S., Olosho, A. I., & Adesibikan, A. A. (2023). Nigerian Biomass for Bioenergy Applications: A Review on the Potential and Challenges. In *Journal of Renewable Materials* (Vol. 11, Issue 12). DOI: 10.32604/jrm.2023.043915

Adeli, K., Nachtane, M., Faik, A., Rachid, A., Tarfaoui, M., & Saifaoui, D. (2024). A deep learning-enhanced framework for sustainable hydrogen production from solar and wind energy in the Moroccan Sahara: Coastal regions focus. *Energy Conversion and Management*, *302*, 118084. Advance online publication. DOI: 10.1016/j.enconman.2024.118084

Adun, H., Ishaku, H. P., & Ogungbemi, A. T. (2022). Towards Renewable energy targets for the Middle East and North African region: A decarbonization assessment of energy-water nexus. *Journal of Cleaner Production*, *374*, 133944. Advance online publication. DOI: 10.1016/j.jclepro.2022.133944

Aghahosseini, A., Bogdanov, D., & Breyer, C. (2020). Towards sustainable development in the MENA region: Analysing the feasibility of a 100% renewable electricity system in 2030. *Energy Strategy Reviews*, *28*, 100466. Advance online publication. DOI: 10.1016/j.esr.2020.100466

Ait Errouhi, A., Choukai, O., Oumimoun, Z., & El Mokhi, C. (2022). Energy efficiency measures and technical-economic study of a photovoltaic self-consumption installation at ENSA Kenitra, Morocco. *Energy Harvesting and Systems*, *9*(2), 193–201. Advance online publication. DOI: 10.1515/ehs-2021-0081

Al-Najjar, H., El-Khozondar, H. J., Pfeifer, C., & Al Afif, R. (2022). Hybrid grid-tie electrification analysis of bio-shared renewable energy systems for domestic application. *Sustainable Cities and Society*, *77*, 103538. Advance online publication. DOI: 10.1016/j.scs.2021.103538

Alaoui, C. (2019). Review and assessment of offshore renewable energy resources in morocco' coastline. In *Cogent Engineering* (Vol. 6, Issue 1). DOI: 10.1080/23311916.2019.1654659

Almulla, Y., Ramirez, C., Joyce, B., Huber-Lee, A., & Fuso-Nerini, F. (2022). From participatory process to robust decision-making: An Agriculture-water-energy nexus analysis for the Souss-Massa basin in Morocco. *Energy for Sustainable Development, 70*, 314–338. Advance online publication. DOI: 10.1016/j.esd.2022.08.009

Ammari, M., Chentouf, M., Ammari, M., & Ben Allal, L. (2022). Assessing National Progress in Achieving the Sustainable Development Goals: A Case Study of Morocco. *Sustainability (Basel), 14*(23), 15582. Advance online publication. DOI: 10.3390/su142315582

Amorim, W. C. S., Cupertino, A. F., Pereira, H. A., & Mendes, V. F. (2024). On sizing of battery energy storage systems for PV plants power smoothing. *Electric Power Systems Research, 229*, 110114. Advance online publication. DOI: 10.1016/j.epsr.2024.110114

Amrani, S. E., Alami Merrouni, A., Touili, S., & Dekhissi, H. (2023). A multi-scenario site suitability analysis to assess the installation of large scale photovoltaic-hydrogen production units. Case study: Eastern Morocco. *Energy Conversion and Management, 295*, 117615. Advance online publication. DOI: 10.1016/j.enconman.2023.117615

Armijo, J., & Philibert, C. (2020). Flexible production of green hydrogen and ammonia from variable solar and wind energy: Case study of Chile and Argentina. *International Journal of Hydrogen Energy, 45*(3), 1541–1558. Advance online publication. DOI: 10.1016/j.ijhydene.2019.11.028

Asiaban, S., Kayedpour, N., Samani, A. E., Bozalakov, D., De Kooning, J. D. M., Crevecoeur, G., & Vandevelde, L. (2021). Wind and solar intermittency and the associated integration challenges: A comprehensive review including the status in the belgian power system. In *Energies* (Vol. 14, Issue 9). DOI: 10.3390/en14092630

Ayad, H., Sari-Hassoun, S. E., Usman, M., & Ahmad, P. (2023). The impact of economic uncertainty, economic growth and energy consumption on environmental degradation in MENA countries: Fresh insights from multiple thresholds NARDL approach. *Environmental Science and Pollution Research International, 30*(1), 1806–1824. Advance online publication. DOI: 10.1007/s11356-022-22256-w PMID: 35921013

Badran, A. (2017). Climate change and water science policy in management. In *Water, Energy and Food Sustainability in the Middle East: The Sustainability Triangle*. DOI: 10.1007/978-3-319-48920-9_1

Bahetta, S., Dahhou, N., & Hasnaoui, R. (2021). Comparative effectiveness of environmental regulation instruments: Case of the Moroccan electricity mix. *AIMS Energy*, *9*(5), 1097–1112. Advance online publication. DOI: 10.3934/energy.2021050

Bahou, S. (2023). Techno-economic assessment of a hydrogen refuelling station powered by an on-grid photovoltaic solar system: A case study in Morocco. *International Journal of Hydrogen Energy*, *48*(61), 23363–23372. Advance online publication. DOI: 10.1016/j.ijhydene.2023.03.220

Bakkari, M., Bossoufi, B., El Kafazi, I., Bouderbala, M., & Karim, M. (2024). A review of wind energy potential in Morocco: New challenges and perspectives. In *Wind Engineering* (Vol. 48, Issue 1). DOI: 10.1177/0309524X231200582

Belaid, F., Boukrami, E., & Amine, R. (2021). Renewable Energy in the MENA Region: Key Challenges and Lessons Learned. In *Lecture Notes in Energy* (Vol. 82). DOI: 10.1007/978-3-030-71403-1_1

Belmahdi, B., Louzazni, M., & El Bouardi, A. (2022). Comparative optimization of global solar radiation forecasting using machine learning and time series models. *Environmental Science and Pollution Research International*, *29*(10), 14871–14888. Advance online publication. DOI: 10.1007/s11356-021-16760-8 PMID: 34625894

Bensouda, M., & Benali, M. (2023). The Role of Institutional Pressure and Dynamic Capabilities in Promoting Energy Efficiency Practices: Evidence from the Moroccan Manufacturing Sector. *International Journal of Energy Economics and Policy*, *13*(1), 352–361. Advance online publication. DOI: 10.32479/ijeep.13697

Bigerna, S. (2024). Connectedness analysis of oil price shocks, inflation, and exchange rate for the MENA region countries. *Resources Policy*, *88*, 104344. Advance online publication. DOI: 10.1016/j.resourpol.2023.104344

Boharb, A., Allouhi, A., El-houari, H., El Markhi, H., Jamil, A., & Kousksou, T. (2022). Energy audit method applied to tertiary buildings: Case study of a University campus. *AIMS Energy*, *10*(3), 506–532. Advance online publication. DOI: 10.3934/energy.2022025

Boullouz, S., Faquir, S., & Yahyaouy, A. (2022). A New Approach to Manage the Energy Flow in a Hybrid Renewable Energy System Based on Green Hydrogen. *Lecture Notes in Networks and Systems, 455 LNNS*. DOI: 10.1007/978-3-031-02447-4_58

Bouyghrissi, S., Murshed, M., Jindal, A., Berjaoui, A., Mahmood, H., & Khanniba, M. (2022). The importance of facilitating renewable energy transition for abating CO2 emissions in Morocco. *Environmental Science and Pollution Research International*, *29*(14), 20752–20767. Advance online publication. DOI: 10.1007/s11356-021-17179-x PMID: 34741744

Cao, J., Wu, H., & Zou, Y. (2024). Advancing sustainable development: The impact of energy transition on accelerating technological progress. *Economic Change and Restructuring*, *57*(2), 70. Advance online publication. DOI: 10.1007/s10644-024-09664-3

Cárdenas, B., Swinfen-Styles, L., Rouse, J., Hoskin, A., Xu, W., & Garvey, S. D. (2021). Energy storage capacity vs. renewable penetration: A study for the UK. *Renewable Energy*, *171*, 849–867. Advance online publication. DOI: 10.1016/j.renene.2021.02.149

Chennaif, M., Maaouane, M., Zahboune, H., Elhafyani, M., & Zouggar, S. (2022). Tri-objective techno-economic sizing optimization of Off-grid and On-grid renewable energy systems using Electric system Cascade Extended analysis and system Advisor Model. *Applied Energy*, *305*, 117844. Advance online publication. DOI: 10.1016/j.apenergy.2021.117844

Cherkasova, E. G. (2023a). SPAIN'S ENERGY SECURITY IN THE CONTEXT OF THE ALGERIAN-MOROCCAN CONFLICT. *World Economy and International Relations*, *67*(9), 48–57. Advance online publication. DOI: 10.20542/0131-2227-2023-67-9-48-57

Cherkasova, E. G. (2023b). SPAIN'S ENERGY SECURITY IN THE CONTEXT OF THE ALGERIAN-MOROCCAN CONFLICT. *World Economy and International Relations*, *67*(9), 48–57. Advance online publication. DOI: 10.20542/0131-2227-2023-67-9-48-57

Chidolue, O., Ngozichukwu, B., Ibekwe, K. I., Illojianya, V. I., Fafure, A. V., & Daudu, C. D. (2024). Decarbonization strategies in energy-intensive industries: Cases from Canada, USA, and Africa. *International Journal of Science and Research Archive*, *11*(1), 361–370. Advance online publication. DOI: 10.30574/ijsra.2024.11.1.0065

Choukri, K., Naddami, A., & Hayani, S. (2017a). Renewable energy in emergent countries: Lessons from energy transition in Morocco. *Energy, Sustainability and Society*, *7*(1), 25. Advance online publication. DOI: 10.1186/s13705-017-0131-2

Choukri, K., Naddami, A., & Hayani, S. (2018). Evaluation of the reserve capacity in a grid supplied by intermittent energy sources. *IET Renewable Power Generation*, *12*(4), 399–406. Advance online publication. DOI: 10.1049/iet-rpg.2017.0119

Choukri, K., Naddami, A., & Hayani, S. M. (2017b). Deep analysis of wind variability and smoothing effect in Moroccan wind farms. *Wind Engineering*, *41*(4), 272–281. Advance online publication. DOI: 10.1177/0309524X17709731

Cloete, S., & Hirth, L. (2020). Flexible power and hydrogen production: Finding synergy between CCS and variable renewables. *Energy*, *192*, 116671. Advance online publication. DOI: 10.1016/j.energy.2019.116671

Dadashi, Z., Mahmoudi, A., & Rashidi, S. (2022). Capacity and strategies of energy production from renewable sources in Arab countries until 2030: a review from renewable energy potentials to environmental issues. In *Environmental Science and Pollution Research* (Vol. 29, Issue 32). DOI: 10.1007/s11356-022-20544-z

de Arce, R., Mahía, R., Medina, E., & Escribano, G. (2012). A simulation of the economic impact of renewable energy development in Morocco. *Energy Policy*, *46*, 335–345. Advance online publication. DOI: 10.1016/j.enpol.2012.03.068

Deininge, F., & Gren, A. (2022). Green jobs for women can combat the climate crisis and boost equality. *Development and a Changing Climate*.

Derouez, F., & Ifa, A. (2024). Sustainable Food Security: Balancing Desalination, Climate Change, and Population Growth in Five Arab Countries Using ARDL and VECM. *Sustainability (Basel)*, *16*(6), 2302. Advance online publication. DOI: 10.3390/su16062302

Dispenza, G., Sergi, F., Napoli, G., Antonucci, V., & Andaloro, L. (2019). Evaluation of hydrogen production cost in different real case studies. *Journal of Energy Storage*, *24*, 100757. Advance online publication. DOI: 10.1016/j.est.2019.100757

Dunlap, A., & Laratte, L. (2022). European Green Deal necropolitics: Exploring 'green' energy transition, degrowth & infrastructural colonization. *Political Geography*, *97*, 102640. Advance online publication. DOI: 10.1016/j.polgeo.2022.102640

El-Emam, R. S., Ezzat, M. F., & Khalid, F. (2022). Assessment of hydrogen as a potential energy storage for urban areas' PV-assisted energy systems - Case study. *International Journal of Hydrogen Energy*, *47*(62), 26209–26222. Advance online publication. DOI: 10.1016/j.ijhydene.2022.01.107

El Hafdaoui, H., Khallaayoun, A., & Ouazzani, K. (2024). Long-term low carbon strategy of Morocco: A review of future scenarios and energy measures. In *Results in Engineering* (Vol. 21). DOI: 10.1016/j.rineng.2023.101724

El Hathat, Z., Sreedharan, V. R., Venkatesh, V. G., Zouadi, T., Arunmozhi, M., & Shi, Y. (2023). Modelling and analyzing the GHG emissions in the VUCA world: Evidence from tomato production in Morocco. *Journal of Cleaner Production*, *382*, 134862. Advance online publication. DOI: 10.1016/j.jclepro.2022.134862

El Majaty, S., Touzani, A., & Kasseh, Y. (2023). Results and perspectives of the application of an energy management system based on ISO 50001 in administrative buildings - case of Morocco. *Materials Today: Proceedings*, *72*, 3233–3237. Advance online publication. DOI: 10.1016/j.matpr.2022.07.094

Elmansour, J., Hajjaji, A., Belhora, F., & Hendrick, P. (2022). Location of seawater pumped storage hydropower plants: Case of Morocco. *Materials Today: Proceedings*, *66*, 45–57. Advance online publication. DOI: 10.1016/j.matpr.2022.03.145

Ersoy, S. R., Terrapon-Pfaff, J., Ribbe, L., & Merrouni, A. A. (2021). Water scenarios modelling for renewable energy development in Southern Morocco. *Journal of Sustainable Development of Energy. Water and Environment Systems*, *9*(1), 0. Advance online publication. DOI: 10.13044/j.sdewes.d8.0335

Firdous, A., Prakash Barala, C., Mathuria, P., & Bhakar, R. (2024). Extended power to hydrogen operations for enhanced grid flexibility in low carbon systems. *Energy Conversion and Management*, *301*, 117982. DOI: 10.1016/j.enconman.2023.117982

Fragkos, P. (2023). Assessing the energy system impacts of Morocco's nationally determined contribution and low-emission pathways. In *Energy Strategy Reviews* (Vol. 47). DOI: 10.1016/j.esr.2023.101081

Gado, M. G., & Hassan, H. (2023). Potential of prospective plans in MENA countries for green hydrogen generation driven by solar and wind power sources. *Solar Energy*, *263*, 111942. Advance online publication. DOI: 10.1016/j.solener.2023.111942

Galimova, T., Fasihi, M., Bogdanov, D., & Breyer, C. (2023). Impact of international transportation chains on cost of green e-hydrogen: Global cost of hydrogen and consequences for Germany and Finland. *Applied Energy*, *347*, 121369. Advance online publication. DOI: 10.1016/j.apenergy.2023.121369

Ge, P., Hu, Q., Wu, Q., Dou, X., Wu, Z., & Ding, Y. (2020). Increasing operational flexibility of integrated energy systems by introducing power to hydrogen. *IET Renewable Power Generation*, *14*(3), 372–380. Advance online publication. DOI: 10.1049/iet-rpg.2019.0663

Ghaemi, S., Li, X., & Mulder, M. (2023). Economic feasibility of green hydrogen in providing flexibility to medium-voltage distribution grids in the presence of local-heat systems. *Applied Energy*, *331*, 120408. Advance online publication. DOI: 10.1016/j.apenergy.2022.120408

Glanville, P., Fridlyand, A., Sutherland, B., Liszka, M., Zhao, Y., Bingham, L., & Jorgensen, K. (2022). Impact of Hydrogen/Natural Gas Blends on Partially Premixed Combustion Equipment: NOx Emission and Operational Performance. *Energies*, *15*(5), 1706. Advance online publication. DOI: 10.3390/en15051706

Gunawan, T. A., & Monaghan, R. F. D. (2022). Techno-econo-environmental comparisons of zero- and low-emission heavy-duty trucks. *Applied Energy*, *308*, 118327. Advance online publication. DOI: 10.1016/j.apenergy.2021.118327

Hafner, M., Raimondi, P. P., & Bonometti, B. (2023). Low-Carbon Energy Strategies in MENA Countries. In *Perspectives on Development in the Middle East and North Africa (MENA) Region: Vol. Part F1210*. DOI: 10.1007/978-3-031-30705-8_4

Haidi, T., & Cheddadi, B. (2022). Wind energy integration in Africa: Development, impacts and barriers. *International Journal of Electrical and Computer Engineering*, *12*(5), 4614. Advance online publication. DOI: 10.11591/ijece.v12i5.pp4614-4622

Haines, M. B., Moore, S., & Adornetto, T. (2023). Suspending democratic (dis) belief: Nonliberal energy polities of solar power in Morocco and Tanzania. *Energy Research & Social Science*, *96*, 102942. Advance online publication. DOI: 10.1016/j.erss.2023.102942

Hajar, A. L., Ahmed, N., & Karim, C. (2023). Deployment of Safety Predictive Analytics to Prevent Workplace Incidents and Promote Event Reduction: A Machine Learning Approach Towards a Data-Driven Safety System. *Proceedings - 2023 3rd International Conference on Digital Data Processing, DDP 2023*. DOI: 10.1109/DDP60485.2023.00042

Hamza, K., Hamid, M., & Mouaad, B. (2022). A Review of Future Fuel Cell Electric Vehicles and Challenges Related to Morocco. *WSEAS Transactions on Power Systems*, *17*, 339–353. Advance online publication. DOI: 10.37394/232016.2022.17.34

Hassan, Q., Algburi, S., Sameen, A. Z., Salman, H. M., & Jaszczur, M. (2023). A review of hybrid renewable energy systems: Solar and wind-powered solutions: Challenges, opportunities, and policy implications. *Results in Engineering*, *20*, 101621. DOI: 10.1016/j.rineng.2023.101621

Hejazi, M., Santos Da Silva, S. R., Miralles-Wilhelm, F., Kim, S., Kyle, P., Liu, Y., Vernon, C., Delgado, A., Edmonds, J., & Clarke, L. (2023). Impacts of water scarcity on agricultural production and electricity generation in the Middle East and North Africa. *Frontiers in Environmental Science*, *11*, 1082930. Advance online publication. DOI: 10.3389/fenvs.2023.1082930

Hossin, M. A., Akorsu, P. K., & Abudu, H. (2023). Regional energy and carbon emissions concerns: Towards cleaner energy transition in Middle East and North Africa. *Social Sciences & Humanities Open*, *8*(1), 100705. Advance online publication. DOI: 10.1016/j.ssaho.2023.100705

Idllalène, S. (2021). The role of environmental Waqf in addressing climate change in the MENA region: A comparative law analysis. In *Climate Change Law and Policy in the Middle East and North Africa Region*.

Idrissi Kaitouni, S., Mghazli, M. O., Nait-Taour, A., Es-Sakali, N., El Mankibi, M., Ahachad, M., & Brigui, J. (2023). Empirical validation and analysis of the energy performance of an ecological Net Zero Energy Building (NZEB) in Benguerir-Morocco. *E3S Web of Conferences, 396*. DOI: 10.1051/e3sconf/202339604023

IRENA. (2021). Morocco and IRENA Partner to Boost Renewables and Green Hydrogen Development. *Press Release*.

Ismagilov, A. F., Khamitov, I. G., & Georgievsky, B. V. (2019). Comparative analysis of oil and gas basins belonging to the MENA region. *Neftyanoe Khozyaystvo - Oil Industry, 2019*(January). DOI: 10.24887/0028-2448-2019-1-20-23

Jalbout, E., Genge, L., & Muesgens, F. (2022). H2Europe: an analysis of long-term hydrogen import-potentials from the MENA region. *International Conference on the European Energy Market, EEM, 2022-September*. DOI: 10.1109/EEM54602.2022.9921055

Jensen, I. G., Wiese, F., Bramstoft, R., & Münster, M. (2020). Potential role of renewable gas in the transition of electricity and district heating systems. *Energy Strategy Reviews*, *27*, 100446. Advance online publication. DOI: 10.1016/j.esr.2019.100446

K, S. (2021). Climate Change Challenge 26th Conference of Parties (COP26) climate summit is crucial but may be disappointing? *Global Journal of Ecology*. DOI: 10.17352/gje.000051

Kasseh, Y., Touzani, A., & El Majaty, S. (2023a). Exemplarity of the State for the Energy Efficiency of Buildings Institutional - Case of Morocco. *Nature Environment and Pollution Technology*, *22*(1), 169–177. Advance online publication. DOI: 10.46488/NEPT.2023.v22i01.015

Kasseh, Y., Touzani, A., & El Majaty, S. (2023b). What public lighting governance model should be deployed in Moroccan cities for sustainable and efficient energy management? *Materials Today: Proceedings*, *72*, 3244–3252. Advance online publication. DOI: 10.1016/j.matpr.2022.07.123

Kharrich, M., Mohammed, O. H., & Akherraz, M. (2019). Assessment of renewable energy sources in Morocco using economical feasibility technique. *International Journal of Renewable Energy Research*, *9*(4). Advance online publication. DOI: 10.20508/ijrer.v9i4.10181.g7791

Khouya, A. (2020). Levelized costs of energy and hydrogen of wind farms and concentrated photovoltaic thermal systems. A case study in Morocco. *International Journal of Hydrogen Energy*, *45*(56), 31632–31650. Advance online publication. DOI: 10.1016/j.ijhydene.2020.08.240

Kim, S., Park, J., Chung, W., Adams, D., & Lee, J. H. (2024). Techno-economic analysis for design and management of international green hydrogen supply chain under uncertainty: An integrated temporal planning approach. *Energy Conversion and Management*, *301*, 118010. DOI: 10.1016/j.enconman.2023.118010

Koirala, B., Hers, S., Morales-España, G., Özdemir, Ö., Sijm, J., & Weeda, M. (2021). Integrated electricity, hydrogen and methane system modelling framework: Application to the Dutch Infrastructure Outlook 2050. *Applied Energy*, *289*, 116713. Advance online publication. DOI: 10.1016/j.apenergy.2021.116713

Kousksou, T., Allouhi, A., Belattar, M., Jamil, A., El Rhafiki, T., Arid, A., & Zeraouli, Y. (2015). Renewable energy potential and national policy directions for sustainable development in Morocco. In *Renewable and Sustainable Energy Reviews* (Vol. 47). DOI: 10.1016/j.rser.2015.02.056

Kouzzi, S., Ganaoui, M. El, Kheiri, A., & Bouardi, A. El. (2022). Local materials, between regional diversity and energy performance to build efficient buildings in a transborder context (morocco-spain). *E3S Web of Conferences, 353*. DOI: 10.1051/e3sconf/202235301003

Lebrouhi, B. E., Lamrani, B., Zeraouli, Y., & Kousksou, T. (2024). Key challenges to ensure Morocco's sustainable transition to a green hydrogen economy. In *International Journal of Hydrogen Energy* (Vol. 49). DOI: 10.1016/j.ijhydene.2023.09.178

Lee, B., & Im, S. (2024). Energy, exergy, and exergoeconomic analyses of plastic waste-to-energy integrated gasification combined cycles with and without heat recovery at a gasifier. *Applied Energy*, *355*, 122243. Advance online publication. DOI: 10.1016/j.apenergy.2023.122243

Li, S., Sun, F., He, H., & Chen, Y. (2017). Optimization for a Grid-connected Hybrid PV-wind-retired HEV Battery Microgrid System. *Energy Procedia*, *105*, 1634–1643. DOI: 10.1016/j.egypro.2017.03.532

Lindner, R. (2023). Green hydrogen partnerships with the Global South. Advancing an energy justice perspective on "tomorrow's oil.". *Sustainable Development (Bradford)*, *31*(2), 1038–1053. Advance online publication. DOI: 10.1002/sd.2439

Lopez, G., Galimova, T., Fasihi, M., Bogdanov, D., & Breyer, C. (2023). Towards defossilised steel: Supply chain options for a green European steel industry. *Energy*, *273*, 127236. Advance online publication. DOI: 10.1016/j.energy.2023.127236

Lu, Q., Yang, Y., Xie, P., Chen, Y., Wu, J., & Lei, Z. (2024). Market Mechanism Designing and Scheduling Strategies of Auxiliary Services for Frequency Control Adapting to Energy Storage Participation. *Dianwang Jishu/Power System Technology, 48*(3). DOI: 10.13335/j.1000-3673.pst.2023.0104

Mahlooji, M., Gaudard, L., Ristic, B., & Madani, K. (2020). The importance of considering resource availability restrictions in energy planning: What is the footprint of electricity generation in the Middle East and North Africa (MENA)? *The Science of the Total Environment*, *717*, 135035. Advance online publication. DOI: 10.1016/j.scitotenv.2019.135035 PMID: 31839284

Makan, A., Gouraizim, M., & Fadili, A. (2022). Sustainability assessment of wastewater treatment systems using cardinal weights and PROMETHEE method: Case study of Morocco. *Environmental Science and Pollution Research International*, *29*(13), 19803–19815. Advance online publication. DOI: 10.1007/s11356-021-17215-w PMID: 34718984

Meliani, M., El, A., El, I., Moumen, A., & Mahmoudi, M. (2021). Materials Today : Proceedings Smart grid implementation in Morocco : Case study. *Materials Today: Proceedings*, •••, 45.

Meliani, M., El Barkany, A., El Abbassi, I., & Mahmoudi, M. (2022). Smart Grid Challenges in Morocco and an Energy Demand Forecasting with Time Series. *International Journal of Engineering Research in Africa*, *61*, 195–215. Advance online publication. DOI: 10.4028/p-2gufv6

Meliani, M., El Barkany, A., El Abbassi, I., Moumen Darcherif, A., & Mahmoudi, M. (2021). Smart grid implementation in Morocco: Case study. *Materials Today: Proceedings*, *45*, 7675–7679. Advance online publication. DOI: 10.1016/j.matpr.2021.03.176

Mena-Cervantes, V. Y., Hernández-Altamirano, R., García-Solares, M. S., & Arreola-Valerio, E. (2023). Biodiesel in Circular Economy. In *Biofuels in Circular Economy*. DOI: 10.1007/978-981-19-5837-3_14

Mewafy, A., Ismael, I., Kaddah, S. S., Hu, W., Chen, Z., & Abulanwar, S. (2024). Optimal design of multiuse hybrid microgrids power by green hydrogen–ammonia. *Renewable & Sustainable Energy Reviews*, *192*, 114174. DOI: 10.1016/j.rser.2023.114174

Mohammadshahi, S. S., Boulaire, F. A., Love, J., Gorji, S. A., & Mackinnon, I. D. R. (2022). A flexible analytical model for operational investigation of solar hydrogen plants. *International Journal of Hydrogen Energy*, *47*(2), 782–808. Advance online publication. DOI: 10.1016/j.ijhydene.2021.10.072

Mohseni, S., Brent, A. C., & Burmester, D. (2020). A comparison of metaheuristics for the optimal capacity planning of an isolated, battery-less, hydrogen-based micro-grid. *Applied Energy*, *259*, 114224. Advance online publication. DOI: 10.1016/j.apenergy.2019.114224

Morocco's Quest for Stronger and Inclusive Growth. (2023). *Morocco's Quest for Stronger and Inclusive Growth.*, DOI: 10.5089/9798400225406.071

Morton, E. M., Deetjen, T. A., & Goodarzi, S. (2023). Optimizing hydrogen production capacity and day ahead market bidding for a wind farm in Texas. *International Journal of Hydrogen Energy*, *48*(46), 17420–17433. Advance online publication. DOI: 10.1016/j.ijhydene.2022.12.354

Nakach, I., Mouhat, O., Shamass, R., & El Mennaouy, F. (2023). Review of strategies for sustainable energy in Morocco. *Polityka Energetyczna*, *26*(2), 65–104. Advance online publication. DOI: 10.33223/epj/163373

Neumann, J., da Rocha, R. C., Debiagi, P., Scholtissek, A., Dammel, F., Stephan, P., & Hasse, C. (2023). Techno-economic assessment of long-distance supply chains of energy carriers: Comparing hydrogen and iron for carbon-free electricity generation. *Applications in Energy and Combustion Science*, *14*, 100128. Advance online publication. DOI: 10.1016/j.jaecs.2023.100128

Nordio, M., Wassie, S. A., Van Sint Annaland, M., Pacheco Tanaka, D. A., Viviente Sole, J. L., & Gallucci, F. (2021). Techno-economic evaluation on a hybrid technology for low hydrogen concentration separation and purification from natural gas grid. *International Journal of Hydrogen Energy*, *46*(45), 23417–23435. Advance online publication. DOI: 10.1016/j.ijhydene.2020.05.009

Norouzi, N. (2023). A framework of sustainable economics: Determining driving forces of the sustainable development in the MENA region. In *Energy Transition in the African Economy Post 2050*. DOI: 10.4018/978-1-7998-8638-9.ch005

Odenweller, A., Ueckerdt, F., Nemet, G. F., Jensterle, M., & Luderer, G. (2022). Probabilistic feasibility space of scaling up green hydrogen supply. *Nature Energy*, *7*(9), 854–865. Advance online publication. DOI: 10.1038/s41560-022-01097-4

Odunaiya, O. G., Soyombo, O. T., Okoli, C. E., Usiagu, G. S., & Ekemezie, I. O.Olusegun Gbenga OdunaiyaOluwatobi Timothy SoyomboChinelo Emilia Okoli-Glory Siwe UsiaguIfeanyi Onyedika Ekemezie. (2024). Renewable energy policies and wildlife conservation: A review of potential conflicts and coexistence strategies. *Open Access Research Journal of Multidisciplinary Studies*, *7*(1), 111–120. Advance online publication. DOI: 10.53022/oarjms.2024.7.1.0015

Okpanachi, E., Ambe-Uva, T., & Fassih, A. (2022). Energy regime reconfiguration and just transitions in the Global South: Lessons for West Africa from Morocco's comparative experience. *Futures*, *139*, 102934. Advance online publication. DOI: 10.1016/j.futures.2022.102934

Okundamiya, M. S. (2021). Integration of photovoltaic and hydrogen fuel cell system for sustainable energy harvesting of a university ICT infrastructure with an irregular electric grid. *Energy Conversion and Management*, *250*, 114928. Advance online publication. DOI: 10.1016/j.enconman.2021.114928

Ouchani, F., Jbaihi, O., Alami Merrouni, A., Ghennioui, A., & Maaroufi, M. (2022). Geographic Information System-based Multi-Criteria Decision-Making analysis for assessing prospective locations of Pumped Hydro Energy Storage plants in Morocco: Towards efficient management of variable renewables. *Journal of Energy Storage*, *55*, 105751. Advance online publication. DOI: 10.1016/j.est.2022.105751

Ouhammou, B., Gargab, F. Z., El idrissi kaitouni, S., Smouh, S., El mrabet, R., Aggour, M., Jamil, A., & Kousksou, T. (2023). Energy saving potential diagnosis for Moroccan university campuses. *AIMS Energy*, *11*(3). DOI: 10.3934/energy.2023030

Park, J., Hwan Ryu, K., Kim, C. H., Chul Cho, W., Kim, M. J., Hun Lee, J., Cho, H. S., & Lee, J. H. (2023). Green hydrogen to tackle the power curtailment: Meteorological data-based capacity factor and techno-economic analysis. *Applied Energy*, *340*, 121016. Advance online publication. DOI: 10.1016/j.apenergy.2023.121016

Plank, F., Daum, B., Muntschick, J., Knodt, M., Hasse, C., Ott, I., & Niemann, A. (2023). Hydrogen: Fueling EU-Morocco Energy Cooperation? *Middle East Policy*, *30*(3), 37–52. Advance online publication. DOI: 10.1111/mepo.12699

Quarton, C. J., & Samsatli, S. (2020). Should we inject hydrogen into gas grids? Practicalities and whole-system value chain optimisation. *Applied Energy*, *275*, 115172. Advance online publication. DOI: 10.1016/j.apenergy.2020.115172

Raab, M., & Dietrich, R. U. (2023). Techno-economic assessment of different aviation fuel supply pathways including LH2 and LCH4 and the influence of the carbon source. *Energy Conversion and Management*, *293*, 117483. Advance online publication. DOI: 10.1016/j.enconman.2023.117483

Rabiee, A., Keane, A., & Soroudi, A. (2021). Green hydrogen: A new flexibility source for security constrained scheduling of power systems with renewable energies. *International Journal of Hydrogen Energy*, *46*(37), 19270–19284. Advance online publication. DOI: 10.1016/j.ijhydene.2021.03.080

Rafik, M., Bahnasse, A., Khiat, A., Bouattane, O., Ouajji, H., & khaili, M. E. (2019). Towards a smart energy sharing in micro smart grid adopting SDN approach. *Procedia Computer Science*, *151*, 717–724. Advance online publication. DOI: 10.1016/j.procs.2019.04.095

Rahmani, K., Ahriz, A., & Bouaziz, N. (2022). Development of a New Residential Energy Management Approach for Retrofit and Transition, Based on Hybrid Energy Sources. *Sustainability (Basel)*, *14*(7), 4069. Advance online publication. DOI: 10.3390/su14074069

Refaat, A., Elbaz, A., Khalifa, A.-E., Mohamed Elsakka, M., Kalas, A., & Hegazy Elfar, M. (2024). Performance evaluation of a novel self-tuning particle swarm optimization algorithm-based maximum power point tracker for porton exchange membrane fuel cells under different operating conditions. *Energy Conversion and Management*, *301*, 118014. DOI: 10.1016/j.enconman.2023.118014

Romeo, L. M., Cavana, M., Bailera, M., Leone, P., Peña, B., & Lisbona, P. (2022a). Non-stoichiometric methanation as strategy to overcome the limitations of green hydrogen injection into the natural gas grid. *Applied Energy*, *309*, 118462. Advance online publication. DOI: 10.1016/j.apenergy.2021.118462

Romeo, L. M., Cavana, M., Bailera, M., Leone, P., Peña, B., & Lisbona, P. (2022b). Non-stoichiometric methanation as strategy to overcome the limitations of green hydrogen injection into the natural gas grid. *Applied Energy*, *309*, 118462. Advance online publication. DOI: 10.1016/j.apenergy.2021.118462

Rosenstiel, A., Monnerie, N., Dersch, J., Roeb, M., Pitz-paal, R., & Sattler, C. (2021). Electrochemical hydrogen production powered by pv/csp hybrid power plants: A modelling approach for cost optimal system design. *Energies*, *14*(12), 3437. Advance online publication. DOI: 10.3390/en14123437

Ruhnau, O., & Schiele, J. (2023). Flexible green hydrogen: The effect of relaxing simultaneity requirements for project design, economics, and power sector emissions. *Energy Policy*, *182*, 113763. Advance online publication. DOI: 10.1016/j.enpol.2023.113763

Shahrtash, F. M. (2023). Sustainable Development, Electricity Generation and Renewable Energy substitution in Middle East Countries and Cooperation of Iran. *Sustainable Environment*, *8*(1), p57. Advance online publication. DOI: 10.22158/se.v8n1p57

Slaoui, Z., Faquir, S., Zidani, K. A., & Yahyaouy, A. (2023). Green Hydrogen for Industry: A Proposed Approach to Manage the Flow of Energy in Industrial Sector in Morocco. *Lecture Notes in Networks and Systems*, *578*, 119–126. Advance online publication. DOI: 10.1007/978-981-19-7660-5_11

Slimani, J., Kadrani, A., El Harraki, I., & Ezzahid, E. H. (2021). Renewable Energy Development in Morocco: Reflections on Optimal Choices through Long-term Bottom-up Modeling. *International Conference on Electrical, Computer, and Energy Technologies, ICECET 2021*. DOI: 10.1109/ICECET52533.2021.9698630

Smith, M. (2020). Morocco aims for global green hydrogen role. In *Petroleum Economist*. Issue December-January.

Stamatakis, E., Perwög, E., Garyfallos, E., Millán, M. S., Zoulias, E., & Chalkiadakis, N. (2022). Hydrogen in Grid Balancing: The European Market Potential for Pressurized Alkaline Electrolyzers. *Energies*, *15*(2), 637. Advance online publication. DOI: 10.3390/en15020637

Stanley, S., Ryan, L., & Flynn, D. (2023). Strategies to increase grid flexibility for an isolated system with over 80% renewable electricity in 2030. *International Conference on the European Energy Market, EEM, 2023-June*. DOI: 10.1109/EEM58374.2023.10161913

Stöckl, F., Schill, W. P., & Zerrahn, A. (2021). Optimal supply chains and power sector benefits of green hydrogen. *Scientific Reports*, *11*(1), 14191. Advance online publication. DOI: 10.1038/s41598-021-92511-6 PMID: 34244545

Szemat-Vielma, W., Scheibz, J., Kasraoui, N., & Al-Omar, F. (2023). *Sun Powered Green Hydrogen - A Comparative Analysis from the Kingdoms Of Morocco and Saudi Arabia*. DOI: 10.2118/214375-MS

Szima, S., Arnaiz del Pozo, C., Cloete, S., Chiesa, P., Jiménez Alvaro, Á., Cormos, A. M., & Amini, S. (2021). Finding synergy between renewables and coal: Flexible power and hydrogen production from advanced IGCC plants with integrated CO2 capture. *Energy Conversion and Management*, *231*, 113866. Advance online publication. DOI: 10.1016/j.enconman.2021.113866

Timilsina, G. R., & Deluque Curiel, I. F. (2023). Subsidy removal, regional trade and CO2 mitigation in the electricity sector in the Middle East and North Africa region. *Energy Policy*, *177*, 113557. Advance online publication. DOI: 10.1016/j.enpol.2023.113557

Tleubergenova, A., Abuov, Y., Danenova, S., Khoyashov, N., Togay, A., & Lee, W. (2023). Resource assessment for green hydrogen production in Kazakhstan. *International Journal of Hydrogen Energy*, *48*(43), 16232–16245. Advance online publication. DOI: 10.1016/j.ijhydene.2023.01.113

Topalović, Z., Haas, R., & Sayer, M. (2024). Economic benefits of PHS and Li-ion storage. Study cases: Austria and Bosnia and Herzegovina. *Applied Energy*, *362*, 122988. Advance online publication. DOI: 10.1016/j.apenergy.2024.122988

Touili, S., Bouaichi, A., Alami Merrouni, A., Amrani, A., El Amrani, A., El Hassouani, Y., & Messaoudi, C. (2022). Performance analysis and economic competiveness of 3 different PV technologies for hydrogen production under the impact of arid climatic conditions of Morocco. *International Journal of Hydrogen Energy*, *47*(74), 31596–31613. Advance online publication. DOI: 10.1016/j.ijhydene.2022.07.088

Utomo, O., Abeysekera, M., & Ugalde-Loo, C. E. (2021). Optimal operation of a hydrogen storage and fuel cell coupled integrated energy system. *Sustainability (Basel)*, *13*(6), 3525. Advance online publication. DOI: 10.3390/su13063525

Valverde, L., Rosa, F., Bordons, C., & Guerra, J. (2016). Energy Management Strategies in hydrogen Smart-Grids: A laboratory experience. *International Journal of Hydrogen Energy*, *41*(31), 13715–13725. Advance online publication. DOI: 10.1016/j.ijhydene.2016.05.279

Wang, D., Muratori, M., Eichman, J., Wei, M., Saxena, S., & Zhang, C. (2018a). Quantifying the flexibility of hydrogen production systems to support large-scale renewable energy integration. *Journal of Power Sources*, *399*, 383–391. Advance online publication. DOI: 10.1016/j.jpowsour.2018.07.101

Wang, D., Muratori, M., Eichman, J., Wei, M., Saxena, S., & Zhang, C. (2018b). Quantifying the flexibility of hydrogen production systems to support large-scale renewable energy integration. *Journal of Power Sources*, *399*, 383–391. Advance online publication. DOI: 10.1016/j.jpowsour.2018.07.101

Weimann, L., Gabrielli, P., Boldrini, A., Kramer, G. J., & Gazzani, M. (2021). Optimal hydrogen production in a wind-dominated zero-emission energy system. *Advances in Applied Energy, 3*, 100032. Advance online publication. DOI: 10.1016/j.adapen.2021.100032

Wodon, Q. (2023). Education to climate change: A missing ingredient in climate action plans for MENA countries? *Orient (Leverkusen), 64*(1).

Xu, D., & Hussain, J. (2023). Globalization, institutions, and environmental quality in Middle East and North African countries. *Environmental Science and Pollution Research International, 30*(26), 68951–68968. Advance online publication. DOI: 10.1007/s11356-023-27348-9 PMID: 37129812

Xu, H., & Shi, B. (2022). Design and System Evaluation of Mixed Waste Plastic Gasification Process Based on Integrated Gasification Combined Cycle System. *Processes (Basel, Switzerland), 10*(3), 499. Advance online publication. DOI: 10.3390/pr10030499

Yan, Y., Ibrahim, R. L., Al-Faryan, M. A. S., & Oke, D. M. (2023). Embracing Eco-Digitalization and Green Finance Policies for Sustainable Environment: Do the Engagements of Multinational Corporations Make or Mar the Target for Selected MENA Countries? *Sustainability (Basel), 15*(15), 12046. Advance online publication. DOI: 10.3390/su151512046

Young, K. E. (2015). The Limits of Gulf Arab Aid: Energy Markets and Foreign Policy. *EUCERS "Reflections" Working Paper Series, 1.*

Zakaria, O., Oumaima, C., Achraf, E. K., Ahmed, A. E., & Driss, Z. (2024). Energy Efficiency Measures and Technical-Economic Study of a Photovoltaic Self-consumption Installation in Kenitra, Morocco. *Lecture Notes in Electrical Engineering, 1141 LNEE.* DOI: 10.1007/978-981-97-0126-1_28

Zarkik, A., & Ouhnini, A. (2022). Integrated transition toward sustainability: The case of the water-energy-food nexus in Morocco. In *Food Security and Climate-Smart Food Systems*. Building Resilience for the Global South., DOI: 10.1007/978-3-030-92738-7_8

Zhao, H., Kamp, L. M., & Lukszo, Z. (2022). The potential of green ammonia production to reduce renewable power curtailment and encourage the energy transition in China. *International Journal of Hydrogen Energy, 47*(44), 18935–18954. Advance online publication. DOI: 10.1016/j.ijhydene.2022.04.088

Zhou, Y., Panteli, M., Moreno, R., & Mancarella, P. (2018). System-level assessment of reliability and resilience provision from microgrids. *Applied Energy*, *230*, 374–392. Advance online publication. DOI: 10.1016/j.apenergy.2018.08.054

ENDNOTES

1 ANRE.MA

2 ANRE.MA

3 ANRE.MA

4 ANRE.MA

5 ANRE.MA

6 Lafarge Hlocim (2018). Renewable energies available on:https://www.lafargeholcim.ma/sites/morocco/files/atoms/files/rdd-_2016_pages.pdf

7 Eco Life (2014). Lafarge Morocco meets 85% of its electricity needs with wind power. Published on 12/26/2014. Available on:https://lavieeco.com/news/economie/lafarge-maroc-satisfait-85-de-ses-besoins-en-electricite-par-leolien-32295.html

8 OCP Group. Energy brochure dated 01/11/2016. Available on the website:https://www.ocpgroup.ma/sites/default/files/alldocs/Plaquette-Energie.pdf

9 SONASID. Sustainable development. Available on the website:http://www.sonasid.ma/Developpement-durable/Environnement#.WpaWMR3FLIU

10 The Economist (2016). Steel industry: Sonasid finalizes a deal with Nareva Holding. Edition No: 4742 dated 01/04/2016. Available on:https://leconomiste.com/article/995923-siderurgie-sonasid-finalise-un-deal-avec-nareva-holding

11 Medias24 (2016). Maghreb Steel will switch to wind energy in 2018. Published on 02/12/2016. Available on:https://www.medias24.com/MAROC/ECONOMIE/ENTREPRISES/169070-Maghreb-Steel-va-passer-al-energie-eolienne-en-2018.html

12 Cements from Morocco. The daily environment. Available on:https://www.cimentsdumaroc.com/fr/lenvironnement-au-quotidien

13 Managem Group (2018). Annual Report. Available on:https://www.managemgroup.com/content/search?SearchText=%C3%A9olien

Compilation of References

Abbass, M., & Akhai, S. (2025). Eco-friendly geopolymer bricks synthesized from alkali-activated sugarcane bagasse ash and rice husk ash. *Industrial Crops and Products*, *233*, 121410. DOI: 10.1016/j.indcrop.2025.121410

Abbass, M., & Akhai, S. (2025). Exploring the sustainability potential of geopolymer concrete with coal bottom ash and basalt fibres. *Multiscale and Multidisciplinary Modeling. Experiments and Design*, *8*(3), 199.

Abdelbasir, M. S., & Esmail Shalan, A. (2019). Intriguing Properties and Applications of Functional Magnetic Materials. Functional Materials.

Abdeljaber, O., Avci, O., Kiranyaz, M. S., Boashash, B., Sodano, H., & Inman, D. J. (2018). 1-D CNNs for structural damage detection: Verification on a structural health monitoring benchmark data. *Neurocomputing*, *275*, 1308–1317. DOI: 10.1016/j.neucom.2017.09.069

Abdel-Karim, N., Preston, E., Moura, J., & Coleman, T. (2018a). Variable energy resource capacity contributions consistent with reserve margin and reliability. *IEEE Power and Energy Society General Meeting, 2018-January*. DOI: 10.1109/PESGM.2017.8274569

Abderahmane, A., & Daniel, L. (2023). Measurement Criteria for the Magnetic Characterization of Magnetic Materials. *IEEE Transactions on Instrumentation and Measurement*, *72*, 1–18. DOI: 10.1109/TIM.2023.3322510

Abeyratne, S. A., & Monfared, R. P. (2016). Blockchain ready manufacturing supply chain using distributed ledger. *International Journal of Research in Engineering and Technology*, *5*(9), 1–10.

Ada, N.. (2021). Blockchain technology for enhancing traceability and efficiency in automobile supply chain—A case study. *Sustainability*, *13*(24), 13667. Advance online publication. DOI: 10.3390/su132413667

Adeel, Aabir., Muhammad, Yasin, Naz., Shazia, Shukrullah. (2022). Synthesis of Carbon Nanotubes via Plasma Arc Discharge Method. Advances in bioinformatics and biomedical engineering book series, DOI: 10.4018/978-1-7998-8398-2.ch005

Adeleke, A. A., Petrus, N., Ayuba, S., Yahya, A. M., Ikubanni, P. P., Okafor, I. S., Emmanuel, S. S., Olosho, A. I., & Adesibikan, A. A. (2023). Nigerian Biomass for Bioenergy Applications: A Review on the Potential and Challenges. In *Journal of Renewable Materials* (Vol. 11, Issue 12). DOI: 10.32604/jrm.2023.043915

Adeli, K., Nachtane, M., Faik, A., Rachid, A., Tarfaoui, M., & Saifaoui, D. (2024). A deep learning-enhanced framework for sustainable hydrogen production from solar and wind energy in the Moroccan Sahara: Coastal regions focus. *Energy Conversion and Management, 302*, 118084. Advance online publication. DOI: 10.1016/j.enconman.2024.118084

Adun, H., Ishaku, H. P., & Ogungbemi, A. T. (2022). Towards Renewable energy targets for the Middle East and North African region: A decarbonization assessment of energy-water nexus. *Journal of Cleaner Production, 374*, 133944. Advance online publication. DOI: 10.1016/j.jclepro.2022.133944

Aghahosseini, A., Bogdanov, D., & Breyer, C. (2020). Towards sustainable development in the MENA region: Analysing the feasibility of a 100% renewable electricity system in 2030. *Energy Strategy Reviews, 28*, 100466. Advance online publication. DOI: 10.1016/j.esr.2020.100466

Aguilar, G. V. (2019). *Sol-Gel Method - Design and Synthesis of New Materials with Interesting Physical*. Chemical and Biological Properties.

Aiswarya, P., David, Cortie., Erin, Smyth., Anton, P., Le, Brun., Francesca, Iacopi. (2024). Epitaxial graphene growth on cubic silicon carbide on silicon with high temperature neutron reflectometry: An operando study. *RSC Advances, 14*(5), 3232–3240. Advance online publication. DOI: 10.1039/D3RA08289J

Ait Errouhi, A., Choukai, O., Oumimoun, Z., & El Mokhi, C. (2022). Energy efficiency measures and technical-economic study of a photovoltaic self-consumption installation at ENSA Kenitra, Morocco. *Energy Harvesting and Systems, 9*(2), 193–201. Advance online publication. DOI: 10.1515/ehs-2021-0081

Akarsu, B., & Serdar Genç, M. (2022). Optimization of electricity and hydrogen production with hybrid renewable energy systems. *Fuel, 324*, 124465. DOI: 10.1016/j.fuel.2022.124465

Akhai, S. (2024). Trends and Environmental Impact of Paper Consumption: A Prognostic Scenario for the Indian Market by 2030-A Case Study. In *International Conference on Interdisciplinary Approaches in Civil Engineering for Sustainable Development* (Vol. 464, pp. 11-18). Springer Nature Singapore. DOI: 10.1007/978-981-97-0910-6_2

Akhai, S. (2023). Navigating the Potential Applications and Challenges of Intelligent and Sustainable Manufacturing for a Greener Future. *Evergreen, 10*(4), 2237–2243. DOI: 10.5109/7160899

Akhai, S., Bansal, S. A., & Singh, S. (2020). A critical review of thermal insulators from natural materials for energy saving in buildings. *Journal of Critical Reviews, 7*(19), 278–283.

Akhai, S., Singh, V. P., & John, S. (2016). Human performance in industrial design centers with small unit air conditioning systems. *Journal of Advanced Research in Production Industrial Engineering, 3*(2), 5–11.

Akhai, S., Srivastava, P., Sharma, V., & Bhatia, A. (2021). Investigating weld strength of AA8011-6062 alloys joined via friction-stir welding using the RSM approach. []. IOP Publishing.]. *Journal of Physics: Conference Series, 1950*(1), 012016. DOI: 10.1088/1742-6596/1950/1/012016

Akinosho, T. D., Oyedele, L. O., Bilal, M., Ajayi, A. O., Delgado, M. D., Akinade, O. O., & Ahmed, A. A. (2020). Deep learning in the construction industry: A review of present status and future innovations. *Journal of Building Engineering, 32*, 101827. DOI: 10.1016/j.jobe.2020.101827

Al-Alawy, A. F., Al-Abodi, E. E., & Kadhim, R. M. (2018). Synthesis and Characterization of Magnetic Iron Oxide Nanoparticles by Co-Precipitation Method at Different Conditions. *Journal of Engineering, 24*(10), 60–72. DOI: 10.31026/j.eng.2018.10.05

Alam, S., & Yao, N. (2019). The impact of preprocessing steps on the accuracy of machine learning algorithms in sentiment analysis. *Computational & Mathematical Organization Theory, 25*(3), 319–335. DOI: 10.1007/s10588-018-9266-8

Alaoui, C. (2019). Review and assessment of offshore renewable energy resources in morocco' coastline. In *Cogent Engineering* (Vol. 6, Issue 1). DOI: 10.1080/23311916.2019.1654659

Alex, A., Sugunan, S. K., & George, G. 1 Synthesis of Micro and Nanostructured Materials by Laser-Assisted Methods Towards Energy Production, Conversion, and Storage Applications. In Materials for Energy Production, Conversion, and Storage (pp. 1-20). CRC Press.

Allocca, L., Baltador, C., Morone, A., Attanasio, C., Cirillo, C., & Gambardella, U. (2014). Magnetic properties of ferromagnetic thin films obtained by Magneto-Optic Kerr Effect measurements. *2014 Third Mediterranean Photonics Conference*, 1-3. DOI: 10.1109/MePhoCo.2014.6866477

Almulla, Y., Ramirez, C., Joyce, B., Huber-Lee, A., & Fuso-Nerini, F. (2022). From participatory process to robust decision-making: An Agriculture-water-energy nexus analysis for the Souss-Massa basin in Morocco. *Energy for Sustainable Development*, *70*, 314–338. Advance online publication. DOI: 10.1016/j.esd.2022.08.009

Alnaimat, F., Alhseinat, E., Banat, F., & Mittal, V. (2016). Electromagnetic–mechanical desalination: Mathematical modeling. *Desalination*, *380*, 75–84. DOI: 10.1016/j.desal.2015.11.012

Al-Najjar, H., El-Khozondar, H. J., Pfeifer, C., & Al Afif, R. (2022). Hybrid grid-tie electrification analysis of bio-shared renewable energy systems for domestic application. *Sustainable Cities and Society*, *77*, 103538. Advance online publication. DOI: 10.1016/j.scs.2021.103538

Alomar, M., Arshad, N., Irshad, M. S., Abdelmohsen, S. A., Ahmed, I., Alhoshani, N., & Alqarni, A. S. (2023). Multi-Walled CNTs/BiVO4 Heterostructures for Solar-Driven Evaporation System and Efficient Photocatalytic Activity against Oxytetracycline. *Nanomaterials (Basel, Switzerland)*, *13*(20), 2764. DOI: 10.3390/nano13202764 PMID: 37887915

Alperin, H.A., Belson, H.S., Clark, A.E., Krall, A.D., & Pickart, S.J. (1970). Forecast on Magnetic Materials.

Ammari, M., Chentouf, M., Ammari, M., & Ben Allal, L. (2022). Assessing National Progress in Achieving the Sustainable Development Goals: A Case Study of Morocco. *Sustainability (Basel)*, *14*(23), 15582. Advance online publication. DOI: 10.3390/su142315582

Amorim, W. C. S., Cupertino, A. F., Pereira, H. A., & Mendes, V. F. (2024). On sizing of battery energy storage systems for PV plants power smoothing. *Electric Power Systems Research*, *229*, 110114. Advance online publication. DOI: 10.1016/j.epsr.2024.110114

Amrani, S. E., Alami Merrouni, A., Touili, S., & Dekhissi, H. (2023). A multi-scenario site suitability analysis to assess the installation of large scale photovoltaic-hydrogen production units. Case study: Eastern Morocco. *Energy Conversion and Management*, *295*, 117615. Advance online publication. DOI: 10.1016/j.enconman.2023.117615

Amran, M., Debbarma, S., & Ozbakkaloglu, T. (2021). Fly ash-based eco-friendly geopolymer concrete: A critical review of the long-term durability properties. *Construction & Building Materials*, *270*, 121857. DOI: 10.1016/j.conbuildmat.2020.121857

Andricacos, P. C., & Robertson, N. (1998). Future directions in electroplated materials for thin-film recording heads. *IBM Journal of Research and Development*, *42*(5), 671–680. DOI: 10.1147/rd.425.0671

Angkiriwang, R., Pujawan, I. N., & Santosa, B. (2014). Managing uncertainty through supply chain flexibility: Reactive vs. proactive approaches. *Production & Manufacturing Research*, *2*(1), 50–70. DOI: 10.1080/21693277.2014.882804

Anner, M. (2020). *Abandoned? The impact of Covid-19 on workers and businesses at the bottom of global garment supply chains*. Center for Global Workers' Rights. https://www.workersrights.org/wp-content/uploads/2020/03/Abandoned-Penn-State-WRC-Report.pdf

An, R., Chu, D., Xie, C., Pan, S., & Yang, Z. (2025). Navigating the Landscape of Next-Generation Nonlinear Optical Materials Discovery: Opportunities and Challenges Driven by Artificial Intelligence. *Accounts of Materials Research*. Advance online publication. DOI: 10.1021/accountsmr.5c00245

Anton, S. R., & Sodano, H. A. (2007). A review of power harvesting using piezoelectric materials (2003–2006). *Smart Materials and Structures*, *16*(3), R1–R21. DOI: 10.1088/0964-1726/16/3/R01

An, Y., Chatzi, E., Sim, S., Laflamme, S., Blachowski, B., & Ou, J. (2019). Recent progress and future trends on damage identification methods for bridge structures. *Structural Control and Health Monitoring*, *26*(10). Advance online publication. DOI: 10.1002/stc.2416

Armijo, J., & Philibert, C. (2020). Flexible production of green hydrogen and ammonia from variable solar and wind energy: Case study of Chile and Argentina. *International Journal of Hydrogen Energy*, *45*(3), 1541–1558. Advance online publication. DOI: 10.1016/j.ijhydene.2019.11.028

Ashwell, M. L. (2017). The digital transformation of intelligence analysis. *Journal of Financial Crime*, *24*(3), 393–411. DOI: 10.1108/JFC-03-2017-0020

Asiaban, S., Kayedpour, N., Samani, A. E., Bozalakov, D., De Kooning, J. D. M., Crevecoeur, G., & Vandevelde, L. (2021). Wind and solar intermittency and the associated integration challenges: A comprehensive review including the status in the belgian power system. In *Energies* (Vol. 14, Issue 9). DOI: 10.3390/en14092630

Aşıkoğlu, A. (2023). Cost analysis of insulation materials used to increase energy performance in buildings with Net Present Value method. *Journal of Sustainable Construction Materials and Technologies*, *8*(2), 134–145. DOI: 10.47481/jscmt.1270831

Atta, N.F., Galal, A., & El-Ads, E.H. (2015). Graphene — A Platform for Sensor and Biosensor Applications.

Ayad, H., Sari-Hassoun, S. E., Usman, M., & Ahmad, P. (2023). The impact of economic uncertainty, economic growth and energy consumption on environmental degradation in MENA countries: Fresh insights from multiple thresholds NARDL approach. *Environmental Science and Pollution Research International*, *30*(1), 1806–1824. Advance online publication. DOI: 10.1007/s11356-022-22256-w PMID: 35921013

Azam, A., Ahmed, A., Wang, H., Wang, Y., & Zhang, Z. (2021). Knowledge structure and research progress in wind power generation (WPG) from 2005 to 2020 using CiteSpace based scientometric analysis. *Journal of Cleaner Production*, *295*, 126496. DOI: 10.1016/j.jclepro.2021.126496

Azzerboni, B., Asti, G., Pareti, L., & Ghidini, M. (2007). *Magnetic Nanostructures in Modern Technology: Spintronics*. Magnetic MEMS and Recording.

Badran, A. (2017). Climate change and water science policy in management. In *Water, Energy and Food Sustainability in the Middle East: The Sustainability Triangle*. DOI: 10.1007/978-3-319-48920-9_1

Bahetta, S., Dahhou, N., & Hasnaoui, R. (2021). Comparative effectiveness of environmental regulation instruments: Case of the Moroccan electricity mix. *AIMS Energy*, *9*(5), 1097–1112. Advance online publication. DOI: 10.3934/energy.2021050

Bahou, S. (2023). Techno-economic assessment of a hydrogen refuelling station powered by an on-grid photovoltaic solar system: A case study in Morocco. *International Journal of Hydrogen Energy*, *48*(61), 23363–23372. Advance online publication. DOI: 10.1016/j.ijhydene.2023.03.220

Bai, Y., Guo, W., Wang, J., Xu, Z., Wang, S., Zhao, Q., & Zhou, J. (2022). Geopolymer bricks prepared by MSWI fly ash and other solid wastes: Moulding pressure and curing method optimisation. *Chemosphere*, *307*, 135987. DOI: 10.1016/j.chemosphere.2022.135987 PMID: 35970219

Bakar, N. A., Peszynski, K., & Sundram, V. P. K. (2016). Abridgment of traditional procurement and e-procurement: Definitions, tools and benefits. *Journal of Emerging Economies and Islamic Research*, *4*(1), 74–91. DOI: 10.24191/jeeir.v4i1.6284

Bakier, Y. M., Ghali, M., & Zahra, W. K. (2020). Highly sensitive fluorescent detection of pyridine using small size carbon quantum dots derived from folic acid. *Journal of Physics. D, Applied Physics*, *53*(40), 405103. DOI: 10.1088/1361-6463/ab985e

Bakkari, M., Bossoufi, B., El Kafazi, I., Bouderbala, M., & Karim, M. (2024). A review of wind energy potential in Morocco: New challenges and perspectives. In *Wind Engineering* (Vol. 48, Issue 1). DOI: 10.1177/0309524X231200582

Bandorf, R., Bloche, A., Ortner, K., Lüthje, H., & Jung, T. D. (2007). High Rate Deposition of Magnetic Material by Gas Flow Sputtering. *Plasma Processes and Polymers*, *4*(S1), 4. DOI: 10.1002/ppap.200730504

Bansal, K., Singh, J., & Dhaliwal, A. S. (2022, February). Synthesis and characterization of Graphene Oxide and its reduction with different reducing agents. [). IOP Publishing.]. *IOP Conference Series. Materials Science and Engineering*, *1225*(1), 012050. DOI: 10.1088/1757-899X/1225/1/012050

Bao, Y., Chen, Z., Wei, S., Xu, Y., Tang, Z., & Li, H. (2019). The State of the Art of Data Science and Engineering in Structural Health Monitoring. *Engineering (Beijing)*, *5*(2), 234–242. DOI: 10.1016/j.eng.2018.11.027

Barnard, A.S., & Snook, I.K. (2008). Thermal stability of graphene edge structure and graphene nanoflakes. *The Journal of chemical physics, 128 9*, 094707 .

Bartlett, A. P., Belanger, R., Amman, M. S., & Venus, D. (2019). In situ determination of the anisotropy field in ferromagnetic films using magnetic susceptibility measurements by MOKE. *Journal of Magnetism and Magnetic Materials*, *477*, 49–54. DOI: 10.1016/j.jmmm.2019.01.037

Basma, B. M., Piétroy, D., Boukhari, M. I., Chahbi, Z., Blanchet, T., Chatelon, J. P., Capraro, S., & Rousseau, J. J. (2024). On thermal characterization method of integrated magnetic components. *Scientific Reports*, *14*(1), 14. DOI: 10.1038/s41598-024-54588-7 PMID: 38418493

Behrens, S. (2011). Preparation of functional magnetic nanocomposites and hybrid materials: Recent progress and future directions. *Nanoscale*, *3*(3), 877–892. DOI: 10.1039/C0NR00634C PMID: 21165500

Belaid, F., Boukrami, E., & Amine, R. (2021). Renewable Energy in the MENA Region: Key Challenges and Lessons Learned. In *Lecture Notes in Energy* (Vol. 82). DOI: 10.1007/978-3-030-71403-1_1

Belmahdi, B., Louzazni, M., & El Bouardi, A. (2022). Comparative optimization of global solar radiation forecasting using machine learning and time series models. *Environmental Science and Pollution Research International*, *29*(10), 14871–14888. Advance online publication. DOI: 10.1007/s11356-021-16760-8 PMID: 34625894

Benckendorff, P., & Zehrer, A. (2013). A NETWORK ANALYSIS OF TOURISM RESEARCH. *Annals of Tourism Research*, *43*, 121–149. DOI: 10.1016/j.annals.2013.04.005

Bensouda, M., & Benali, M. (2023). The Role of Institutional Pressure and Dynamic Capabilities in Promoting Energy Efficiency Practices: Evidence from the Moroccan Manufacturing Sector. *International Journal of Energy Economics and Policy*, *13*(1), 352–361. Advance online publication. DOI: 10.32479/ijeep.13697

Bhagat, M., Singh, S., Suresh, S., Arisutha, S., Verma, S., & Kansal, S. K. (2023). Utilization of value-added products from fly ash: An industrial waste. In *Advanced Materials from Recycled Waste* (pp. 239–254). Elsevier. DOI: 10.1016/B978-0-323-85604-1.00003-2

Bienhaus, F., & Haddud, A. (2018). Procurement 4.0: Factors influencing the digitisation of procurement and supply chains. *Business Process Management Journal*, *24*(4), 965–984. DOI: 10.1108/BPMJ-06-2017-0139

Bigerna, S. (2024). Connectedness analysis of oil price shocks, inflation, and exchange rate for the MENA region countries. *Resources Policy*, *88*, 104344. Advance online publication. DOI: 10.1016/j.resourpol.2023.104344

Biswas, C., & Lee, Y. H. (2011). Graphene Versus Carbon Nanotubes in Electronic Devices. *Advanced Functional Materials*, *21*(20), 21. DOI: 10.1002/adfm.201101241

Boharb, A., Allouhi, A., El-houari, H., El Markhi, H., Jamil, A., & Kousksou, T. (2022). Energy audit method applied to tertiary buildings: Case study of a University campus. *AIMS Energy*, *10*(3), 506–532. Advance online publication. DOI: 10.3934/energy.2022025

Bolotin, K. I., Ghahari, F., Shulman, M. D., Stormer, H. L., & Kim, P. (2009). Observation of the fractional quantum Hall effect in graphene. *Nature*, *462*(7270), 196–199. DOI: 10.1038/nature08582 PMID: 19881489

Boston Consulting Group. (2023). *How to mitigate supply chain risk*. https://www.bcg.com/publications/2023/how-to-mitigate-supply-chain-risk

Boullouz, S., Faquir, S., & Yahyaouy, A. (2022). A New Approach to Manage the Energy Flow in a Hybrid Renewable Energy System Based on Green Hydrogen. *Lecture Notes in Networks and Systems, 455 LNNS*. DOI: 10.1007/978-3-031-02447-4_58

Bourgoin, M. (2003). *Études en magnétohydrodynamique, application à l'effet dynamo* (Doctoral dissertation, Ecole normale supérieure de lyon-ENS LYON).

Bouyghrissi, S., Murshed, M., Jindal, A., Berjaoui, A., Mahmood, H., & Khanniba, M. (2022). The importance of facilitating renewable energy transition for abating CO2 emissions in Morocco. *Environmental Science and Pollution Research International, 29*(14), 20752–20767. Advance online publication. DOI: 10.1007/s11356-021-17179-x PMID: 34741744

Bowler, A. L., Pound, M. P., & Watson, N. J. (2022). A review of ultrasonic sensing and machine learning methods to monitor industrial processes. *Ultrasonics, 124*, 106776. DOI: 10.1016/j.ultras.2022.106776 PMID: 35653984

Brandon-Jones, A., & Kauppi, K. (2018). Examining the antecedents of the technology acceptance model within e-procurement. *International Journal of Operations & Production Management, 38*(1), 22–42. DOI: 10.1108/IJOPM-06-2015-0346

Broz, M. E., Cook, R. F., & Whitney, D. L. (2006). Microhardness, toughness, and modulus of Mohs scale minerals. *The American Mineralogist, 91*(1), 135–142. DOI: 10.2138/am.2006.1844

Burgei, W. A., Pechan, M. J., & Jaeger, H. (2003). A simple vibrating sample magnetometer for use in a materials physics course. *American Journal of Physics, 71*(8), 825–828. DOI: 10.1119/1.1572149

Burn, S., Hoang, M., Zarzo, D., Olewniak, F., Campos, E., Bolto, B., & Barron, O. (2015). Desalination techniques—A review of the opportunities for desalination in agriculture. *Desalination, 364*, 2–16. DOI: 10.1016/j.desal.2015.01.041

Cai, R., Yang, H., He, J., & Zhu, W. (2009). The effects of magnetic fields on water molecular hydrogen bonds. *Journal of Molecular Structure, 938*(1-3), 15–19. DOI: 10.1016/j.molstruc.2009.08.037

Cao, J., Wu, H., & Zou, Y. (2024). Advancing sustainable development: The impact of energy transition on accelerating technological progress. *Economic Change and Restructuring, 57*(2), 70. Advance online publication. DOI: 10.1007/s10644-024-09664-3

Cárdenas, B., Swinfen-Styles, L., Rouse, J., Hoskin, A., Xu, W., & Garvey, S. D. (2021). Energy storage capacity vs. renewable penetration: A study for the UK. *Renewable Energy*, *171*, 849–867. Advance online publication. DOI: 10.1016/j.renene.2021.02.149

Caspani, S., Magalhães, R., Araújo, J. P., & Sousa, C. T. (2020). Magnetic Nanomaterials as Contrast Agents for MRI. *Materials (Basel)*, *13*(11), 13. DOI: 10.3390/ma13112586 PMID: 32517085

Centenaro, E., Lany, M., Meyer, T., & Gasparin, E. (2019). *Structural Health Monitoring Solutions for Offshore Platforms*. OMC-2019-1031.

Chadge, R. B., Sunheriya, N., Giri, J. P., & Shrivastava, N. (2022). Experimental investigation of different electrical configurations and topologies for Photovoltaic system. *Materials Today: Proceedings*, *57*, 316–320. DOI: 10.1016/j.matpr.2022.03.097

Chae, S. H., & Lee, Y. H. (2014). Carbon nanotubes and graphene towards soft electronics. *Nano Convergence*, *1*(1), 1. DOI: 10.1186/s40580-014-0015-5 PMID: 28936384

Chander, S. (2024). *Harnessing Nanotechnology for Advancements in Magnetism: Current Trends and Future Prospects. International Journal of Innovative Science and Research Technology*. IJISRT.

Chandrashekhar, A. U. (2013). Studies on potential usage of fly ash in steel fiber reinforced concrete for rigid pavements (Doctoral dissertation, National Institute of Technology Karnataka, Surathkal).

Chapman, J., Truong, V. K., Elbourne, A., Gangadoo, S., Cheeseman, S., Rajapaksha, P., Latham, K., Crawford, R. J., & Cozzolino, D. (2020). Combining Chemometrics and Sensors: Toward New Applications in Monitoring and Environmental Analysis. *Chemical Reviews*, *120*(13), 6048–6069. DOI: 10.1021/acs.chemrev.9b00616 PMID: 32364371

Chauhan, A. (2024). *Exploring the Principles of Optics Through Modern Technologies*. DOI: 10.22541/au.172660512.22821656/v1

Cha, Y.-J., Ali, R., Lewis, J., & Büyük ztürk, O. (2024). Deep learning-based structural health monitoring. *Automation in Construction*, *161*, 105328. DOI: 10.1016/j.autcon.2024.105328

Cheng, W., Ping, Y., Zhang, Y., Chuang, K., & Liu, Y. (2013). Magnetic resonance imaging (MRI) contrast agents for tumor diagnosis. *Journal of Healthcare Engineering*, *4*(1), 23–45. DOI: 10.1260/2040-2295.4.1.23 PMID: 23502248

Chen, J., Zhu, J., Zhang, D., Lattery, D. M., Li, M., Wang, J., & Wang, X. (2016). Time-Resolved Magneto-Optical Kerr Effect of Magnetic Thin Films for Ultrafast Thermal Characterization. *The Journal of Physical Chemistry Letters*, *7*(13), 2328–2332. DOI: 10.1021/acs.jpclett.6b00945 PMID: 27269127

Chennaif, M., Maaouane, M., Zahboune, H., Elhafyani, M., & Zouggar, S. (2022). Tri-objective techno-economic sizing optimization of Off-grid and On-grid renewable energy systems using Electric system Cascade Extended analysis and system Advisor Model. *Applied Energy*, *305*, 117844. Advance online publication. DOI: 10.1016/j.apenergy.2021.117844

Cherkasova, E. G. (2023a). SPAIN'S ENERGY SECURITY IN THE CONTEXT OF THE ALGERIAN-MOROCCAN CONFLICT. *World Economy and International Relations*, *67*(9), 48–57. Advance online publication. DOI: 10.20542/0131-2227-2023-67-9-48-57

Chidolue, O., Ngozichukwu, B., Ibekwe, K. I., Illojianya, V. I., Fafure, A. V., & Daudu, C. D. (2024). Decarbonization strategies in energy-intensive industries: Cases from Canada, USA, and Africa. *International Journal of Science and Research Archive*, *11*(1), 361–370. Advance online publication. DOI: 10.30574/ijsra.2024.11.1.0065

Choukri, K., Naddami, A., & Hayani, S. (2017a). Renewable energy in emergent countries: Lessons from energy transition in Morocco. *Energy, Sustainability and Society*, *7*(1), 25. Advance online publication. DOI: 10.1186/s13705-017-0131-2

Choukri, K., Naddami, A., & Hayani, S. (2018). Evaluation of the reserve capacity in a grid supplied by intermittent energy sources. *IET Renewable Power Generation*, *12*(4), 399–406. Advance online publication. DOI: 10.1049/iet-rpg.2017.0119

Choukri, K., Naddami, A., & Hayani, S. M. (2017b). Deep analysis of wind variability and smoothing effect in Moroccan wind farms. *Wind Engineering*, *41*(4), 272–281. Advance online publication. DOI: 10.1177/0309524X17709731

Chu, Z., Xu, B., & Liang, J. (2023). Direct Application of Carbon Nanotubes (CNTs) Grown by Chemical Vapor Deposition (CVD) for Integrated Circuits (ICs) Interconnection: Challenges and Developments. *Nanomaterials (Basel, Switzerland)*, *13*(20), 2791. DOI: 10.3390/nano13202791 PMID: 37887942

Cid-Samamed, A., Correa-Duarte, M. Á., Mariño-López, A., & Diniz, M. S. (2023). Exposure to oxidized multi-walled CNTs can lead to oxidative stress in the asian freshwater clam Corbicula fluminea (Müller, 1774). *International Journal of Molecular Sciences*, *24*(22), 16122. DOI: 10.3390/ijms242216122 PMID: 38003314

Cloete, S., & Hirth, L. (2020). Flexible power and hydrogen production: Finding synergy between CCS and variable renewables. *Energy*, *192*, 116671. Advance online publication. DOI: 10.1016/j.energy.2019.116671

Coey, J. M. (2010). *Magnetism and magnetic materials*. Cambridge university press.

Costa, A. A., Arantes, A., & Tavares, L. V. (2013). Evidence of the impacts of public e-procurement: The Portuguese experience. *Journal of Purchasing and Supply Management*, *19*(4), 238–246. DOI: 10.1016/j.pursup.2013.07.004

Coufal, H., Dhar, L., & Mee, C. D. (2006). Materials for Magnetic Data Storage: The Ongoing Quest for Superior Magnetic Materials. *MRS Bulletin*, *31*(5), 374–378. DOI: 10.1557/mrs2006.96

Cruickshank, D.B. (2011). Microwave Materials for Wireless Applications.

Cuc, S. (2023). Unlocking the potential of blockchain technology in the textile and fashion industry. *FinTech*, *2*(2), 311–326. DOI: 10.3390/fintech2020018

Culbert, J., Hobert, A., Jahn, N., Haupka, N., Schmidt, M., Donner, P., & Mayr, P. (2024). *Reference Coverage Analysis of OpenAlex compared to Web of Science and Scopus* (No. arXiv:2401.16359). arXiv. https://arxiv.org/abs/2401.16359

da Costa, T. P., Quinteiro, P., Arroja, L., & Dias, A. C. (2022). Environmental performance of different end-of-life alternatives of wood fly ash by a consequential perspective. *Sustainable Materials and Technologies*, *32*, e00411. DOI: 10.1016/j.susmat.2022.e00411

Dadashi, Z., Mahmoudi, A., & Rashidi, S. (2022). Capacity and strategies of energy production from renewable sources in Arab countries until 2030: a review from renewable energy potentials to environmental issues. In *Environmental Science and Pollution Research* (Vol. 29, Issue 32). DOI: 10.1007/s11356-022-20544-z

Dai, D., Zhang, Y., Yang, S., Kong, W., Yang, J., & Zhang, J. (2024). Recent Advances in Functional Materials for Optical Data Storage. *Molecules (Basel, Switzerland)*, *29*(1), 254. DOI: 10.3390/molecules29010254 PMID: 38202837

Dai, L., Chang, D. W., Baek, J., & Lu, W. (2012). Carbon nanomaterials for advanced energy conversion and storage. *Small*, *8*(8), 1130–1166. DOI: 10.1002/smll.201101594 PMID: 22383334

Dan, A. K., Bhattacharjee, D., Ghosh, S., Behera, S. S., Bindhani, B. K., Das, D., & Parhi, P. K. (2021). Prospective utilization of coal fly ash for making advanced materials. In *Clean Coal Technologies: Beneficiation, Utilization, Transport Phenomena and Prospective* (pp. 511–531). Springer International Publishing. DOI: 10.1007/978-3-030-68502-7_20

Dariyal, P., Sehrawat, M., Dhakate, S. R., & Singh, B. P. (2023). Carbon nanotubes and graphene for conducting wires. In *Emerging applications of carbon nanotubes and graphene* (pp. 105–129). CRC Press. DOI: 10.1201/9781003231943-6

Darmawan, C. C., Ye, L., Samani, M. K., Fu, Y., & Liu, J. (2017). Graphene-CNT hybrid material as potential thermal solution in electronics applications. *2017 IMAPS Nordic Conference on Microelectronics Packaging (NordPac)*, 190-193. DOI: 10.1109/NORDPAC.2017.7993191

Das, A., & Wazeer, A. (2022). Graphene and carbon nanotubes (CNTs)-based biosensor for life sciences applications. In *Next Generation Smart Nano-Bio-Devices* (pp. 61–79). Springer Nature Singapore.

Das, D. K., Patnaik, P., Nayak, S. K., & Barala, M. (2023). Assessing the Effect of Electronic Pseudopotentials and Relativistic Treatments on the Structural and Electrical Properties of GaN: A DFT Study. *Physics and Chemistry of Solid State*, *24*(4), 714–721. DOI: 10.15330/pcss.24.4.714-721

Das, D. K., Patnaik, P., Nayak, S. K., & Barala, M. (2024). Electronic and half-metallic behavior of copper-doped silicon carbide armchair single-walled nanotubes using density functional theory. *International Journal of Computational Materials Science and Engineering*, •••, 2450011.

de Arce, R., Mahía, R., Medina, E , & Escribano, G. (2012). A simulation of the economic impact of renewable energy development in Morocco. *Energy Policy*, *46*, 335–345. Advance online publication. DOI: 10.1016/j.enpol.2012.03.068

De, S., & Son, M. (2023). REVUE BIBLIOGRAPHIQUE ET ÉTUDE DE CAS SUR LE DESSALEMENT D'EAU DE MER ET SON IMPACT SUR LE MILIEU MARIN.

Deaver, B.S., Bucelot, T.J., & Finley, J.J. (1978). SQUIDs for measuring the magnetic properties of materials.

Deininge, F., & Gren, A. (2022). Green jobs for women can combat the climate crisis and boost equality. *Development and a Changing Climate*.

Derouez, F., & Ifa, A. (2024). Sustainable Food Security: Balancing Desalination, Climate Change, and Population Growth in Five Arab Countries Using ARDL and VECM. *Sustainability (Basel)*, *16*(6), 2302. Advance online publication. DOI: 10.3390/su16062302

Devi, M., Wang, H., Moon, S., Sharma, S., & Strauss, V. (2023). Laser-Carbonization–A powerful tool for micro-fabrication of patterned electronic carbons. *Advanced Materials*, *35*(38), 2211054. DOI: 10.1002/adma.202211054 PMID: 36841955

Dhiman, S., Seema, S., & Akhai, S. (2025). Performance evaluation of OPC 43 concrete incorporating micro silica and recycled ceramic tile waste as partial cement and coarse aggregate replacements. *Asian Journal of Civil Engineering*, 1-19.

Diez, A., Khoa, N. L. D., Makki Alamdari, M., Wang, Y., Chen, F., & Runcie, P. (2016). A clustering approach for structural health monitoring on bridges. *Journal of Civil Structural Health Monitoring*, *6*(3), 429–445. DOI: 10.1007/s13349-016-0160-0

Dispenza, G., Sergi, F., Napoli, G., Antonucci, V., & Andaloro, L. (2019). Evaluation of hydrogen production cost in different real case studies. *Journal of Energy Storage*, *24*, 100757. Advance online publication. DOI: 10.1016/j.est.2019.100757

Dong, G., & Liu, H. (Eds.). (2018). *Feature engineering for machine learning and data analytics*. CRC press.

Dong, K., Zhang, S., & Wang, Q. (2015). A new class of ion-ion interaction: Z-bond. *Science China. Chemistry*, *58*(3), 495–500. DOI: 10.1007/s11426-014-5147-2

Duan, W. H., Quek, S. T., & Wang, Q. (2005). Free vibration analysis of piezoelectric coupled thin and thick annular plate. *Journal of Sound and Vibration*, *281*(1-2), 119 ?"139. DOI: 10.1016/j.jsv.2004.01.009

Duan, X.-F., Zhou, L.-P., Li, H.-R., Hu, S.-J., Zheng, W., Xu, X., Zhang, R., Chen, X., Guo, X.-Q., & Sun, Q.-F. (2023). Excited-Multimer Mediated Supramolecular Upconversion on Multicomponent Lanthanide-Organic Assemblies. *Journal of the American Chemical Society*, *145*(42), 23121–23130. DOI: 10.1021/jacs.3c06775 PMID: 37844009

Du, J., Wang, Q., & Song, T. (2018). Improvement of the measurement method of rock magnetic properties based on the scanning SQUID microscope with an in-situ magnetization/demagnetization field. *2018 IEEE International Magnetic Conference (INTERMAG)*, 1-1. DOI: 10.1109/INTMAG.2018.8508800

Dunlap, A., & Laratte, L. (2022). European Green Deal necropolitics: Exploring 'green' energy transition, degrowth & infrastructural colonization. *Political Geography, 97*, 102640. Advance online publication. DOI: 10.1016/j.polgeo.2022.102640

Dwivedi, A., & Jain, M. K. (2014). Fly ash–waste management and overview: A review. *Recent Research in Science and Technology, 6*(1).

El Hafdaoui, H., Khallaayoun, A., & Ouazzani, K. (2024). Long-term low carbon strategy of Morocco: A review of future scenarios and energy measures. In *Results in Engineering* (Vol. 21). DOI: 10.1016/j.rineng.2023.101724

El Hathat, Z., Sreedharan, V. R., Venkatesh, V. G., Zouadi, T., Arunmozhi, M., & Shi, Y. (2023). Modelling and analyzing the GHG emissions in the VUCA world: Evidence from tomato production in Morocco. *Journal of Cleaner Production, 382*, 134862. Advance online publication. DOI: 10.1016/j.jclepro.2022.134862

El Majaty, S., Touzani, A., & Kasseh, Y. (2023). Results and perspectives of the application of an energy management system based on ISO 50001 in administrative buildings - case of Morocco. *Materials Today: Proceedings, 72*, 3233–3237. Advance online publication. DOI: 10.1016/j.matpr.2022.07.094

El Mir, H., King, S., Skote, M., & Perinpanayagam, S. (2023). *Machine Learning Requirements for the Airworthiness of Structural Health Monitoring Systems in Aircraft.* 1–13. https://www.icaf.aero/icaf2023/proceedings/documents/128.pdf

El-Emam, R. S., Ezzat, M. F., & Khalid, F. (2022). Assessment of hydrogen as a potential energy storage for urban areas' PV-assisted energy systems - Case study. *International Journal of Hydrogen Energy, 47*(62), 26209–26222. Advance online publication. DOI: 10.1016/j.ijhydene.2022.01.107

Elmansour, J., Hajjaji, A., Belhora, F., & Hendrick, P. (2022). Location of seawater pumped storage hydropower plants: Case of Morocco. *Materials Today: Proceedings, 66*, 45–57. Advance online publication. DOI: 10.1016/j.matpr.2022.03.145

El-Shamy, A. M., Farag, H. K., & Saad, W. (2017). Comparative study of removal of heavy metals from industrial wastewater using clay and activated carbon in batch and continuous flow systems. *Egyptian Journal of Chemistry, 60*(6), 1165–1175.

Eltouny, K., Gomaa, M., & Liang, X. (2023). Unsupervised Learning Methods for Data-Driven Vibration-Based Structural Health Monitoring: A Review. *Sensors (Basel), 23*(6), 3290. DOI: 10.3390/s23063290 PMID: 36992001

Ersoy, S. R., Terrapon-Pfaff, J., Ribbe, L., & Merrouni, A. A. (2021). Water scenarios modelling for renewable energy development in Southern Morocco. *Journal of Sustainable Development of Energy. Water and Environment Systems*, *9*(1), 0. Advance online publication. DOI: 10.13044/j.sdewes.d8.0335

Esplandiu, M. J., Baeza, M., Olivé-Monllau, R., Céspedes, F., & Bartrolí, J. (2011). *Development of tunable nanocomposites made from carbon nanotubes for electrochemical applications*. InTech.

Evans, M., & Guthrie, W. S. (2023, May). Reducing the Carbon Footprint of Concrete by Expanding the Use of Supplementary Cementitious Materials. In *2023 Intermountain Engineering, Technology and Computing (IETC)* (pp. 253-256). IEEE. DOI: 10.1109/IETC57902.2023.10152065

Evans, P. G., & Isaacs, E. D. (2006). Magnetic x-ray microdiffraction. *Journal of Physics. D, Applied Physics*, *39*(15), R245–R263. DOI: 10.1088/0022-3727/39/15/R01

Fadhillahanafi, N.M., Leong, K.Y., & Risby, M.S. (2014). STABILITY AND THERMAL CONDUCTIVITY CHARACTERISTICS OF CARBON NANOTUBE BASED NANOFLUIDS.

Fan, W., Zhang, L., & Liu, T. (2017). Graphene-CNT Hybrids for Environmental Applications.

Fan, W., & Qiao, P. (2011). Vibration-based Damage Identification Methods: A Review and Comparative Study. *Structural Health Monitoring*, *10*(1), 83–111. DOI: 10.1177/1475921710365419

Farhana, S., & Noordin, M.F. (2016). Technology and performance: Carbon nanotube (CNT) field effect transistor (FET) in VLSI circuit design. *2016 IEEE 7th Annual Ubiquitous Computing, Electronics & Mobile Communication Conference (UEMCON)*, 1-4.

Fatorachian, H., & Kazemi, H. (2021). Impact of Industry 4.0 on supply chain performance. *Production Planning and Control*, *32*(1), 63–81. DOI: 10.1080/09537287.2020.1712487

Fedosyuk, V. M. (2018). MAGNETIC NANOMATERIALS AND NANOSTRUCTURES. TRENDS OF DEVELOPMENT. Proceedings of the National Academy of Sciences of Belarus, Physical-Technical Series.

Ferreira, F.V., Cividanes, L.D., Brito, F.S., Menezes, B.R., Franceschi, W., Nunes Simonetti, E.A., & Thim, G.P. (2016). Functionalizing Graphene and Carbon Nanotubes: A Review. *Functionalizing Graphene and Carbon Nanotubes*.

Figueiredo, E., & Brownjohn, J. (2022). Three decades of statistical pattern recognition paradigm for SHM of bridges. *Structural Health Monitoring*, *21*(6), 3018–3054. DOI: 10.1177/14759217221075241

Filchakova, M., & Saik, V. (2021). *Single-walled carbon nanotubes: structure, properties, applications, and health & safety*. Tuball.

Fiorillo, F. (2004). Measurement and characterization of magnetic materials.

Fiorillo, F. (2010). Measurements of magnetic materials. *Metrologia*, *47*(2), S114–S142. DOI: 10.1088/0026-1394/47/2/S11

Firdous, A., Prakash Barala, C., Mathuria, P., & Bhakar, R. (2024). Extended power to hydrogen operations for enhanced grid flexibility in low carbon systems. *Energy Conversion and Management*, *301*, 117982. DOI: 10.1016/j.enconman.2023.117982

Fischer, P., Sanz-Hernández, D., Streubel, R., & Fernández-Pacheco, A. (2020). Launching a new dimension with 3D magnetic nanostructures. *APL Materials*, *8*(1), 010701. DOI: 10.1063/1.5134474

Flahaut, E., Evariste, L., Liné, C., Mottier, A., Mouchet, F., Barret, M., Pinelli, E., Galibert, A. M., Soula, B., Larue, C., & Gauthier, L. (2020). (Invited) Environmental Impact of Carbon Nanotubes & Graphene and Related Materials (GRMs). *ECS Meeting Abstracts, MA2021-01*, 591-591.

Flah, M., Nunez, I., Ben Chaabene, W., & Nehdi, M. L. (2021). Machine Learning Algorithms in Civil Structural Health Monitoring: A Systematic Review. *Archives of Computational Methods in Engineering*, *28*(4), 2621–2643. DOI: 10.1007/s11831-020-09471-9

Fleet, E. F., Chatraphorn, S., Wellstood, F. C., & Eylem, C. (2001). Determination of magnetic properties using a room-temperature scanning SQUID microscope. *IEEE Transactions on Applied Superconductivity*, *11*(1), 1180–1183. DOI: 10.1109/77.919559

Florio, G. (2022). Structural features of magnetic materials.

Fouaidi, M., Belfallah, K., Jamal, M., & Belouaggadia, N. (2023). Transient analysis of functionally graded graphene oxide powders-reinforced porous composite beams resting on elastic foundations using the reproducing kernel particle meshless method. *Engineering Analysis with Boundary Elements*, *146*, 460 ?"472. DOI: 10.1016/j.enganabound.2022.10.029

Fouaidi, M., Jamal, M., Zaite, A., & Belouaggadia, N. (2021). Bending analysis of functionally graded graphene oxide powder-reinforced composite beams using a meshfree method. *Aerospace Science and Technology*, *110*, 106479. DOI: 10.1016/j.ast.2020.106479

Fragkos, P. (2023). Assessing the energy system impacts of Morocco's nationally determined contribution and low-emission pathways. In *Energy Strategy Reviews* (Vol. 47). DOI: 10.1016/j.esr.2023.101081

Franco, V. R., & Gutfleisch, O. (2012). Magnetic Materials for Energy Applications. *Journal of the Minerals Metals & Materials Society*, *64*(7), 750–751. DOI: 10.1007/s11837-012-0348-7

Fu, X., Mol, W., Yuan, S., Xu, L., Song, J., Dong, K., & Jin, F. (2018). Measurement System of Ferromagnetic Film Magnetic Properties Based on Mazneto-optical Kerr Effect. 2018 37th Chinese Control Conference (CCC), 7370-7373.

Gabor, N.M., Xu, X., & Alden, J.S. (2009). Optoelectronics in Carbon Nanotube Photodiodes and Graphene Hetero-Interface Devices.

Gado, M. G., & Hassan, H. (2023). Potential of prospective plans in MENA countries for green hydrogen generation driven by solar and wind power sources. *Solar Energy*, *263*, 111942. Advance online publication. DOI: 10.1016/j.solener.2023.111942

Galimova, T., Fasihi, M., Bogdanov, D., & Breyer, C. (2023). Impact of international transportation chains on cost of green e-hydrogen: Global cost of hydrogen and consequences for Germany and Finland. *Applied Energy*, *347*, 121369. Advance online publication. DOI: 10.1016/j.apenergy.2023.121369

Garg, P., Gupta, B., Chauhan, A. K., Sivarajah, U., Gupta, S., & Modgil, S. (2021). Measuring the perceived benefits of implementing blockchain technology in the banking sector. *Technological Forecasting and Social Change*, *163*, 120407. Advance online publication. DOI: 10.1016/j.techfore.2020.120407

Garnett, J. C. (2019). Prediction of Mohs hardness with machine learning methods using compositional features. In *Machine Learning in Chemistry: Data-Driven Algorithms, Learning Systems, and Predictions* (pp. 23-48). American Chemical Society. DOI: 10.1021/bk-2019-1326.ch002

Ge, P., Hu, Q., Wu, Q., Dou, X., Wu, Z., & Ding, Y. (2020). Increasing operational flexibility of integrated energy systems by introducing power to hydrogen. *IET Renewable Power Generation*, *14*(3), 372–380. Advance online publication. DOI: 10.1049/iet-rpg.2019.0663

Geraldes, C., & Laurent, S. (2009). Classification and basic properties of contrast agents for magnetic resonance imaging. *Contrast Media & Molecular Imaging*, *4*(1), 1–23. DOI: 10.1002/cmmi.265 PMID: 19156706

Gerberich, W. W., Ballarini, R., Hintsala, E. D., Mishra, M., Molinari, J. F., & Szlufarska, I. (2015). Toward demystifying the Mohs hardness scale. *Journal of the American Ceramic Society*, *98*(9), 2681–2688. DOI: 10.1111/jace.13753

Ghadimi, P., Wang, C., Lim, M. K., & Heavey, C. (2019). Intelligent sustainable supplier selection using multi-agent technology: Theory and application for Industry 4.0 supply chains. *Computers & Industrial Engineering*, *127*, 588–600. DOI: 10.1016/j.cie.2018.10.050

Ghaemi, S., Li, X., & Mulder, M. (2023). Economic feasibility of green hydrogen in providing flexibility to medium-voltage distribution grids in the presence of local-heat systems. *Applied Energy*, *331*, 120408. Advance online publication. DOI: 10.1016/j.apenergy.2022.120408

Gharehbaghi, V. R., Noroozinejad Farsangi, E., Noori, M., Yang, T. Y., Li, S., Nguyen, A., Málaga-Chuquitaype, C., Gardoni, P., & Mirjalili, S. (2022). A Critical Review on Structural Health Monitoring: Definitions, Methods, and Perspectives. *Archives of Computational Methods in Engineering*, *29*(4), 2209–2235. DOI: 10.1007/s11831-021-09665-9

Gholami, A., Ameri, M., Zandi, M., & Gavagsaz Ghoachani, R. (2022). Electrical, thermal and optical modeling of photovoltaic systems: Step-by-step guide and comparative review study. *Sustainable Energy Technologies and Assessments*, *49*, 101711. DOI: 10.1016/j.seta.2021.101711

Giri, J., Ray, A., Dasgupta, S., Datta, D., & Bahadur, D. (2003). Investigation on Tc tuned nano particles of magnetic oxides for hyperthermia applications. *Bio-Medical Materials and Engineering*, *13*(4), 387–399. PMID: 14646053

Glanville, P., Fridlyand, A., Sutherland, B., Liszka, M., Zhao, Y., Bingham, L., & Jorgensen, K. (2022). Impact of Hydrogen/Natural Gas Blends on Partially Premixed Combustion Equipment: NOx Emission and Operational Performance. *Energies*, *15*(5), 1706. Advance online publication. DOI: 10.3390/en15051706

Glas, A. H., & Kleemann, F. C. (2016). The impact of Industry 4.0 on procurement and supply management: A conceptual and qualitative analysis. *International Journal of Business Management and Invention*, *5*(6), 55–66. http://www.ijbmi.org

Godwin, H. A., Nameth, C., Avery, D., Bergeson, L. L., Bernard, D., Beryt, E., Boyes, W. K., Brown, S. C., Clippinger, A. J., Cohen, Y., Doa, M. J., Hendren, C. O., Holden, P. A., Houck, K. A., Kane, A. B., Klaessig, F. C., Kodas, T. T., Landsiedel, R., Lynch, I., & Nel, A. E. (2015). Nanomaterial categorization for assessing risk potential to facilitate regulatory decision-making. *ACS Nano*, *9*(4), 3409–3417. DOI: 10.1021/acsnano.5b00941 PMID: 25791861

Goli, P. (2014). Thermal Management Using Graphene and Carbon-Nanotubes.

Gomez-Cabrera, A., & Escamilla-Ambrosio, P. J. (2022). Review of Machine-Learning Techniques Applied to Structural Health Monitoring Systems for Building and Bridge Structures. *Applied Sciences (Basel, Switzerland)*, *12*(21), 10754. DOI: 10.3390/app122110754

Gosling, J. H., Makarovsky, O., Wang, F., Cottam, N. D., Greenaway, M. T., Patanè, A., . . . Fromhold, T. M. (2020). Universal mobility characteristics of graphene originating from electron/hole scattering by ionised impurities. arXiv preprint arXiv:2005.07961.

Gosling, J. H., Makarovsky, O., Wang, F., Cottam, N. D., Greenaway, M. T., Patanè, A., Wildman, R. D., Tuck, C. J., Turyanska, L., & Fromhold, T. M. (2021). Universal mobility characteristics of graphene originating from charge scattering by ionised impurities. *Communications on Physics*, *4*(1), 30. DOI: 10.1038/s42005-021-00518-2

Grömping, U. (2009). Variable importance assessment in regression: Linear regression versus random forest. *The American Statistician*, *63*(4), 308–319. DOI: 10.1198/tast.2009.08199

Gude, V. G. (2016). Desalination and sustainability–an appraisal and current perspective. *Water Research*, *89*, 87–106. DOI: 10.1016/j.watres.2015.11.012 PMID: 26641014

Guerpinar, T., Guadiana, G., Ioannidis, P. A., Straub, N., & Henke, M. (2021). The current state of blockchain applications in supply chain management. In *ACM International Conference Proceeding Series* (pp. 168–175). Association for Computing Machinery. https://doi.org/DOI: 10.1145/3460537.3460568

Gunawan, T. A., & Monaghan, R. F. D. (2022). Techno-econo-environmental comparisons of zero- and low-emission heavy-duty trucks. *Applied Energy*, *308*, 118327. Advance online publication. DOI: 10.1016/j.apenergy.2021.118327

Guo, S. J., Xu, Y. S., Cheng, R., Zhou, J., & Chen, X. (2022). Thermal Hall effect in insulating quantum materials. *Innovation (Cambridge (Mass.))*, *3*(5), 3. DOI: 10.1016/j.xinn.2022.100290 PMID: 36039089

Gutfleisch, O., Willard, M. A., Brück, E., Chen, C. H., Sankar, S. G., & Liu, J. P. (2011). Magnetic Materials and Devices for the 21st Century: Stronger, Lighter, and More Energy Efficient. *Advanced Materials*, *23*(7), 23. DOI: 10.1002/adma.201002180 PMID: 21294168

Hafner, M., Raimondi, P. P., & Bonometti, B. (2023). Low-Carbon Energy Strategies in MENA Countries. In *Perspectives on Development in the Middle East and North Africa (MENA) Region: Vol. Part F1210*. DOI: 10.1007/978-3-031-30705-8_4

Hahn, G. J. (2020). Industry 4.0: A supply chain innovation perspective. *International Journal of Production Research*, *58*(5), 1425–1441. DOI: 10.1080/00207543.2019.1641642

Haidi, T., & Cheddadi, B. (2022). Wind energy integration in Africa: Development, impacts and barriers. *International Journal of Electrical and Computer Engineering*, *12*(5), 4614. Advance online publication. DOI: 10.11591/ijece.v12i5.pp4614-4622

Haines, M. B., Moore, S., & Adornetto, T. (2023). Suspending democratic (dis) belief: Nonliberal energy polities of solar power in Morocco and Tanzania. *Energy Research & Social Science*, *96*, 102942. Advance online publication. DOI: 10.1016/j.erss.2023.102942

Haines, W. (1984). VSM profiling of CoCr films: A new analytical technique. *IEEE Transactions on Magnetics*, *20*(5), 812–814. DOI: 10.1109/TMAG.1984.1063365

Hajalilou, A., Tavakoli, M., & Parvini, E. (2023). Type and Characteristics of Magnetic Materials.

Hajar, A. L., Ahmed, N., & Karim, C. (2023). Deployment of Safety Predictive Analytics to Prevent Workplace Incidents and Promote Event Reduction: A Machine Learning Approach Towards a Data-Driven Safety System. *Proceedings - 2023 3rd International Conference on Digital Data Processing, DDP 2023*. DOI: 10.1109/DDP60485.2023.00042

Hamza, K., Hamid, M., & Mouaad, B. (2022). A Review of Future Fuel Cell Electric Vehicles and Challenges Related to Morocco. *WSEAS Transactions on Power Systems*, *17*, 339–353. Advance online publication. DOI: 10.37394/232016.2022.17.34

Handfield, R., Jeong, S., & Choi, T. (2019). Emerging procurement technology: Data analytics and cognitive analytics. *International Journal of Physical Distribution & Logistics Management*, *49*(10), 972–1002. DOI: 10.1108/IJPDLM-11-2017-0348

Hany, E., Fouad, N., Abdel-Wahab, M., & Sadek, E. (2021). Investigating the mechanical and thermal properties of compressed earth bricks made by eco-friendly stabilization materials as partial or full replacement of cement. *Construction & Building Materials*, *281*, 122535. DOI: 10.1016/j.conbuildmat.2021.122535

Hao, D., Qi, L., Tairab, A. M., Ahmed, A., Azam, A., Luo, D., Pan, Y., Zhang, Z., & Yan, J. (2022). Solar energy harvesting technologies for PV self-powered applications: A comprehensive review. *Renewable Energy*, *188*, 678–697. DOI: 10.1016/j.renene.2022.02.066

Hao, Y., Zhang, S., Fang, B., Sun, F., Liu, H., & Li, H. (2022). A Review of Smart Materials for the Boost of Soft Actuators, Soft Sensors, and Robotics Applications. *Chinese Journal of Mechanical Engineering*, *35*(1), 1–16. DOI: 10.1186/s10033-022-00707-2

Harris, V. G. (2010). The role of magnetic materials in RF, microwave, and mm-wave devices: the quest for self-biased materials. *Proceedings of the IEEE 2010 National Aerospace & Electronics Conference*, 1-96. DOI: 10.1109/NAECON.2010.5712910

Hartman, H., & Williams, J. J. (Eds.). (2024). *Materials: An environmental primer*. Routledge. DOI: 10.4324/9781003486923

Hasnain, S., & Nayak, A. K. (2019). *Functionalization of Carbon Nanotubes*. Carbon Nanotubes for Targeted Drug Delivery. DOI: 10.1007/978-981-15-0910-0_4

Hasnain, S., & Nayak, A. K. (2019). *Regulatory Considerations of Carbon Nanotubes*. Carbon Nanotubes for Targeted Drug Delivery. DOI: 10.1007/978-981-15-0910-0_15

Hasnat, A., Moheman, A., Usmani, M. A., Bhawani, S. A., & Alotaibi, K. M. (2024). Carbon Nanotube and Graphene-Reinforced Vegetable Oil-Based Nanocomposites. In *Vegetable Oil-Based Composites: Processing, Properties and Applications* (pp. 179–190). Springer Nature Singapore. DOI: 10.1007/978-981-99-9959-0_10

Hassan, Q., Algburi, S., Sameen, A. Z., Salman, H. M., & Jaszczur, M. (2023). A review of hybrid renewable energy systems: Solar and wind-powered solutions: Challenges, opportunities, and policy implications. *Results in Engineering*, *20*, 101621. DOI: 10.1016/j.rineng.2023.101621

Hejazi, M., Santos Da Silva, S. R., Miralles-Wilhelm, F., Kim, S., Kyle, P., Liu, Y., Vernon, C., Delgado, A., Edmonds, J., & Clarke, L. (2023). Impacts of water scarcity on agricultural production and electricity generation in the Middle East and North Africa. *Frontiers in Environmental Science*, *11*, 1082930. Advance online publication. DOI: 10.3389/fenvs.2023.1082930

Hernández, A. R. R., Cruz, A. G., & Campos-Delgado, J. (2022). Chemical Vapor Deposition Synthesis of Graphene on Copper Foils. In Graphene-A Wonder Material for Scientists and Engineers. IntechOpen

Heshmati, M., & Amini, Y. (2020). An electromechanical finite element model for new CNTs-reinforced harvesters subjected to harmonic and random base excitations. *Iranian Journal of Science and Technology. Transaction of Mechanical Engineering*, *44*(1), 163 ?"181. DOI: 10.1007/s40997-018-0254-x

He, Y., Zhou, Z., Pan, Y., Chong, F., Wu, B., Xiao, K., & Li, H. (2024). Review of data security within energy blockchain: A comprehensive analysis of storage, management, and utilization. *High-Confidence Computing. High-Confidence Computing*, *100233*(3), 100233. Advance online publication. DOI: 10.1016/j.hcc.2024.100233

Hirohata, A., Sukegawa, H., Yanagihara, H., Žutić, I., Seki, T., Mizukami, S., & Swaminathan, R. (2015). Roadmap for Emerging Materials for Spintronic Device Applications. *IEEE Transactions on Magnetics*, *51*(10), 1–11. DOI: 10.1109/TMAG.2015.2457393

Hoffart, D. J., Tiburcio, J., de la Torre, A., Knight, L. K., & Loeb, S. J. (2008). Cooperative ion-ion interactions in the formation of interpenetrated molecules. *Angewandte Chemie International Edition*, *47*(1), 97–101. DOI: 10.1002/anie.200703019 PMID: 18008294

Holm, A. P., Pecharsky, V. K., Gschneidner, K.Jr, Rink, R., & Jirmanus, M. (2004). X-ray powder diffractometer for in situ structural studies in magnetic fields from 0 to 35 kOe between 2.2 and 315 K. *The Review of Scientific Instruments*, *75*(4), 1081–1088. DOI: 10.1063/1.1667253

Holmes, H. R., DeRouin, A. J., Wright, S., Riedemann, T. M., Lograsso, T. A., Rajachar, R. M., & Ghee Ong, K. (2014). Biodegradation and biocompatibility of mechanically active magnetoelastic materials. *Smart Materials and Structures*, *23*(9), 23. DOI: 10.1088/0964-1726/23/9/095036

Hoshi, Y., Naoe, M., & Yamanaka, S. (1982). High-rate, low-temperature sputtering method of facing-targets type and its application for deposition of magnetic films. *Electronics and Communications in Japan (Part I Communications)*, *65*(6), 106–112. DOI: 10.1002/ecja.4410650614

Hossin, M. A., Akorsu, P. K., & Abudu, H. (2023). Regional energy and carbon emissions concerns: Towards cleaner energy transition in Middle East and North Africa. *Social Sciences & Humanities Open*, *8*(1), 100705. Advance online publication. DOI: 10.1016/j.ssaho.2023.100705

Huang, Y.-M., Singh, K. J., Liu, A.-C., Lin, C.-C., Chen, Z., Wang, K., Lin, Y., Liu, Z., Wu, T., & Kuo, H.-C. (2020). Advances in Quantum-Dot-Based Displays. *Nanomaterials (Basel, Switzerland)*, *10*(7), 1327. DOI: 10.3390/nano10071327 PMID: 32640754

Hu, L., Zhang, R., & Chen, Q. (2014). Synthesis and assembly of nanomaterials under magnetic fields. *Nanoscale*, *6*(23), 14064–14105. DOI: 10.1039/C4NR05108D PMID: 25338267

Hu, M., Yue, N., & Groves, R. M. (2024). Damage Classification of a Bolted Connection using Guided Waves and Explainable Artificial Intelligence. *Procedia Structural Integrity*, *52*, 224–233. DOI: 10.1016/j.prostr.2023.12.023

Idllalène, S. (2021). The role of environmental Waqf in addressing climate change in the MENA region: A comparative law analysis. In *Climate Change Law and Policy in the Middle East and North Africa Region*.

Idrissi Kaitouni, S., Mghazli, M. O., Nait-Taour, A., Es-Sakali, N., El Mankibi, M., Ahachad, M., & Brigui, J. (2023). Empirical validation and analysis of the energy performance of an ecological Net Zero Energy Building (NZEB) in Benguerir-Morocco. *E3S Web of Conferences, 396*. DOI: 10.1051/e3sconf/202339604023

Idumah, C. I., & Hassan, A. (2016). Emerging trends in graphene carbon based polymer nanocomposites and applications. *Reviews in Chemical Engineering*, *32*(2), 223–264. DOI: 10.1515/revce-2015-0038

Imris, P., & Schenk, D. (2007). *Method and device for the desalination of in particular seawater with the aid of alternating fields of identical frequencies* (European Patent No. EP1828060A1). European Patent Office. https://patents.google.com/patent/EP1828060A1/en

Institute of Electrical and Electronics Engineers. (2019). *Proceedings of 2019 8th International Conference on Industrial Technology and Management: ICITM 2019, March 2-4, 2019, Cambridge, UK.*

IRENA. (2021). Morocco and IRENA Partner to Boost Renewables and Green Hydrogen Development. *Press Release.*

Islam, M. S., Kurawaki, J., Kusumoto, Y., Abdulla-Al-Mamun, M., & Mukhlish, M. Z. (2011). Hydrothermal Novel Synthesis of Neck-structured Hyperthermia-suitable Magnetic (Fe 3 O 4, γ-Fe 2 O 3 and α-Fe 2 O 3) Nanoparticles. *Journal of Scientific Research*, *4*, 99. DOI: 10.3329/jsr.v4i1.8727

Ismagilov, A. F., Khamitov, I. G., & Georgievsky, B. V. (2019). Comparative analysis of oil and gas basins belonging to the MENA region. *Neftyanoe Khozyaystvo - Oil Industry, 2019*(January). DOI: 10.24887/0028-2448-2019-1-20-23

Issa, B.A., & M. Obaidat, I. (2019). Magnetic Nanoparticles as MRI Contrast Agents. Magnetic Resonance Imaging.

Ivan, S. (2021, June 22). Filippo, Giannazzo., Rositsa, Yakimova. (2021). Epitaxial Graphene on 4H-SiC (0001) as a Versatile Platform for Materials Growth: Mini-Review. *Applied Sciences (Basel, Switzerland), 11*(13), 5784. Advance online publication. DOI: 10.3390/app11135784

Jahani, N., Sepehri, A., Vandchali, H. R., & Tirkolaee, E. B. (2021). Application of industry 4.0 in the procurement processes of supply chains: A systematic literature review. *Sustainability (Basel), 13*(14), 7520. Advance online publication. DOI: 10.3390/su13147520

Jalbout, E., Genge, L., & Muesgens, F. (2022). H2Europe: an analysis of long-term hydrogen import-potentials from the MENA region. *International Conference on the European Energy Market, EEM, 2022-September*. DOI: 10.1109/EEM54602.2022.9921055

Jamwal, A., Agrawal, R., Sharma, M., & Giallanza, A. (2021). Industry 4.0 technologies for manufacturing sustainability: A systematic review and future research directions. *Applied Sciences (Basel, Switzerland), 11*(12), 5725. DOI: 10.3390/app11125725

Jardine, A. K. S., Lin, D., & Banjevic, D. (2006). A review on machinery diagnostics and prognostics implementing condition-based maintenance. *Mechanical Systems and Signal Processing*, *20*(7), 1483–1510. DOI: 10.1016/j.ymssp.2005.09.012

Jensen, I. G., Wiese, F., Bramstoft, R., & Münster, M. (2020). Potential role of renewable gas in the transition of electricity and district heating systems. *Energy Strategy Reviews*, *27*, 100446. Advance online publication. DOI: 10.1016/j.esr.2019.100446

Jensen, J. A., Nikolov, S. I., Gammelmark, K. L., & Pedersen, M. H. (2006). Synthetic aperture ultrasound imaging. *Ultrasonics*, *44*, e5–e15. DOI: 10.1016/j.ultras.2006.07.017 PMID: 16959281

Jones, G. A. (1978). Magnetic contrast in the scanning electron microscope: An appraisal of techniques and their applications. *Journal of Magnetism and Magnetic Materials*, *8*(4), 263–285. DOI: 10.1016/0304-8853(78)90096-3

Jonnala, S. N., Gogoi, D., Devi, S., Kumar, M., & Kumar, C. (2024). A comprehensive study of building materials and bricks for residential construction. *Construction & Building Materials, 425*, 135931. DOI: 10.1016/j.conbuildmat.2024.135931

Jumas, D., Mohd-Rahim, F. A., Ariani, V., & Florisa, A. N. (2023). Obtaining professional skills in the public construction procurement system: A challenge from the viewpoint of Indonesian government worker. *Journal of Design and Built Environment, 23*(1), 48–58. DOI: 10.22452/jdbe.vol23no1.4

K, S. (2021). Climate Change Challenge 26th Conference of Parties (COP26) climate summit is crucial but may be disappointing? *Global Journal of Ecology*. DOI: 10.17352/gje.000051

Kachhi, B., Singh, K.K., Singh, A.K., & Sharma, D.K. (2021). Role of Carbon Nanotubes as Energy Storage Materials.

Kademaunga, C. K., & Phiri, J. (2019). Factors affecting successful implementation of electronic procurement in government institutions based on the technology acceptance model. *Open Journal of Business and Management, 7*(4), 1705–1714. DOI: 10.4236/ojbm.2019.74118

Karmakar, A., Ghosh, P., Banerjee, P. S., & De, D. (2023). ChainSure: Agent free insurance system using blockchain for Healthcare 4.0. *Intelligent Systems with Applications, 17*, 200177. Advance online publication. DOI: 10.1016/j.iswa.2023.200177

Karpen, W. L., Becker, R. J., & Arnold, W. (1998). Characterization of Electric and Magnetic Material Properties with Eddy-Current Measurements. *Nondestructive Testing and Evaluation, 15*(2), 93–107. DOI: 10.1080/10589759908952867

Kasseh, Y., Touzani, A., & El Majaty, S. (2023a). Exemplarity of the State for the Energy Efficiency of Buildings Institutional - Case of Morocco. *Nature Environment and Pollution Technology, 22*(1), 169–177. Advance online publication. DOI: 10.46488/NEPT.2023.v22i01.015

Kasseh, Y., Touzani, A., & El Majaty, S. (2023b). What public lighting governance model should be deployed in Moroccan cities for sustainable and efficient energy management? *Materials Today: Proceedings, 72*, 3244–3252. Advance online publication. DOI: 10.1016/j.matpr.2022.07.123

Katsikouli, P., Wilde, A. S., Dragoni, N., & Høgh-Jensen, H. (2021). On the benefits and challenges of blockchains for managing food supply chains. *Journal of the Science of Food and Agriculture, 101*(6), 2175–2181. DOI: 10.1002/jsfa.10883 PMID: 33063342

Katzmarek, D. A., Mancini, A., Maier, S. A., & Iacopi, F. (2023). Direct synthesis of nanopatterned epitaxial graphene on silicon carbide. *Nanotechnology*, *34*(40), 405302. DOI: 10.1088/1361-6528/ace369 PMID: 37399800

Khanam, J., Ahmed, M.F., Zaman, S , Sharmin, N., & Ahmed, S. (2022). Synthesis of nano-sized magnetic iron oxide by a simple and facile co-precipitation method.

Khan, J., & Jaafar, M. (2021). Reduction efficiencies of natural substances for reduced graphene oxide synthesis. *Journal of Materials Science*, *56*(33), 18477–18492. DOI: 10.1007/s10853-021-06492-y

Khan, M. U., & Shaida, M. A. (2023). Reduction mechanism of graphene oxide including various parameters affecting the C/O ratio. *Materials Today. Communications*, *36*, 106577. DOI: 10.1016/j.mtcomm.2023.106577

Kharrich, M., Mohammed, O. H., & Akherraz, M. (2019). Assessment of renewable energy sources in Morocco using economical feasibility technique. *International Journal of Renewable Energy Research*, *9*(4). Advance online publication. DOI: 10.20508/ijrer.v9i4.10181.g7791

Khouya, A. (2020). Levelized costs of energy and hydrogen of wind farms and concentrated photovoltaic thermal systems. A case study in Morocco. *International Journal of Hydrogen Energy*, *45*(56), 31632–31650. Advance online publication. DOI: 10.1016/j.ijhydene.2020.08.240

Kim, E. J., Desforges, A., Speyer, L., Ghanbaja, J., Gleize, J., Estellé, P., & Vigolo, B. (2017). Graphene for Water-Based Nanofluid Preparation: Effect of Chemical Modifications on Dispersion and Stability. *Journal of Nanofluids*, *6*(3), 603–613. DOI: 10.1166/jon.2017.1353

Kim, H., & Kim, T. (2002). Measurement of hardness on traditional ceramics. *Journal of the European Ceramic Society*, *22*(9-10), 1437–1445. DOI: 10.1016/S0955-2219(01)00457-5

Kim, M. C., & Zhu, Y. (2018). Scientometrics of Scientometrics: Mapping Historical Footprint and Emerging Technologies in Scientometrics. In Jibu, M., & Osabe, Y. (Eds.), *Scientometrics*. InTech., DOI: 10.5772/intechopen.77951

Kim, S., Park, J., Chung, W., Adams, D., & Lee, J. H. (2024). Techno-economic analysis for design and management of international green hydrogen supply chain under uncertainty: An integrated temporal planning approach. *Energy Conversion and Management*, *301*, 118010. DOI: 10.1016/j.enconman.2023.118010

Kimura, F., & Kimura, T. (2018). Magnetically textured powders—An alternative to single-crystal and powder X-ray diffraction methods. *CrystEngComm*, *20*(7), 861–872. DOI: 10.1039/C7CE01305A

Kinloch, I. A., Suhr, J., Lou, J., Young, R. J., & Ajayan, P. M. (2018). Composites with carbon nanotubes and graphene: An outlook. *Science*, *362*(6414), 547–553. DOI: 10.1126/science.aat7439 PMID: 30385571

Kiranyaz, S., Avci, O., Abdeljaber, O., Ince, T., Gabbouj, M., & Inman, D. J. (2021). 1D convolutional neural networks and applications: A survey. *Mechanical Systems and Signal Processing*, *151*, 107398. DOI: 10.1016/j.ymssp.2020.107398

Kirby, A. (2023). Exploratory Bibliometrics: Using VOSviewer as a Preliminary Research Tool. *Publications/MDPI*, *11*(1), 10. DOI: 10.3390/publications11010010

Kitanovski, A. (2020). Energy Applications of Magnetocaloric Materials. *Advanced Energy Materials*, •••, 10.

Kohashi, T. (2018). Magnetization Analysis by Spin-Polarized Scanning Electron Microscopy. *Scanning*, *2018*, 2018. DOI: 10.1155/2018/2420747 PMID: 29507649

Koirala, B., Hers, S., Morales-España, G., Özdemir, Ö., Sijm, J., & Weeda, M. (2021). Integrated electricity, hydrogen and methane system modelling framework: Application to the Dutch Infrastructure Outlook 2050. *Applied Energy*, *289*, 116713. Advance online publication. DOI: 10.1016/j.apenergy.2021.116713

Kollipara, P. S., Li, J., & Zheng, Y. (2020). Optical Patterning of Two-Dimensional Materials. *Research*, *2020*, 2020/6581250. Advance online publication. DOI: 10.34133/2020/6581250 PMID: 32043085

Kong, J. (2023, August). Electrical Control in the Chemical Vapor Deposition Synthesis of Graphene and Related Materials. In Electrochemical Society Meeting Abstracts 243 (No. 13, pp. 1302-1302). The Electrochemical Society, Inc. DOI: 10.1149/MA2023-01131302mtgabs

Kongkajun, N., Laitila, E. A., Ineure, P., Prakaypan, W., Cherdhirunkorn, B., & Chakartnarodom, P. (2020). Soil-cement bricks produced from local clay brick waste and soft sludge from fiber cement production. *Case Studies in Construction Materials*, *13*, e00448. DOI: 10.1016/j.cscm.2020.e00448

Kornak, R., Niznansky, D., Haimann, K., Tylus, W., & Maruszewski, K. (2005). Synthesis of magnetic nanoparticles via the sol-gel technique. *Materials Science Poland*.

Korucuk, S., Tirkolaee, E. B., Aytekin, A., Karabasevic, D., & Karamaşa, Ç. (2023). Agile supply chain management based on critical success factors and most ideal risk reduction strategy in the era of Industry 4.0: Application to plastic industry. *Operations Management Research : Advancing Practice Through Research, 16*(4), 1698–1719. DOI: 10.1007/s12063-023-00360-5

Kousksou, T., Allouhi, A., Belattar, M., Jamil, A., El Rhafiki, T., Arid, A., & Zeraouli, Y. (2015). Renewable energy potential and national policy directions for sustainable development in Morocco. In *Renewable and Sustainable Energy Reviews* (Vol. 47). DOI: 10.1016/j.rser.2015.02.056

Kouzzi, S., Ganaoui, M. El, Kheiri, A., & Bouardi, A. El. (2022). Local materials, between regional diversity and energy performance to build efficient buildings in a transborder context (morocco-spain). *E3S Web of Conferences, 353*. DOI: 10.1051/e3sconf/202235301003

Kramer, G. (2016). Electronic public procurement as a tool of reform in South African public procurement. *African Public Procurement Law Journal, 3*(1), 1–43. DOI: 10.14803/3-1-16

Krings, A., Boglietti, A., Cavagnino, A., & Sprague, S. (2017). Soft Magnetic Material Status and Trends in Electric Machines. *IEEE Transactions on Industrial Electronics, 64*(3), 2405–2414. DOI: 10.1109/TIE.2016.2613844

Krings, A., & Soulard, J. (2015). Experimental characterization of magnetic materials for electrical machine applications. *2015 IEEE Workshop on Electrical Machines Design, Control and Diagnosis (WEMDCD)*, 85-89. DOI: 10.1109/WEMDCD.2015.7194514

Kryder, M. H. (2003). Future trends in magnetic storage technology. Joint NAPMRC 2003. *Digest of Technical Papers*, •••, 68.

Kumar, P. (2023). Carbon Nanotubes and Graphene in Photovoltaics. In Emerging Applications of Carbon Nanotubes and Graphene (pp. 131-159). CRC Press. DOI: 10.1201/9781003231943-7

Kumar, A., & Kumar, S. (2023). Light-controlled receptors for environmentally and biologically relevant anions. *Chemical Engineering Journal, 474*, 145493. DOI: 10.1016/j.cej.2023.145493

Kumaravel, A., & Sathyamoorthi, S. (2024). Carbon nanotubes-based sensors to detect environmental contaminants. In *Nanotechnology-Based Sensors for Detection of Environmental Pollution* (pp. 349–370). Elsevier. DOI: 10.1016/B978-0-443-14118-8.00018-8

Kumar, C. S., & Mohammad, F. (2011). Magnetic nanomaterials for hyperthermia-based therapy and controlled drug delivery. *Advanced Drug Delivery Reviews, 63*(9), 789–808. DOI: 10.1016/j.addr.2011.03.008 PMID: 21447363

Kumar, H., Wadhwa, A. S., Akhai, S., & Kaushik, A. (2024). Parametric analysis, modeling and optimization of the process parameters in electric discharge machining of aluminium metal matrix composite. *Engineering Research Express, 6*(2), 025542. DOI: 10.1088/2631-8695/ad4ba9

Kumar, H., Wadhwa, A. S., Akhai, S., & Kaushik, A. (2024). Parametric optimization of the machining performance of Al-SiCp composite using combination of response surface methodology and desirability function. *Engineering Research Express, 6*(2), 025505. DOI: 10.1088/2631-8695/ad38ff

Kumar, P., & Akhai, S. (2022). Effective energy management in smart buildings using VRV/VRF systems. In *Additive Manufacturing in Industry 4.0* (pp. 27–35). CRC Press. DOI: 10.1201/9781003360001-2

Kumbhakar, P., Chowde Gowda, C., & Tiwary, C. S. (2021). Advance Optical Properties and Emerging Applications of 2D Materials. *Frontiers in Materials, 8*, 721514. Advance online publication. DOI: 10.3389/fmats.2021.721514

Kwon, H., Yang, Y., Kim, G., Gim, D., & Ha, M. (2024). Anisotropy in magnetic materials for sensors and actuators in soft robotic systems. *Nanoscale, 16*(14), 6778–6819. DOI: 10.1039/D3NR05737B PMID: 38502047

Laaz, H., Rhimini, H., Nissabouri, S., & Mekkaoui, M. (2025). Signal Processing Analysis of Lamb Wave Propagation in Stainless Steel Plates with Thickness Defects: Implications for AI-Based Non-Destructive Testing. *2025 5th International Conference on Innovative Research in Applied Science, Engineering and Technology (IRASET)*, 1–10. DOI: 10.1109/IRASET64571.2025.11008279

Laaz, H., Mekkaoui, M., Nissabouri, S., & Rhimini, H. (2026). Investigation of Lamb Wave Excitation and Modal Behavior in Thin Plates for Non-Destructive Testing. In Barkaoui, A., Mabrouki, T., Degiovanni, A., Ben Zineb, T., & El Otmani, R. (Eds.), *Advances in Mechanics, Materials, and Energy* (pp. 117–136). Springer Nature Switzerland., DOI: 10.1007/978-3-032-12980-2_11

Lau, C. N., Xia, F., & Cao, L. (2020). Emergent quantum materials. *MRS Bulletin, 45*(5), 340–347. DOI: 10.1557/mrs.2020.125

Le, V.T., Ngo, C.L., Le, Q.T., Ngo, T.T., Nguyen, D.N., & Vu, M.T. (2013). Surface modification and functionalization of carbon nanotube with some organic compounds. *Advances in Natural Sciences: Nanoscience and Nanotechnology, 4*.

Lebrouhi, B. E., Lamrani, B., Zeraouli, Y., & Kousksou, T. (2024). Key challenges to ensure Morocco's sustainable transition to a green hydrogen economy. In *International Journal of Hydrogen Energy* (Vol. 49). DOI: 10.1016/j.ijhydene.2023.09.178

Lee, B., & Im, S. (2024). Energy, exergy, and exergoeconomic analyses of plastic waste-to-energy integrated gasification combined cycles with and without heat recovery at a gasifier. *Applied Energy*, *355*, 122243. Advance online publication. DOI: 10.1016/j.apenergy.2023.122243

Lee, I., & Lee, K. (2015). The Internet of Things (IoT): Applications, investments, and challenges for enterprises. *Business Horizons*, *58*(4), 431–440. DOI: 10.1016/j.bushor.2015.03.008

Lee, J. H., Avsar, A., Jung, J., Tan, J. Y., Watanabe, K., Taniguchi, T., Natarajan, S., Eda, G., Adam, S., Castro Neto, A. H., & Özyilmaz, B. (2015). Van der Waals force: A dominant factor for reactivity of graphene. *Nano Letters*, *15*(1), 319–325. DOI: 10.1021/nl5036012 PMID: 25493357

Libra, M., Petrik, T., Poulek, V., Tyukhov, I. I., & Kourim, P. (2021). Changes in the Efficiency of Photovoltaic Energy Conversion in Temperature Range With Extreme Limits. *IEEE Journal of Photovoltaics*, *11*(6), 1479–1484. DOI: 10.1109/JPHOTOV.2021.3108484

Li, H., Yang, S., Zha, Z., Fei, B., & Wang, X. (2023). Hygrothermal properties analysis of bamboo building envelope with different insulation systems in five climate zones. *Buildings (Basel, Switzerland)*, *13*(5), 1214. DOI: 10.3390/buildings13051214

Li, J., Li, X., Liang, S., Zhang, Y., Ye, Q., Zhang, L., Pan, C., Lv, R., Li, Q., Xiao, K., Yan, Y., Chen, Y., Hu, J., Hou, H., & Yang, J. (2021). Preparation of water-permeable bricks derived from fly ash by autoclaving. *Construction & Building Materials*, *271*, 121556. DOI: 10.1016/j.conbuildmat.2020.121556

Li, J., Maiti, A., Springer, M., & Gray, T. (2020). Blockchain for supply chain quality management: Challenges and opportunities in context of open manufacturing and industrial internet of things. *International Journal of Computer Integrated Manufacturing*, *33*(12), 1321–1355. DOI: 10.1080/0951192X.2020.1815853

Li, L., Kong, W., & Chen, F. (2022). Femtosecond laser-inscribed optical waveguides in dielectric crystals: A concise review and recent advances. *Advanced Photonics*, *4*(02). Advance online publication. DOI: 10.1117/1.AP.4.2.024002

Lindner, R. (2023). Green hydrogen partnerships with the Global South. Advancing an energy justice perspective on "tomorrow's oil.". *Sustainable Development (Bradford)*, *31*(2), 1038–1053. Advance online publication. DOI: 10.1002/sd.2439

Lindskog, H., Vegas, N., Babb, J., & Brege, S. (n.d.). Will electronic procurement change the public sector's purchasing behaviour? In *9th Annual ISOneWorld Conference: Issues in Information Systems*.

Lin, H., Gao, Y., Wang, X., Nan, T., Liu, M., Lou, J., Yang, G., Zhou, Z., Yang, X., Wu, J., Li, M., Hu, Z., & Sun, N. X. (2016). Integrated Magnetics and Multiferroics for Compact and Power-Efficient Sensing, Memory, Power, RF, and Microwave Electronics. *IEEE Transactions on Magnetics*, *52*(7), 1–8. DOI: 10.1109/TMAG.2016.2514982

Lin, L., Jiang, W., Xu, X., & Xu, P. (2020). A critical review of the application of electromagnetic fields for scaling control in water systems: Mechanisms, characterization, and operation. *NPJ Clean Water*, *3*(1), 25. DOI: 10.1038/s41545-020-0071-9

Lin, X., Chen, A., & Wang, S. (2011). Research progress in biocompatibility of magnetic iron oxide nanoparticles. *Kexue Tongbao*, *56*(26), 2223–2228. DOI: 10.1360/972010-2264

Liotine, M., & Ginocchio, D. (2019). The supply blockchain: Integrating blockchain technology within supply chain operations. In *Technology in supply chain management and logistics: Current practice and future applications* (pp. 57–89). Elsevier., DOI: 10.1016/B978-0-12-815956-9.00004-1

Li, Q., Guo, Z., Zhao, X., Zhang, T., Chen, J., & Wei, Y. (2020). One-pot synthesis of 2,2′-dipicolylamine derived highly photoluminescent nitrogen-doped carbon quantum dots for Fe 3+ detection and fingermark detection. *Nanotechnology*, *31*(33), 335501. DOI: 10.1088/1361-6528/ab8f4b PMID: 32357348

Li, Q., Luo, J., Li, Z., Rummeli, M. H., & Liu, L. (2024). Numerical investigation on the effect of gas-phase dynamics on graphene growth in chemical vapor deposition. *Journal of Applied Physics*, *135*(12), 125302. DOI: 10.1063/5.0199471

Li, S., Sun, F., He, H., & Chen, Y. (2017). Optimization for a Grid-connected Hybrid PV-wind-retired HEV Battery Microgrid System. *Energy Procedia*, *105*, 1634–1643. DOI: 10.1016/j.egypro.2017.03.532

Liu, S., Yan, L., Huang, J., Zhang, Q., & Zhou, B. (2022). Controlling upconversion in emerging multilayer core–shell nanostructures: From fundamentals to frontier applications. *Chemical Society Reviews*, *51*(5), 1729–1765. DOI: 10.1039/D1CS00753J PMID: 35188156

Liu, Y. (2002). Solving the Structural Mysteries of Magnetic Materials in TEM. *Microscopy and Microanalysis*, *8*(S02), 402–403. DOI: 10.1017/S1431927602100778

Li, Y., Zhu, H. Y., Hon, C. J., Jiang, Y., & Li, Y. (2013). Hydrothermal & Solvothermal Synthesis of Nanoscale Magnetic Materials (SCI). *Huaxue Jinzhan*, *25*, 276.

Li, Z., Ma, X., & Xin, H. (2017). Feature engineering of machine-learning chemisorption models for catalyst design. *Catalysis Today*, *280*, 232–238. DOI: 10.1016/j.cattod.2016.04.013

Li, Z., Sun, Y., Ge, S., Zhu, F., Yin, F., Gu, L., Yang, F., Hu, P., Chen, G., Wang, K., & Volinsky, A. A. (2023). An Overview of Synthesis and Structural Regulation of Magnetic Nanomaterials Prepared by Chemical Coprecipitation. *Metals*, *13*(1), 152. DOI: 10.3390/met13010152

Lizé, E., Rébillat, M., Mechbal, N., & Bolzmacher, C. (2018). Optimal dual-PZT sizing and network design for baseline-free SHM of complex anisotropic composite structures. *Smart Materials and Structures*, *27*(11), 115018. DOI: 10.1088/1361-665X/aad534

Lopez, G., Galimova, T., Fasihi, M., Bogdanov, D., & Breyer, C. (2023). Towards defossilised steel: Supply chain options for a green European steel industry. *Energy*, *273*, 127236. Advance online publication. DOI: 10.1016/j.energy.2023.127236

Lu, Q., Yang, Y., Xie, P., Chen, Y., Wu, J., & Lei, Z. (2024). Market Mechanism Designing and Scheduling Strategies of Auxiliary Services for Frequency Control Adapting to Energy Storage Participation. *Dianwang Jishu/Power System Technology, 48*(3). DOI: 10.13335/j.1000-3673.pst.2023.0104

Luckey, D., Fritz, H., Legatiuk, D., Peralta Abadía, J. J., Walther, C., & Smarsly, K. (2022). Explainable Artificial Intelligence to Advance Structural Health Monitoring. In Cury, A., Ribeiro, D., Ubertini, F., & Todd, M. D. (Eds.), *Structural Health Monitoring Based on Data Science Techniques* (Vol. 21, pp. 331–346). Springer International Publishing., DOI: 10.1007/978-3-030-81716-9_16

Luryi, S., Xu, J., & Zaslavsky, A. (2013). Future Trends in Microelectronics: Frontiers and Innovations.

M., Stetsenko., Rashad, Abaszade. (2023). X-ray phase analysis of carbon nanotubes obtained by the arc discharge method. DOI: 10.61640/ujeas.2023.0503

Machado, F.M., Lima, E.C., Jauris, I.M., & Adebayo, M.A. (2015). Carbon Nanomaterials for Environmental Applications.

Mahalingam, S., Omar, A., Manap, A., & Abd Rahim, N. (2023). Synthesis and applications of carbon-polymer composites and nanocomposite functional materials. In Functional Materials from Carbon, Inorganic, and Organic Sources (pp. 71-105). Woodhead Publishing. DOI: 10.1016/B978-0-323-85788-8.00020-3

Mahbubi, K., Zaki, A., & Nugroho, G. (2024). Bibliometric and Scientometric Trends in Structural Health Monitoring Using Fiber-Optic Sensors: A Comprehensive Review. *Journal of Civil and Hydraulic Engineering*, *2*(1), 51–64. DOI: 10.56578/jche020104

Mahesh, V., & Mangalasseri, A. S. (2024). Agglomeration effects of CNTs on the energy harvesting performance of multifield interactive magneto-electro-elastic/ nanocomposite unimorph smart beam. *Mechanics Based Design of Structures and Machines*, *52*(3), 1324 ?"1350. DOI: 10.1080/15397734.2022.2144886

Mahlooji, M., Gaudard, L., Ristic, B., & Madani, K. (2020). The importance of considering resoûrce availability restrictions in energy planning: What is the footprint of electricity generation in the Middle East and North Africa (MENA)? *The Science of the Total Environment*, *717*, 135035. Advance online publication. DOI: 10.1016/j.scitotenv.2019.135035 PMID: 31839284

Mahmoudpour, M., Ding, S., Lyu, Z., Ebrahimi, G., Du, D., Ezzati Nazhad Dolatabadi, J., Torbati, M., & Lin, Y. (2021). Aptamer functionalized nanomaterials for biomedical applications: Recent advances and new horizons. *Nano Today*, *39*, 101177. DOI: 10.1016/j.nantod.2021.101177

Mahyuni, L. P., Adrian, R., Darma, G. S., Krisnawijaya, N. N. K., Dewi, I. G. A. A. P., & Permana, G. P. L. (2020). Mapping the potentials of blockchain in improving supply chain performance. *Cogent Business & Management*, *7*(1), 1788329. Advance online publication. DOI: 10.1080/23311975.2020.1788329

Maiti, D., Tong, X., Mou, X., & Yang, K. (2019). Carbon-Based Nanomaterials for Biomedical Applications: A Recent Study. *Frontiers in Pharmacology, 9*. manpure, S., Singh, G., & Jewariya, M. (2023). Carbon Nanotubes and Graphene: The Novel Materials for Terahertz Detection. In Emerging Applications of Carbon Nanotubes and Graphene (pp. 275-288). CRC Press.

Majidi, S., Zeinali Sehrig, F., Farkhani, S. M., Soleymani Golоujeh, M., & Akbarzadeh, A. (2016). Current methods for synthesis of magnetic nanoparticles. *Artificial Cells, Nanomedicine, and Biotechnology*, *44*(2), 722–734. DOI: 10.3109/21691401.2014.982802 PMID: 25435409

Makan, A., Gouraizim, M., & Fadili, A. (2022). Sustainability assessment of wastewater treatment systems using cardinal weights and PROMETHEE method: Case study of Morocco. *Environmental Science and Pollution Research International*, *29*(13), 19803–19815. Advance online publication. DOI: 10.1007/s11356-021-17215-w PMID: 34718984

Malekloo, A., Ozer, E., AlHamaydeh, M., & Girolami, M. (2022). Machine learning and structural health monitoring overview with emerging technology and high-dimensional data source highlights. *Structural Health Monitoring*, *21*(4), 1906–1955. DOI: 10.1177/14759217211036880

Mallick, P., Das, D. K., Mohanty, D., Patnaik, P., Mohapatra, R. K., & Satpathy, S. K. (2024). Introduction to Smart Materials in Micro-and Nanoscale. In Smart Micro-and Nanomaterials for Drug Delivery (pp. 1-17). CRC Press. DOI: 10.1201/9781003468424-1

Mallmann, E. J., Sombra, A. S., Góes, J. C., & Fechine, P. B. (2013). Yttrium Iron Garnet: Properties and Applications Review. *Diffusion and Defect Data, Solid State Data. Part B, Solid State Phenomena*, *202*, 65–96. DOI: 10.4028/www.scientific.net/SSP.202.65

Malozemoff, A. P., Maguire, J. F., Gamble, B. B., & Kalsi, S. S. (2002). Power applications of high-temperature superconductors: Status and perspectives. *IEEE Transactions on Applied Superconductivity*, *12*(1), 778–781. DOI: 10.1109/TASC.2002.1018517

Margine, E. R., Bocquet, M., & Blase, X. (2008). Thermal stability of graphene and nanotube covalent functionalization. *Nano Letters*, *8*(10), 3315–3319. DOI: 10.1021/nl801718f PMID: 18767882

Masi, J.V., & Masi, S.D. (1997). Sol-gel magnetic materials. Proceedings: Electrical Insulation Conference and Electrical Manufacturing and Coil Winding Conference, 565-570.

Matizamhuka, W. R. (2018). The Impact of Magnetic Materials in Renewable Energy-Related Technologies in the 21st Century Industrial Revolution: The Case of South Africa. *Advances in Materials Science and Engineering*, *2018*(1), 3149412. DOI: 10.1155/2018/3149412

McHenry, M. E., & Laughlin, D. E. (2000). Nano-Scale Materials Development for Future Magnetic Applications. *ChemInform*, •••, 31.

Mehtab, A., Ahmed, J., Alshehri, S. M., Mao, Y., & Ahmad, T. (2022). Rare earth doped metal oxide nanoparticles for photocatalysis: A perspective. *Nanotechnology*, *33*(14), 142001. DOI: 10.1088/1361-6528/ac43e7 PMID: 34915455

Meliani, M., El Barkany, A., El Abbassi, I., & Mahmoudi, M. (2022). Smart Grid Challenges in Morocco and an Energy Demand Forecasting with Time Series. *International Journal of Engineering Research in Africa*, *61*, 195–215. Advance online publication. DOI: 10.4028/p-2gufv6

Meliani, M., El Barkany, A., El Abbassi, I., Moumen Darcherif, A., & Mahmoudi, M. (2021). Smart grid implementation in Morocco: Case study. *Materials Today: Proceedings*, *45*, 7675–7679. Advance online publication. DOI: 10.1016/j.matpr.2021.03.176

Meliani, M., El, A., El, I., Moumen, A., & Mahmoudi, M. (2021). Materials Today : Proceedings Smart grid implementation in Morocco : Case study. *Materials Today: Proceedings*, •••, 45.

Menaa, F., Fatemeh, Y., Vashist, S. K., Iqbal, H., Sharts, O. N., & Menaa, B. (2021). Graphene, an Interesting Nanocarbon Allotrope for Biosensing Applications: Advances, Insights, and Prospects. *Biomedical Engineering and Computational Biology*, *12*, 12. DOI: 10.1177/1179597220983821 PMID: 33716517

Mena-Cervantes, V. Y., Hernández-Altamirano, R., García-Solares, M. S., & Arreola-Valerio, E. (2023). Biodiesel in Circular Economy. In *Biofuels in Circular Economy*. DOI: 10.1007/978-981-19-5837-3_14

Menezes, B. R., Rodrigues, K. F., Fonseca, B. C., Ribas, R. G., Montanheiro, T. L., & Thim, G. P. (2019). Recent advances in the use of carbon nanotubes as smart biomaterials. *Journal of Materials Chemistry. B*, *7*(9), 1343–1360. DOI: 10.1039/C8TB02419G PMID: 32255006

Mészáros, I. (2007). Development of a Novel Vibrating Sample Magnetometer. *Materials Science Forum*, *537-538*, 413–418. DOI: 10.4028/www.scientific.net/MSF.537-538.413

Mewafy, A., Ismael, I., Kaddah, S. S., Hu, W., Chen, Z., & Abulanwar, S. (2024). Optimal design of multiuse hybrid microgrids power by green hydrogen–ammonia. *Renewable & Sustainable Energy Reviews*, *192*, 114174. DOI: 10.1016/j.rser.2023.114174

Mghaiouini, R., Saifaoui, D., & Salah, M. (2022). New ecological and economic process to save energy by electromagnetic fields. *Materials Today: Proceedings*, *66*, 11–16. DOI: 10.1016/j.matpr.2022.02.568

Mhd Razali, M. N., Ruzlan, N. N., & Sufian, A. H. (2023, August). Carbon Nanotube-Reinforced Polymer Composites for Biomedical Applications. In *Innovative Manufacturing, Mechatronics & Materials Forum* (pp. 255–264). Springer Nature Singapore.

Mikkonen, R. (2002). Highlights of SC power applications in Europe. *IEEE Transactions on Applied Superconductivity*, *12*(1), 782–787. DOI: 10.1109/TASC.2002.1018518

Mishra, A. N., Devaraj, S., & Vaidyanathan, G. (2013). Capability hierarchy in electronic procurement and procurement process performance: An empirical analysis. *Journal of Operations Management*, *31*(6), 376–390. DOI: 10.1016/j.jom.2013.07.011

Mitra, M., & Gopalakrishnan, S. (2016). Guided wave based structural health monitoring: A review. *Smart Materials and Structures*, *25*(5), 053001. DOI: 10.1088/0964-1726/25/5/053001

Mohammadshahi, S. S., Boulaire, F. A., Love, J., Gorji, S. A., & Mackinnon, I. D. R. (2022). A flexible analytical model for operational investigation of solar hydrogen plants. *International Journal of Hydrogen Energy*, *47*(2), 782–808. Advance online publication. DOI: 10.1016/j.ijhydene.2021.10.072

Mohseni, S., Brent, A. C., & Burmester, D. (2020). A comparison of metaheuristics for the optimal capacity planning of an isolated, battery-less, hydrogen-based micro-grid. *Applied Energy*, *259*, 114224. Advance online publication. DOI: 10.1016/j.apenergy.2019.114224

Morel, J. C., Mesbah, A., Oggero, M., & Walker, P. (2001). Building houses with local materials: Means to drastically reduce the environmental impact of construction. *Building and Environment*, *36*(10), 1119–1126. DOI: 10.1016/S0360-1323(00)00054-8

Morocco's Quest for Stronger and Inclusive Growth. (2023). *Morocco's Quest for Stronger and Inclusive Growth.*, DOI: 10.5089/9798400225406.071

Morón, C., Cabrera, C., Morón, A., García, A., & González, M. (2015). Magnetic Sensors Based on Amorphous Ferromagnetic Materials: A Review. *Sensors (Basel)*, *15*(11), 28340–28366. DOI: 10.3390/s151128340 PMID: 26569244

Morton, E. M., Deetjen, T. A., & Goodarzi, S. (2023). Optimizing hydrogen production capacity and day ahead market bidding for a wind farm in Texas. *International Journal of Hydrogen Energy*, *48*(46), 17420–17433. Advance online publication. DOI: 10.1016/j.ijhydene.2022.12.354

Mosin, O., & Ignatov, I. (2015). Magnetohydrodynamic cell for magnetic water treatment. *Nanotechnology Research and Practice*, *6*(2), 81–92. DOI: 10.13187/ejnr.2015.6.81

Motaung, J. R., & Sifolo, P. P. S. (2023). Benefits and barriers of digital procurement: Lessons from an airport company. *Sustainability (Basel)*, *15*(5), 4610. Advance online publication. DOI: 10.3390/su15054610

Muhit, M.S., & Hoque, R. (2011). Magnetic and Electric Characterization of Materials for Electrical Machines.

Mukhopadhyay, S. C. (2015). Wearable Sensors for Human Activity Monitoring: A Review. *IEEE Sensors Journal*, *15*(3), 1321–1330. DOI: 10.1109/JSEN.2014.2370945

Murjani, B. O., Kadu, P. S., Bansod, M., Vaidya, S. S., & Yadav, M. D. (2022). Carbon nanotubes in biomedical applications: Current status, promises, and challenges. *Carbon Letters*, *32*(5), 1207–1226. DOI: 10.1007/s42823-022-00364-4 PMID: 40477687

Mushnikov, N. V. (2022). Basics of magnetism and magnetic materials. In *Magnetic Materials and Technologies for Medical Applications* (pp. 23–55). Woodhead Publishing. DOI: 10.1016/B978-0-12-822532-5.00010-8

Naima, S.G., Zohra, S.F., & Samira, S. (2013). Magnetic materials for clean environment energy applications.

Nakach, I., Mouhat, O., Shamass, R., & El Mennaouy, F. (2023). Review of strategies for sustainable energy in Morocco. *Polityka Energetyczna*, *26*(2), 65–104. Advance online publication. DOI: 10.33223/epj/163373

Nakamura, K., Ohta, Y., Itoh, A., & Hayashi, C. (1982). Magnetic properties of thin films prepared by continuous vapor deposition. *IEEE Transactions on Magnetics*, *18*(6), 1077–1079. DOI: 10.1109/TMAG.1982.1062143

Nakamura, K., Yamada, T., Ohta, Y., & Itoh, A. (1982). "GT target", A new high rate sputtering target of magnetic materials. *IEEE Transactions on Magnetics*, *18*(6), 1080–1082. DOI: 10.1109/TMAG.1982.1062177

Namdeo, M., Saxena, S., Tankhiwale, R., Bajpai, M., Mohan, Y. M., & Bajpai, S. K. (2008). Magnetic nanoparticles for drug delivery applications. *Journal of Nanoscience and Nanotechnology*, *8*(7), 3247–3271. DOI: 10.1166/jnn.2008.399 PMID: 19051873

Naoe, M., Yamanaka, S., & Hoshi, Y. (1980). Facing targets type of sputtering method for deposition of magnetic metal films at low temperature and high rate. *IEEE Transactions on Magnetics*, *16*(5), 646–648. DOI: 10.1109/TMAG.1980.1060683

Narayanan, P., Pramanik, R., & Arockiarajan, A. (2024). Hard magnetics and soft materials—A synergy. *Smart Materials and Structures*, *33*(4), 043001. DOI: 10.1088/1361-665X/ad2bd8

Nayak, S. K., Patnaik, P., Das, D. K., & Barala, M. (2024). Electronic and half-metallic behaviour of manganese doped SiC Zigzag nanotubes using density functional theory. *Chemical Physics Impact*, *8*, 100417. DOI: 10.1016/j.chphi.2023.100417

Ndukwu, M. C., Bennamoun, L., Simo-Tagne, M., Ibeh, M. I., Abada, U. C., & Ekop, I. E. (2021). Influence of drying applications on wood, brick and concrete used as building materials: A review. *Journal of Building Pathology and Rehabilitation*, *6*(1), 1–19. DOI: 10.1007/s41024-021-00119-0

Neumann, W., Kirmse, H., Haeusler, I., Zheng, C., & Mogilatenko, A. (2009). Quantitative Structure Analysis of Nano-sized Materials by Transmission Electron Microscopy. MRS Proceedings, 1184. DOI: 10.1557/PROC-1184-GG03-02

Neumann, J., da Rocha, R. C., Debiagi, P., Scholtissek, A., Dammel, F., Stephan, P., & Hasse, C. (2023). Techno-economic assessment of long-distance supply chains of energy carriers: Comparing hydrogen and iron for carbon-free electricity generation. *Applications in Energy and Combustion Science*, *14*, 100128. Advance online publication. DOI: 10.1016/j.jaecs.2023.100128

Nguyen, M.H., Bui, H.T., Pham, V.T., Phan, N.H., Nguyen, T.H., Nguyen, V.C., Le, D.Q., Phan, H.K., & Phan, N.M. (2016). Thermo-mechanical properties of carbon nanotubes and applications in thermal management. *Advances in Natural Sciences: Nanoscience and Nanotechnology, 7*.

Niazi, A., Poddar, P., & Rastogi, A. K. (2000). A precision, low-cost vibrating sample magnetometer. *CURRENT SCIENCE-BANGALORE*, *79*(1), 99–109.

Niazi, K., Parsi, M. K., & Mohammadi, M. (2022). Nonlinear Dynamic Analysis of Hybrid Piezoelectric ??Magnetostrictive Energy ??Harvesting Systems. *Journal of Sensors*, *2022*(1), 8921779. DOI: 10.1155/2022/8921779

Nikov, R. G., Avdeev, G., Dikovska, A. O., Koleva, M. E., & Nedyalkov, N. (2020). Microstructural characterization of nanocomposites produced by laser ablation in a magnetic field. *Journal of Physics: Conference Series*, *1492*(1), 1492. DOI: 10.1088/1742-6596/1492/1/012057

Nimanpure, S., Singh, G., & Jewariya, M. (2023). Carbon Nanotubes and Graphene: The Novel Materials for Terahertz Detection. In Emerging Applications of Carbon Nanotubes and Graphene (pp. 275-288). CRC Press.

Nordio, M., Wassie, S. A., Van Sint Annaland, M., Pacheco Tanaka, D. A., Viviente Sole, J. L., & Gallucci, F. (2021). Techno-economic evaluation on a hybrid technology for low hydrogen concentration separation and purification from natural gas grid. *International Journal of Hydrogen Energy*, *46*(45), 23417–23435. Advance online publication. DOI: 10.1016/j.ijhydene.2020.05.009

Nore, C., Guermond, J. L., Léorat, J., & Luddens, F. (2014, July). Effet dynamo dans des géométries cylindriques. In *Symposium de Génie Électrique 2014*.

Norouzi, N. (2023). A framework of sustainable economics: Determining driving forces of the sustainable development in the MENA region. In *Energy Transition in the African Economy Post 2050*. DOI: 10.4018/978-1-7998-8638-9.ch005

Nur'aeni, R., & Zalsahra, R. (2024). Bibliometric Analysis of Artificial Intelligence. *West Science Interdisciplinary Studies*, *2*(1), 119–128. DOI: 10.58812/wsis.v2i01.563

Odenweller, A., Ueckerdt, F., Nemet, G. F., Jensterle, M., & Luderer, G. (2022). Probabilistic feasibility space of scaling up green hydrogen supply. *Nature Energy*, *7*(9), 854–865. Advance online publication. DOI: 10.1038/s41560-022-01097-4

Odunaiya, O. G., Soyombo, O. T., Okoli, C. E., Usiagu, G. S., & Ekemezie, I. O.Olusegun Gbenga OdunaiyaOluwatobi Timothy SoyomboChinelo Emilia Okoli-Glory Siwe UsiaguIfeanyi Onyedika Ekemezie. (2024). Renewable energy policies and wildlife conservation: A review of potential conflicts and coexistence strategies. *Open Access Research Journal of Multidisciplinary Studies*, *7*(1), 111–120. Advance online publication. DOI: 10.53022/oarjms.2024.7.1.0015

Oh, S., Yang, H., & Kim, S. W. (2014). Managerial capabilities of information technology and firm performance: Role of e-procurement system type. *International Journal of Production Research*, *52*(15), 4488–4506. DOI: 10.1080/00207543.2013.867084

Okpanachi, E., Ambe-Uva, T., & Fassih, A. (2022). Energy regime reconfiguration and just transitions in the Global South: Lessons for West Africa from Morocco's comparative experience. *Futures*, *139*, 102934. Advance online publication. DOI: 10.1016/j.futures.2022.102934

Okundamiya, M. S. (2021). Integration of photovoltaic and hydrogen fuel cell system for sustainable energy harvesting of a university ICT infrastructure with an irregular electric grid. *Energy Conversion and Management*, *250*, 114928. Advance online publication. DOI: 10.1016/j.enconman.2021.114928

Ouchani, F., Jbaihi, O., Alami Merrouni, A., Ghennioui, A., & Maaroufi, M. (2022). Geographic Information System-based Multi-Criteria Decision-Making analysis for assessing prospective locations of Pumped Hydro Energy Storage plants in Morocco: Towards efficient management of variable renewables. *Journal of Energy Storage*, *55*, 105751. Advance online publication. DOI: 10.1016/j.est.2022.105751

Ouhammou, B., Gargab, F. Z., El idrissi kaitouni, S., Smouh, S., El mrabet, R., Aggour, M., Jamil, A., & Kousksou, T. (2023). Energy saving potential diagnosis for Moroccan university campuses. *AIMS Energy*, *11*(3). DOI: 10.3934/energy.2023030

Paipetis, A.S., & Kostopoulos, V. (2013). Carbon nanotube enhanced aerospace composite materials: a new generation of multifunctional hybrid structural composites.

Pamme, N. (2006). Magnetism and microfluidics. *Lab on a Chip*, *6*(1), 24–38. DOI: 10.1039/B513005K PMID: 16372066

Panagopoulos, A. (2022). Brine management (saline water & wastewater effluents): Sustainable utilization and resource recovery strategy through Minimal and Zero Liquid Discharge (MLD & ZLD) desalination systems. *Chemical Engineering and Processing*, *176*, 108944. DOI: 10.1016/j.cep.2022.108944

Panda, P., & Sahu, G. P. (2012). *E-procurement implementation: Critical analysis of success factors' impact on project outcome*. *SSRN*. https://ssrn.com/abstract=2019575 DOI: 10.2139/ssrn.2019575

Pan, H., Qi, L., Zhang, Z., & Yan, J. (2021). Kinetic energy harvesting technologies for applications in land transportation: A comprehensive review. *Applied Energy*, *286*, 116518. DOI: 10.1016/j.apenergy.2021.116518

Park, J., Hwan Ryu, K., Kim, C. H., Chul Cho, W., Kim, M. J., Hun Lee, J., Cho, H. S., & Lee, J. H. (2023). Green hydrogen to tackle the power curtailment: Meteorological data-based capacity factor and techno-economic analysis. *Applied Energy*, *340*, 121016. Advance online publication. DOI: 10.1016/j.apenergy.2023.121016

Park, K., Heo, H., & Yang, D. R. (2018). Feasibility study of a forward osmosis/crystallization/reverse osmosis hybrid process with high-temperature operation: Modeling, experiments, and energy consumption. *Journal of Membrane Science*, *555*, 206–219. DOI: 10.1016/j.memsci.2018.03.031

Park, S.-M., & Jun, B.-H. (2023). Synthesis and Applications of Optical Materials. *Nanomaterials (Basel, Switzerland)*, *13*(2), 297. DOI: 10.3390/nano13020297 PMID: 36678049

Park, S., Srivastava, D., & Cho, K. (2003). Generalized Chemical Reactivity of Curved Surfaces: Carbon Nanotubes. *Nano Letters*, *3*(9), 1273–1277. DOI: 10.1021/nl0342747

Pasquale, M. (2003). Magnetic sensors and actuators. *Journal of Magnetics*, *8*(1), 60–69. DOI: 10.4283/JMAG.2003.8.1.060

Pásztory, Z. (2021). An overview of factors influencing thermal conductivity of building insulation materials. *Journal of Building Engineering*, *44*, 102604. DOI: 10.1016/j.jobe.2021.102604

Pathirage, C. S. N., Li, J., Li, L., Hao, H., Liu, W., & Ni, P. (2018). Structural damage identification based on autoencoder neural networks and deep learning. *Engineering Structures*, *172*, 13–28. DOI: 10.1016/j.engstruct.2018.05.109

Pawar, P., Anumalla, S., & Sharma, S. (2023). Role of carbon nanotubes (CNTs) in transgenic plant development. *Biotechnology and Bioengineering*, *120*(12), 3493–3500. DOI: 10.1002/bit.28550 PMID: 37691181

Pereira, C. R., Christopher, M., & Lago Da Silva, A. (2014). Achieving supply chain resilience: The role of procurement. *Supply Chain Management*, *19*(5/6), 626–642. DOI: 10.1108/SCM-09-2013-0346

Petrescu, L. (2009). MAGNETIC MATERIAL CHARACTERIZATION BY OPEN SAMPLE MEASUREMENTS.

Pham, A. T. T., Wallace, A., Zhang, X., Tohl, D., Fu, H., Chuah, C., Reynolds, K. J., Ramsey, C., & Tang, Y. (2021). Optical-Based Biosensors and Their Portable Healthcare Devices for Detecting and Monitoring Biomarkers in Body Fluids. *Diagnostics (Basel)*, *11*(7), 1285. DOI: 10.3390/diagnostics11071285 PMID: 34359368

Phonphuak, N., & Thiansem, S. (2012). Using charcoal to increase properties and durability of fired test briquettes. *Construction & Building Materials*, *29*, 612–618. DOI: 10.1016/j.conbuildmat.2011.11.018

Piyathanavong, V., Huynh, V. N., Karnjana, J., & Olapiriyakul, S. (2024). Role of project management on Sustainable Supply Chain development through Industry 4.0 technologies and Circular Economy during the COVID-19 pandemic: A multiple case study of Thai metals industry. *Operations Management Research : Advancing Practice Through Research*, *17*(1), 13–37. DOI: 10.1007/s12063-022-00283-7

Plank, F., Daum, B., Muntschick, J., Knodt, M., Hasse, C., Ott, I., & Niemann, A. (2023). Hydrogen: Fueling EU-Morocco Energy Cooperation? *Middle East Policy*, *30*(3), 37–52. Advance online publication. DOI: 10.1111/mepo.12699

Polytechnic, B. (1991). Magnetism: It is Permanent.

Pouponneau, P., Yahia, L., Merhi, Y., Epure, L. M., & Martel, S. (2006). Biocompatibility of Candidate Materials for the Realization of Medical Microdevices. 2006 International Conference of the IEEE Engineering in Medicine and Biology Society, 2362-2365.

Prasert, K., Sanglaow, T., Liangruksa, M., & Sutthibutpong, T. (2024). Effects of stacking layers and different doping elements on the electronic structures and quantum capacitance of graphene: A DFT study. *Journal of Physics and Chemistry of Solids*, *185*, 111758. DOI: 10.1016/j.jpcs.2023.111758

Pulliam, G. R. (1967). Chemical Vapor Growth of Single-Crystal Magnetic Oxide Films. *Journal of Applied Physics*, *38*(3), 1120–1126. DOI: 10.1063/1.1709508

Pulmano, C. E., Estuar, M. R. J. E., De Leon, M. M., Tan, H. C. L., Co, N. A. S., & Tamayo, L. P. V. (2023). Towards the development of a blockchain-based decentralized digital credential system using hyperledger fabric for participatory governance. *Procedia Computer Science*, *219*, 99–106. DOI: 10.1016/j.procs.2023.01.269

Pulugundla, G. (2013). *Numerical modelling of liquid metal flows interacting with strongly inhomogeneous magnetic fields* (Doctoral dissertation, TU Ilmenau).

Qiang, Z., Jiaqi, H., Mengqiang, Z., Weizhong, Q., & Fei, W. (2013). Review on mass production and industrialization of carbon nanotubes.

Qin, M., & Bau, H. H. (2009). Magneto-hydrodynamic flow in electrolyte solutions. In *Comsol conference*.

Qing, X., Liao, Y., Wang, Y., Chen, B., Zhang, F., & Wang, Y. (2022). Machine Learning Based Quantitative Damage Monitoring of Composite Structure. *International Journal of Smart and Nano Materials*, *13*(2), 167–202. DOI: 10.1080/19475411.2022.2054878

Qiu, C., Zhang, Z., Xiao, M., Yang, Y., Zhong, D., & Peng, L. (2017). Scaling carbon nanotube complementary transistors to 5-nm gate lengths. *Science*, *355*(6322), 271–276. DOI: 10.1126/science.aaj1628 PMID: 28104886

Quarton, C. J., & Samsatli, S. (2020). Should we inject hydrogen into gas grids? Practicalities and whole-system value chain optimisation. *Applied Energy*, *275*, 115172. Advance online publication. DOI: 10.1016/j.apenergy.2020.115172

Raab, M., & Dietrich, R. U. (2023). Techno-economic assessment of different aviation fuel supply pathways including LH2 and LCH4 and the influence of the carbon source. *Energy Conversion and Management*, *293*, 117483. Advance online publication. DOI: 10.1016/j.enconman.2023.117483

Rabiee, A., Keane, A., & Soroudi, A. (2021). Green hydrogen: A new flexibility source for security constrained scheduling of power systems with renewable energies. *International Journal of Hydrogen Energy*, *46*(37), 19270–19284. Advance online publication. DOI: 10.1016/j.ijhydene.2021.03.080

Radcliffe, B. (n.d.). *The basics of tariffs and trade barriers*. Investopedia. https://www.investopedia.com/articles/economics/08/tariff-trade-barrier-basics.asp

Rafik, M., Bahnasse, A., Khiat, A., Bouattane, O., Ouajji, H., & khaili, M. E. (2019). Towards a smart energy sharing in micro smart grid adopting SDN approach. *Procedia Computer Science*, *151*, 717–724. Advance online publication. DOI: 10.1016/j.procs.2019.04.095

Rahita, A. C., Zaki, A., Nugroho, G., & Yadi, S. (2024). Internet of Things (IoT) in Structural Health Monitoring: A Decade of Research Trends. *Instrumentation Mesure Métrologie*, *23*(2), 123–139. DOI: 10.18280/i2m.230205

Rahmani, K., Ahriz, A., & Bouaziz, N. (2022). Development of a New Residential Energy Management Approach for Retrofit and Transition, Based on Hybrid Energy Sources. *Sustainability (Basel)*, *14*(7), 4069. Advance online publication. DOI: 10.3390/su14074069

Rahmat, M. N., & Saleh, A. M. (2023). Strength and environmental evaluation of building bricks using industrial waste for livable environments. *Construction & Building Materials*, *403*, 132864. DOI: 10.1016/j.conbuildmat.2023.132864

Rajavel, K., Minitha, C. R., Ranjith, K. S., & Rajendra Kumar, R. T. (2012). Recent progress on the synthesis and applications of carbon based nanostructures. *Recent Patents on Nanotechnology*, *6*(2), 99–104. DOI: 10.2174/187221012800270199 PMID: 22292609

Ramatlo, D. A., Wilke, D. N., & Loveday, P. W. (2023). Digital Twin Hybrid Modeling for Enhancing Guided Wave Ultrasound Inspection Signals in Welded Rails. *Mathematical & Computational Applications*, *28*(2), 58. DOI: 10.3390/mca28020058

Ranjani, M., Jesurani, D., Priyadharshini, M., & Vennila, S. M. (2016). SOL-GEL SYNTHESIS AND CHARACTERIZATION OF ZINC SUBSTITUTED COBALT FERRITE MAGNETIC NANOPARTICLES. *International Journal of Approximate Reasoning*, *4*, 53–58.

Rao, R., Pint, C. L., Islam, A., Weatherup, R. S., Hofmann, S., Meshot, E. R., Wu, F., Zhou, C., Dee, N. T., Amama, P. B., Carpena-Núñez, J., Shi, W., Plata, D. L., Penev, E. S., Yakobson, B. I., Balbuena, P., Bichara, C., Futaba, D. N., Noda, S., & Hart, A. (2018). Carbon Nanotubes and Related Nanomaterials: Critical Advances and Challenges for Synthesis toward Mainstream Commercial Applications. *ACS Nano*, *12*(12), 11756–11784. DOI: 10.1021/acsnano.8b06511 PMID: 30516055

Rashmi, W., Ismail, A. F., Sopyan, I., Jameel, A. T., Yusof, F., Khalid, M., & Mubarak, N. M. (2011). Stability and thermal conductivity enhancement of carbon nanotube nanofluid using gum arabic. *Journal of Experimental Nanoscience*, *6*(6), 567–579. DOI: 10.1080/17458080.2010.487229

Rathee, M., & Singh, N. (2022). Durability properties of copper slag and coal bottom ash based I-shaped geopolymer paver blocks. *Construction & Building Materials*, *347*, 128461. DOI: 10.1016/j.conbuildmat.2022.128461

Rathore, D., Kurchania, R., & Pandey, R. K. (2012). Synthesis and Characterization of Cobalt Ferrite Nanoparticles Grown by Chemical Co-precipitation Method. *Invertis Journal of Science & Technology*, *5*, 132–137.

Reddy, J., Lakshya, A.K., Raj, R., Kumar, L., & Chowdhury, A. (2023). Characterizing Curie and Néel Point Phase Transitions via Thermal Techniques. physica status solidi (b), 260.

Refaat, A., Elbaz, A., Khalifa, A.-E., Mohamed Elsakka, M., Kalas, A., & Hegazy Elfar, M. (2024). Performance evaluation of a novel self-tuning particle swarm optimization algorithm-based maximum power point tracker for porton exchange membrane fuel cells under different operating conditions. *Energy Conversion and Management*, *301*, 118014. DOI: 10.1016/j.enconman.2023.118014

Remane, G., Hanelt, A., Nickerson, R. C., & Kolbe, L. M. (2017). Discovering digital business models in traditional industries. *The Journal of Business Strategy*, *38*(2), 41–51. DOI: 10.1108/JBS-10-2016-0127

Riley-Reid, T. D. (2015). The hidden cost of digitization – Things to consider. *Collection Building*, *34*(3), 89–93. DOI: 10.1108/CB-01-2015-0001

Rodrigues, L., & Jewell, G. W. (2014). Model Specific Characterization of Soft Magnetic Materials for Core Loss Prediction in Electrical Machines. *IEEE Transactions on Magnetics*, *50*(11), 1–4. DOI: 10.1109/TMAG.2014.2328099

Roman, A. V., & Thai, K. V. (2013). Public procurement in the United States. In *Exzellente öffentliche Beschaffung* (pp. 245–258). Springer., DOI: 10.1007/978-3-658-00567-2_14

Romeo, L. M., Cavana, M., Bailera, M., Leone, P., Peña, B., & Lisbona, P. (2022a). Non-stoichiometric methanation as strategy to overcome the limitations of green hydrogen injection into the natural gas grid. *Applied Energy*, *309*, 118462. Advance online publication. DOI: 10.1016/j.apenergy.2021.118462

Rosenstiel, A., Monnerie, N., Dersch, J., Roeb, M., Pitz-paal, R., & Sattler, C. (2021). Electrochemical hydrogen production powered by pv/csp hybrid power plants: A modelling approach for cost optimal system design. *Energies*, *14*(12), 3437. Advance online publication. DOI: 10.3390/en14123437

Roy, R. K. (2001). *Design of experiments using the Taguchi approach: 16 steps to product and process improvement*. John Wiley & Sons.

Ruhnau, O., & Schiele, J. (2023). Flexible green hydrogen: The effect of relaxing simultaneity requirements for project design, economics, and power sector emissions. *Energy Policy*, *182*, 113763. Advance online publication. DOI: 10.1016/j.enpol.2023.113763

Sahoo, S., & Das, S. K. (2009). Fractional quantum Hall effect in graphene. *Indian Journal of Pure and Applied Physics*, *47*, 658–662.

Salah, N., & Barra Ndiaye, E. (2024). Towards an optimization of automatic defect detection by artificial neural network using Lamb waves. *IAES International Journal of Artificial Intelligence (IJ-AI)*, *13*(2), 1459. DOI: 10.11591/ijai.v13.i2.pp1459-1468

Salah, N., Hamza, E., Mustapha, J., & Mansouri, K. (2024). Towards an automatic classification of welding defect by convolutional neural network and robot classifier. *Indonesian Journal of Electrical Engineering and Computer Science*, *33*(3), 1768. DOI: 10.11591/ijeecs.v33.i3.pp1768-1774

Salawu, O. S. (1997). Detection of structural damage through changes in frequency: A review. *Engineering Structures*, *19*(9), 718–723. DOI: 10.1016/S0141-0296(96)00149-6

Sali, N., Biwalkar, S., Swamy, A., Khadilkar, N., & Sahu, S. (2023). Blockchain in food supply chain: Benefits, challenges and review. *International Journal for Research in Applied Science and Engineering Technology*, *11*(4), 4654–4661. DOI: 10.22214/ijraset.2023.50661

Samal, A. K., Panigrahi, M., Ganguly, R. I., & Dash, R. R. (2023). Historical development of construction materials–From Stone Age to Modern Age. In *Development of Geopolymer from Pond Ash-Thermal Power Plant Waste: Novel Constructional Materials for Civil Engineers* (pp. 1-70). Elsevier. DOI: 10.1002/9781394167975.ch1

Sangwan, K. S., Choudhary, K., & Agarwal, S. (2020). Sustainability assessment of sanitary ware supply chain using life cycle assessment framework—a case study. In *Enhancing Future Skills and Entrepreneurship: 3rd Indo-German Conference on Sustainability in Engineering* (pp. 167-179). Springer International Publishing.

Saucedo-Jimenez, D., Medina-Sánchez, I., Padilla-Pérez, D. A., Couder-Castañeda, C., & Sanchez, J. L. L. (2023). Double-walled carbon nanotubes synthesis by pulsed electric arc discharge with spinning anode.

Sen, S., Singh, A., Bera, C., Roy, S., & Kailasam, K. (2022). Recent developments in biomass derived cellulose aerogel materials for thermal insulation application: A review. *Cellulose (London, England)*, *29*(9), 4805–4833. DOI: 10.1007/s10570-022-04586-7

Shabir, M., Hussain, I., Işık, Ö., Razzaq, K., & Mehroush, I. (2023). The role of innovation in environmental-related technologies and institutional quality to drive environmental sustainability. *Frontiers in Environmental Science*, *11*, 1174827. Advance online publication. DOI: 10.3389/fenvs.2023.1174827

Shahrtash, F. M. (2023). Sustainable Development, Electricity Generation and Renewable Energy substitution in Middle East Countries and Cooperation of Iran. *Sustainable Environment*, *8*(1), p57. Advance online publication. DOI: 10.22158/se.v8n1p57

Sharma, P. R., Mehra, N. K., Jain, K., & Jain, N. K. (2016). Biomedical Applications of Carbon Nanotubes: A Critical Review. *Current Drug Delivery*, *13*(6), 796–817. DOI: 10.2174/1567201813666160623091814 PMID: 27339036

Sharma, V., & Akhai, S. (2019). Mechanical Behaviour of Fly Ash Reinforced Aluminum Composite Prepared by Casting. *Journal of Advanced Research in Mechanical Engineering and Technology*, *6*(1&2), 23–26. DOI: 10.24321/2454.8650.201904

Sharma, V., & Akhai, S. (2019). Trends in Utilization of Coal Fly Ash in India: A Review. *Journal of Engineering Design and Analysis*, *2*(1), 12–16.

Shavelkina, M. B., Malikov, M. M., Ivanov, P. P., Borodina, T. I., & Valyano, G. E. (2024). Nanosecond pulsed laser ablation of synthetic graphite in liquids for the synthesis of spherical graphene. *Journal of Laser Applications*, *36*(2), 022021. DOI: 10.2351/7.0001329

Sherman, K. L. (2017). *Continuous flow centrifuge with application of water desalination and purification* (W.O. Patent No. WO2017030598A1). World Intellectual Property Organization. https://patents.google.com/patent/WO2017030598A1/en

Shifa, M., Toor, Z. S., & Tariq, F. (2024). Arc discharge synthesis and multistep purification of multiwall carbon nanotubes. *Nano*, *19*(02), 2450007. DOI: 10.1142/S1793292024500073

Singh, S., Maiti, S., Bisht, R. S., Panigrahi, S. K., & Yadav, S. (2024). Large CO2 reduction and enhanced thermal performance of agro-forestry, construction and demolition waste based fly ash bricks for sustainable construction. *Scientific Reports*, *14*(1), 8368. DOI: 10.1038/s41598-024-59012-8 PMID: 38600139

Siochi, E. J., & Harrison, J. S. (2015). Structural nanocomposites for aerospace applications. *MRS Bulletin*, *40*(10), 829–835. DOI: 10.1557/mrs.2015.228

Slaoui, Z., Faquir, S., Zidani, K. A., & Yahyaouy, A. (2023). Green Hydrogen for Industry: A Proposed Approach to Manage the Flow of Energy in Industrial Sector in Morocco. *Lecture Notes in Networks and Systems*, *578*, 119–126. Advance online publication. DOI: 10.1007/978-981-19-7660-5_11

Slimani, J., Kadrani, A., El Harraki, I., & Ezzahid, E. H. (2021). Renewable Energy Development in Morocco: Reflections on Optimal Choices through Long-term Bottom-up Modeling. *International Conference on Electrical, Computer, and Energy Technologies, ICECET 2021*. DOI: 10.1109/ICECET52533.2021.9698630

Small, H. (1973). Co-citation in the scientific literature: A new measure of the relationship between two documents. *Journal of the American Society for Information Science*, *24*(4), 265–269. DOI: 10.1002/asi.4630240406

Smith, D.J., McCartney, M.R., & Dunin-Borkowski, R.E. (2006). Nanoscale Structural and Magnetic Characterization Using Electron Microscopy.

Smith, M. (2020). Morocco aims for global green hydrogen role. In *Petroleum Economist*. Issue December-January.

Sobhani, E., & Safaei, B. (2023). Dynamic analysis of functionally graded embedded graphene oxide powder nanocomposite conical shells resting on flexible foundations. *Engineering Analysis with Boundary Elements*, *152*, 757 ?"773. DOI: 10.1016/j.enganabound.2023.05.006

Song, J. M., Sung, J., & Park, T. (2019). Applications of blockchain to improve supply chain traceability. *Procedia Computer Science*, *162*, 119–122. DOI: 10.1016/j.procs.2019.11.266

Sowter, G. A. (1987). Soft magnetic materials for audio transformers: History, production, and applications. *Journal of the Audio Engineering Society*, *35*, 760–777.

Spencer, B. F., Ruiz-Sandoval, M. E., & Kurata, N. (2004). Smart sensing technology: Opportunities and challenges. *Structural Control and Health Monitoring*, *11*(4), 349–368. DOI: 10.1002/stc.48

Sreenivasulu, K. V., & Srikanth, V. V. (2016). Fascinating Magnetic Energy Storage Nanomaterials: A Brief Review. *Recent Patents on Nanotechnology*, *11*(2), 116–122. PMID: 28286941

Stamatakis, E., Perwög, E., Garyfallos, E., Millán, M. S., Zoulias, E., & Chalkiadakis, N. (2022). Hydrogen in Grid Balancing: The European Market Potential for Pressurized Alkaline Electrolyzers. *Energies*, *15*(2), 637. Advance online publication. DOI: 10.3390/en15020637

Stanley, S., Ryan, L., & Flynn, D. (2023). Strategies to increase grid flexibility for an isolated system with over 80% renewable electricity in 2030. *International Conference on the European Energy Market, EEM, 2023-June*. DOI: 10.1109/EEM58374.2023.10161913

Stöckl, F., Schill, W. P., & Zerrahn, A. (2021). Optimal supply chains and power sector benefits of green hydrogen. *Scientific Reports*, *11*(1), 14191. Advance online publication. DOI: 10.1038/s41598-021-92511-6 PMID: 34244545

Subhedar, K. M., & Singh, B. P. (2023). Introduction to Carbon Nanotubes and Graphene and their Emerging Applications. In *Emerging Applications of Carbon Nanotubes and Graphene* (pp. 1–22). CRC Press. DOI: 10.1201/9781003231943-1

Sun, H., Burton, H. V., & Huang, H. (2021). Machine learning applications for building structural design and performance assessment: State-of-the-art review. *Journal of Building Engineering*, *33*, 101816. DOI: 10.1016/j.jobe.2020.101816

Sun, J., Wu, M., Jiang, H., Fan, X., & Zhao, T. (2021). Advances in the design and fabrication of high-performance flow battery electrodes for renewable energy storage. *Advances in Applied Energy*, *2*, 100016. DOI: 10.1016/j.adapen.2021.100016

Sun, Z., & Xia, Y. (2022). Study on Structural Health Monitoring and Damage Detection in the Field of Civil Engineering. *Highlights in Science. Engineering and Technology*, *10*, 92–98. DOI: 10.54097/hset.v10i.1231

Szemat-Vielma, W., Scheibz, J., Kasraoui, N., & Al-Omar, F. (2023). *Sun Powered Green Hydrogen - A Comparative Analysis from the Kingdoms Of Morocco and Saudi Arabia*. DOI: 10.2118/214375-MS

Szima, S., Arnaiz del Pozo, C., Cloete, S., Chiesa, P., Jiménez Alvaro, Á., Cormos, A. M., & Amini, S. (2021). Finding synergy between renewables and coal: Flexible power and hydrogen production from advanced IGCC plants with integrated CO2 capture. *Energy Conversion and Management*, *231*, 113866. Advance online publication. DOI: 10.1016/j.enconman.2021.113866

Tabor, D. (1954). Mohs's hardness scale-a physical interpretation. *Proceedings of the Physical Society. Section B*, *67*(3), 249–257. DOI: 10.1088/0370-1301/67/3/310

Tachiki, M., & Kobayashi, T. (1999). Manipulation of Laser Ablation Plume by Magnetic Field Application. *Japanese Journal of Applied Physics*, *38*(6R), 3642. DOI: 10.1143/JJAP.38.3642

Takakura, A., Beppu, K., Nishihara, T., Fukui, A., Kozeki, T., Namazu, T., Miyauchi, Y., & Itami, K. (2019). Strength of carbon nanotubes depends on their chemical structures. *Nature Communications*, *10*(1), 3040. DOI: 10.1038/s41467-019-10959-7 PMID: 31292451

Tang, F., & Wan, X. (2023). Evolution of topological states in magnetic materials by symmetry-breaking. arXiv preprint arXiv:2302.13622.

Tan, K. C. (2001). A framework of supply chain management literature. *European Journal of Purchasing & Supply Management*, *7*(1), 39–48. DOI: 10.1016/S0969-7012(00)00020-4

Tannous, C., & Comstock, R. (2017). Magnetic Information-Storage Materials.

Thai, H.-T. (2022). Machine learning for structural engineering: A state-of-the-art review. *Structures, 38*, 448–491. DOI: 10.1016/j.istruc.2022.02.003

Thakur, A., & Tiwary, A. K. (2021, November). A comparative study on conventional clay bricks and autoclaved aerated concrete blocks. [). IOP Publishing.]. *IOP Conference Series. Earth and Environmental Science*, *889*(1), 012061. DOI: 10.1088/1755-1315/889/1/012061

Tham, A., Siah, K. X., Noor, I. M., Ghazali, I., Tan, S. T., & Tamchek, N. (2024). Fabrication and structural modification of graphene oxide-tetraethyl orthosilicate solution via liquid-phase pulsed laser ablation. *Journal of Laser Applications*, *36*(1), 012023. DOI: 10.2351/7.0001227

Thangamani, G. J., Deshmukh, K., Sadasivuni, K. K., Chidambaram, K., Ahamed, M. B., Ponnamma, D., AlMaadeed, M. A., & Pasha, S. K. (2017). Recent advances in electrochemical biosensor and gas sensors based on graphene and carbon nanotubes (CNT) - A review. *Advanced Materials Letters*, *8*, 196–205. DOI: 10.5185/amlett.2017.7042

Thareja, P., & Akhai, S. (2017). Processing parameters of powder aluminium-fly ash P/M composites. *Journal of advanced research in manufacturing, material science & metallurgical engineering, 4*(3&4), 24-35.

Thareja, P., & Akhai, S. (2016). Processing Aluminum Fly Ash Composites via Parametric Analysis of Stir Casting. *Journal of Advanced Research in Manufacturing. Material Science & Metallurgical Engineering, 3*(3&4), 21–28.

Thornhill, S., Wu, N., Wang, Z., Shi, Q., & Chen, J. (2008). Graphene nanoribbon field-effect transistors. *2008 IEEE International Symposium on Circuits and Systems*, 169-172. DOI: 10.1109/ISCAS.2008.4541381

Timilsina, G. R., & Deluque Curiel, I. F. (2023). Subsidy removal, regional trade and CO2 mitigation in the electricity sector in the Middle East and North Africa region. *Energy Policy, 177*, 113557. Advance online publication. DOI: 10.1016/j.enpol.2023.113557

Tkachenko, N., Tang, K., McCarten, M., Reece, S., Kampmann, D., Hickey, C., ... & Caldecott, B. (2023). Global database of cement production assets and upstream

Tleubergenova, A., Abuov, Y., Danenova, S., Khoyashov, N., Togay, A., & Lee, W. (2023). Resource assessment for green hydrogen production in Kazakhstan. *International Journal of Hydrogen Energy, 48*(43), 16232–16245. Advance online publication. DOI: 10.1016/j.ijhydene.2023.01.113

Topalović, Z., Haas, R., & Sayer, M. (2024). Economic benefits of PHS and Li-ion storage. Study cases: Austria and Bosnia and Herzegovina. *Applied Energy, 362*, 122988. Advance online publication. DOI: 10.1016/j.apenergy.2024.122988

Touili, S., Bouaichi, A., Alami Merrouni, A., Amrani, A., El Amrani, A., El Hassouani, Y., & Messaoudi, C. (2022). Performance analysis and economic competiveness of 3 different PV technologies for hydrogen production under the impact of arid climatic conditions of Morocco. *International Journal of Hydrogen Energy, 47*(74), 31596–31613. Advance online publication. DOI: 10.1016/j.ijhydene.2022.07.088

Tripathi, S. K., Kaur, R., Bharti, S., Gawri, I., & Kaur, J. (2015). Recent Advances in Synthesis, Properties and Applications of Magnetic Oxide Nanomaterials. *Diffusion and Defect Data, Solid State Data. Part B, Solid State Phenomena, 232*, 1–44. DOI: 10.4028/www.scientific.net/SSP.232.1

Trout, S.R. (2001). Material selection of permanent magnets, considering thermal properties correctly. Proceedings: Electrical Insulation Conference and Electrical Manufacturing and Coil Winding Conference (Cat. No.01CH37264), 365-370.

Tung, T. T., Nine, M. J., Krebsz, M., Pasinszki, T., Coghlan, C. J., Tran, D. N., & Losic, D. (2017). Recent Advances in Sensing Applications of Graphene Assemblies and Their Composites. *Advanced Functional Materials*, *27*(46), 27. DOI: 10.1002/adfm.201702891

Tupa, J., Simota, J., & Steiner, F. (2017). Aspects of risk management implementation for Industry 4.0. *Procedia Manufacturing*, *11*, 1223–1230. DOI: 10.1016/j.promfg.2017.07.248

Tyagi, K., Verma, A. K., Gahtori, B., & Dhakate, S. R. (2023). Carbon Nanotubes and Graphene-Based Thermoelectric Materials: A Futuristic Approach for Energy Harvesting. In Emerging Applications of Carbon Nanotubes and Graphene (pp. 205-226). CRC Press.

Unguris, J., Scheinfein, M. R., Celotta, R. J., & Pierce, D. T. (1991). Scanning Electron Microscopy with Polarization Analysis: Studies of Magnetic Microstructure. *Journal of Vacuum Science and Technology*, •••, 239–265.

Ünverdi, Ö. (2018). Nanotribological Properties Of Epitaxial Graphene Grown On C-Terminated Face Of Silicon Carbide Semiconductor. *Eskişehir Teknik Üniversitesi Bilim ve Teknoloji Dergisi B-Teorik Bilimler*, *6*(2), 228–234. DOI: 10.20290/aubtdb.464691

Urmi, R., Banerjee, P., Singh, M., Singh, R., Chhillar, S., Sharma, N., Chandra, A., Singh, N., & Qamar, I. (2024). Revolutionizing biomedicine: Aptamer-based nanomaterials and nanodevices for therapeutic applications. *Biotechnology Reports (Amsterdam, Netherlands)*, *42*, e00843. DOI: 10.1016/j.btre.2024.e00843 PMID: 38881649

Utomo, O., Abeysekera, M., & Ugalde-Loo, C. E. (2021). Optimal operation of a hydrogen storage and fuel cell coupled integrated energy system. *Sustainability (Basel)*, *13*(6), 3525. Advance online publication. DOI: 10.3390/su13063525

Valverde, L., Rosa, F., Bordons, C., & Guerra, J. (2016). Energy Management Strategies in hydrogen Smart-Grids: A laboratory experience. *International Journal of Hydrogen Energy*, *41*(31), 13715–13725. Advance online publication. DOI: 10.1016/j.ijhydene.2016.05.279

Vashist, E.K. (2014). Graphene: Emerging Trends, Prospects and Challenges. *Journal of Nanomaterials & Molecular Nanotechnology, 2014*.

Vasile, C., Engel, T., Risser, M., Muller, C., & Imbs, I.A. (2008). ENERGY EFFICIENT AND ENVIRONMENTAL SAFE MAGNETIC COOLING SYSTEM.

Verhille, G., Plihon, N., Khalilov, R., Pinton, J. F., & Frick, P. (2011, August). Modification of a von Kármán turbulent flow by a magnetic field. In *CFM 2011 - 20th French Congress of Mechanics*. AFM, Maison de la Mécanique, 39/41 rue Louis Blanc - 92400 Courbevoie.

Verma, M., & Malvi, L. (2021). Comparative analysis between properties of fly ash bricks and normal clay bricks. *International Journal of Scientific Modern Research and Technology*, *4*(3), 26–33.

Vettier, C., & Mcwhan, D. B. (1989). Magnetic X-ray diffraction from ferromagnetic materials. *Physica B, Condensed Matter*, *159*(2), 106–110. DOI: 10.1016/0921-4526(89)90027-6

Vijayan, D. S., Mohan, A., Revathy, J., Parthiban, D., & Varatharajan, R. (2021). Evaluation of the impact of thermal performance on various building bricks and blocks: A review. *Environmental Technology & Innovation*, *23*, 101577. DOI: 10.1016/j.eti.2021.101577

Volk, R., Stengel, J., & Schultmann, F. (2014). Building Information Modeling (BIM) for existing buildings—Literature review and future needs. *Automation in Construction*, *38*, 109–127. DOI: 10.1016/j.autcon.2013.10.023

Wang, Y., & Weng, G. J. (2018). Electrical conductivity of carbon nanotube-and graphene-based nanocomposites. Micromechanics and nanomechanics of composite solids, 123-156.

Wang, B., Hu, C., & Dai, L. (2016). Functionalized carbon nanotubes and graphene-based materials for energy storage. *Chemical Communications (Cambridge)*, *52*(100), 14350–14360. DOI: 10.1039/C6CC05581H PMID: 27709194

Wang, C., Yin, L., Zhang, L., Xiang, D., & Gao, R. (2010). Metal Oxide Gas Sensors: Sensitivity and Influencing Factors. *Sensors (Basel)*, *10*(3), 2088–2106. DOI: 10.3390/s100302088 PMID: 22294916

Wang, D., Muratori, M., Eichman, J., Wei, M., Saxena, S., & Zhang, C. (2018a). Quantifying the flexibility of hydrogen production systems to support large-scale renewable energy integration. *Journal of Power Sources*, *399*, 383–391. Advance online publication. DOI: 10.1016/j.jpowsour.2018.07.101

Wang, G., Wang, F., Shen, F., Jiang, T., Chen, Z., & Hu, P. (2020). Experimental and optical performances of a solar CPV device using a linear Fresnel reflector concentrator. *Renewable Energy*, *146*, 2351–2361. DOI: 10.1016/j.renene.2019.08.090

Wang, Q., & Wu, N. (2012). Optimal design of a piezoelectric coupled beam for power harvesting. *Smart Materials and Structures*, *21*(8), 085013. DOI: 10.1088/0964-1726/21/8/085013

Wang, R., & Haller, P. (2022). Applications of wood ash as a construction material in civil engineering: A review. *Biomass Conversion and Biorefinery*, •••, 1–21. PMID: 35342681

Wang, X., Luo, Y., Liao, J., Joselevich, E., & Xu, J. (2024). Selective-Area Growth of Aligned Organic Semiconductor Nanowires and Their Device Integration. *Advanced Functional Materials*, *34*(7), 2308708. Advance online publication. DOI: 10.1002/adfm.202308708

Wang, Y., Pan, C., Chu, W., Vipin, A. K., & Sun, L. (2019). Environmental Remediation Applications of Carbon Nanotubes and Graphene Oxide: Adsorption and Catalysis. *Nanomaterials (Basel, Switzerland)*, *9*(3), 9. DOI: 10.3390/nano9030439 PMID: 30875970

Wang, Y., Wei, H., & Li, Z. (2018). Effect of magnetic field on the physical properties of water. *Results in Physics*, *8*, 262–267. DOI: 10.1016/j.rinp.2017.12.022

Wanner, J., Heinrich, K., & Zschech, P. (2020). *How much AI do you require? Decision factors for adopting AI technology*. https://www.researchgate.net/publication/344350604

Wardly, G. A. (1971). Magnetic Contrast in the Scanning Electron Microscope. *Journal of Applied Physics*, *42*(1), 376–386. DOI: 10.1063/1.1659604

Wei, J., Wei, Q., Ma, Z., & Zeng, H. (2016). Atomic and electronic structures of carbon nanotube covalent connecting with graphene by oxygen molecular. *The European Physical Journal Applied Physics*, *73*(2), 20401. DOI: 10.1051/epjap/2016150444

Wei, L., & Zhang, L. (2022). Atomic Simulations of (8, 0) CNT-Graphene by SCC-DFTB Algorithm. *Nanomaterials (Basel, Switzerland)*, *12*(8), 1361. DOI: 10.3390/nano12081361 PMID: 35458069

Wei, L., & Zhang, L. (2023). Structures and electron states of (4, 4) CNT-graphene hybrid system through DFTB investigation. *Materials Today. Communications*, *35*, 105576. DOI: 10.1016/j.mtcomm.2023.105576

Wei, M., Lee, S. H., Hong, T., Conlon, B., McKenzie, L., Hendron, B., & German, A. (2021). Approaches to cost-effective near-net zero energy new homes with time-of-use value of energy and battery storage. *Advances in Applied Energy*, *2*, 100018. DOI: 10.1016/j.adapen.2021.100018

Weimann, L., Gabrielli, P., Boldrini, A., Kramer, G. J., & Gazzani, M. (2021). Optimal hydrogen production in a wind-dominated zero-emission energy system. *Advances in Applied Energy*, *3*, 100032. Advance online publication. DOI: 10.1016/j.adapen.2021.100032

Weisberg, S. (2005). *Applied linear regression* (Vol. 528). John Wiley & Sons. DOI: 10.1002/0471704091

Welipitiya, D., He, Y. L., Zhang, J., Oden, P. I., Thundat, T., Warmack, R. J., Gobulukoglu, I., Shan, Z., Sellmyer, D. J., & Dowben, P. A. (1996). Fabrication of large arrays of micron-scale magnetic features by selective area organometallic chemical vapor deposition. *Journal of Applied Physics*, *80*(3), 1867–1871. DOI: 10.1063/1.362999

Wen, L., Li, F., & Cheng, H. (2016). Carbon Nanotubes and Graphene for Flexible Electrochemical Energy Storage: From Materials to Devices. *Advanced Materials*, *28*(22), 28. DOI: 10.1002/adma.201504225 PMID: 26748581

Wodon, Q. (2023). Education to climate change: A missing ingredient in climate action plans for MENA countries? *Orient (Leverkusen)*, *64*(1).

Worden, K., & Manson, G. (2007). The application of machine learning to structural health monitoring. *Philosophical Transactions of the Royal Society A: Mathematical, Physical and Engineering Sciences, 365*(1851), 515–537. DOI: 10.1098/rsta.2006.1938

Wujcik, E. K., & Monty, C. N. (2013). Nanotechnology for implantable sensors: Carbon nanotubes and graphene in medicine. *Wiley Interdisciplinary Reviews. Nanomedicine and Nanobiotechnology*, *5*(3), 233–249. DOI: 10.1002/wnan.1213 PMID: 23450525

Wu, M., & Huang, S. (2017). Magnetic nanoparticles in cancer diagnosis, drug delivery and treatment. *Molecular and Clinical Oncology*, *7*(5), 738–746. DOI: 10.3892/mco.2017.1399 PMID: 29075487

Wu, T., Sher, C.-W., Lin, Y., Lee, C.-F., Liang, S., Lu, Y., & Chen, H. (1557). S.-W., Guo, W., Kuo, H.-C., & Chen, Z. (2018). Mini-LED and Micro-LED: Promising Candidates for the Next Generation Display Technology. *Applied Sciences (Basel, Switzerland)*, *8*(9), 1557. Advance online publication. DOI: 10.3390/app8091557

Xie, S., Ai, L., Cui, C., Fu, T., Cheng, X., Qu, F., & Tan, W. (2021). Functional Aptamer-Embedded Nanomaterials for Diagnostics and Therapeutics. *ACS Applied Materials & Interfaces*, *13*(8), 9542–9560. DOI: 10.1021/acsami.0c19562 PMID: 33595277

Xie, X. D., Wu, N., Yuen, K. V., & Wang, Q. (2013). Energy harvesting from high-rise buildings by a piezoelectric coupled cantilever with a proof mass. *International Journal of Engineering Science*, *72*, 98 ?"106. DOI: 10.1016/j.ijengsci.2013.07.004

Xie, Y., Shibata, K., & Mizoguchi, T. (2023). Tensile strength of a transverse grain boundary in a single-walled carbon nanotube. *Journal of the Ceramic Society of Japan*, *131*(10), 621–627. DOI: 10.2109/jcersj2.23062

Xu, D., & Hussain, J. (2023). Globalization, institutions, and environmental quality in Middle East and North African countries. *Environmental Science and Pollution Research International*, *30*(26), 68951–68968. Advance online publication. DOI: 10.1007/s11356-023-27348-9 PMID: 37129812

Xu, H., & Shi, B. (2022). Design and System Evaluation of Mixed Waste Plastic Gasification Process Based on Integrated Gasification Combined Cycle System. *Processes (Basel, Switzerland)*, *10*(3), 499. Advance online publication. DOI: 10.3390/pr10030499

Xu, J., Cao, Z., Zhang, Y., Yuan, Z., Lou, Z., Xu, X., & Wang, X. (2018). A review of functionalized carbon nanotubes and graphene for heavy metal adsorption from water: Preparation, application, and mechanism. *Chemosphere*, *195*, 351–364. DOI: 10.1016/j.chemosphere.2017.12.061 PMID: 29272803

Xu, Z., Inoue, T., Nishina, Y., & Kobayashi, Y. (2022). Stacking order reduction in multilayer graphene by inserting nanospacers. *Journal of Applied Physics*, *132*(17), 174305. DOI: 10.1063/5.0103826

Yadav, V. K., Gacem, A., Choudhary, N., Rai, A., Kumar, P., Yadav, K. K., Abbas, M., Khedher, N. B., Awwad, N. S., Barik, D., & Islam, S. (2022). Status of coal-based thermal power plants, coal fly ash production, utilization in India and their emerging applications. *Minerals (Basel)*, *12*(12), 1503. DOI: 10.3390/min12121503

Yamashita, S. (2012). Carbon nanotube and graphene for photonic and optoelectronic applications. *2012 Photonics Global Conference (PGC)*, 1-3.

Yang, B., Yang, N., Zhao, D., Chen, F., Yuan, X., Kou, B., Hou, Y., & Xie, G. (2023). Numerical Simulation of Graphene Growth by Chemical Vapor Deposition Based on Tesla Valve Structure. *Coatings*, *13*(3), 564. DOI: 10.3390/coatings13030564

Yang, C., Jiang, Y., Hao, S. H., Yan, X. Y., Hong, D. F., & Naranmandura, H. (2022). Aptamers: An emerging navigation tool of therapeutic agents for targeted cancer therapy. *Journal of Materials Chemistry. B*, *10*(1), 20–33. DOI: 10.1039/D1TB02098F PMID: 34881767

Yan, K., Karthick Kannan, P., Doonyapisut, D., Wu, K., Chung, C., & Zhang, J. (2021). Advanced Functional Electroactive and Photoactive Materials for Monitoring the Environmental Pollutants. *Advanced Functional Materials*, *31*(12), 2008227. Advance online publication. DOI: 10.1002/adfm.202008227

Yan, Y., Ibrahim, R. L., Al-Faryan, M. A. S., & Oke, D. M. (2023). Embracing Eco-Digitalization and Green Finance Policies for Sustainable Environment: Do the Engagements of Multinational Corporations Make or Mar the Target for Selected MENA Countries? *Sustainability (Basel)*, *15*(15), 12046. Advance online publication. DOI: 10.3390/su151512046

Yao, C., & Ma, Y. (2021). Superconducting materials: Challenges and opportunities for large-scale applications. *iScience*, *24*(6), 24. DOI: 10.1016/j.isci.2021.102541 PMID: 34136765

Yao, J., Li, P., Li, L., & Yang, M. (2018). Biochemistry and biomedicine of quantum dots: From biodetection to bioimaging, drug discovery, diagnostics, and therapy. *Acta Biomaterialia*, *74*, 36–55. DOI: 10.1016/j.actbio.2018.05.004 PMID: 29734008

Yilmaz, D., & Dal, L. (2015). Hydraulic properties estimation of an experimental urban soil column constructed with waste brick and compost. *International Journal of Pure and Applied Sciences*, *1*, 9–15.

Yin, H., Zhang, Y., Lu, P., Zhang, Y., & Liu, J. (2020). Analysis of heat dissipation characteristics of three-dimensional graphene-carbon nanotube composite structures. *2020 21st International Conference on Electronic Packaging Technology (ICEPT)*, 1-5.

You, F., Wassick, J. M., & Grossmann, I. E. (2008). Risk management for a global supply chain planning under uncertainty: Models and algorithms. *AIChE Journal. American Institute of Chemical Engineers*, *55*(4), 931–946. DOI: 10.1002/aic.11721

Young, K. E. (2015). The Limits of Gulf Arab Aid: Energy Markets and Foreign Policy. *EUCERS "Reflections" Working Paper Series, 1.*

Zadeh-Haghighi, H., & Simon, C. (2022). Magnetic field effects in biology from the perspective of the radical pair mechanism. *Journal of the Royal Society, Interface*, *19*(193), 20220325. DOI: 10.1098/rsif.2022.0325 PMID: 35919980

Zakaria, O., Oumaima, C., Achraf, E. K., Ahmed, A. E., & Driss, Z. (2024). Energy Efficiency Measures and Technical-Economic Study of a Photovoltaic Self-consumption Installation in Kenitra, Morocco. *Lecture Notes in Electrical Engineering, 1141 LNEE*. DOI: 10.1007/978-981-97-0126-1_28

Zalite, I., Heidemane, G., Kuznetsova, L. V., & Maiorov, M. (2015). Hydrothermal Synthesis of Cobalt Ferrite Nanosized Powders. *IOP Conference Series. Materials Science and Engineering*, *77*, 012011. DOI: 10.1088/1757-899X/77/1/012011

Zanane, F. Z., Sadki, K., Drissi, L. B., Saidi, E. H., & Bousmina, M. (2022). Graphene multilayers nanoribbons with chirality from molecular dynamics. *Materials Science and Engineering B*, *286*, 115982. DOI: 10.1016/j.mseb.2022.115982

Zarbin, A. J., & Oliveira, M. R. (2013). Nanoestruturas de carbono (nanotubos, grafeno): Quo Vadis? *Química Nova*, *36*(10), 1533–1539. DOI: 10.1590/S0100-40422013001000009

Zarkik, A., & Ouhnini, A. (2022). Integrated transition toward sustainability: The case of the water-energy-food nexus in Morocco. In *Food Security and Climate-Smart Food Systems*. Building Resilience for the Global South., DOI: 10.1007/978-3-030-92738-7_8

Zelaya, C. V. G. (2019, April). Towards explaining the effects of data preprocessing on machine learning. In *2019 IEEE 35th international conference on data engineering (ICDE)* (pp. 2086-2090). IEEE.

Zeng, X., Xiao, Y., Ji, X., & Wang, G. (2021). Mineral identification based on deep learning that combines image and mohs hardness. *Minerals (Basel)*, *11*(5), 506. DOI: 10.3390/min11050506

Zhang, H., Zhou, T., Xie, G., Cao, J., & Yang, Z. (2014). Thermal transport in folded zigzag and armchair graphene nanoribbons. *Applied Physics Letters*, *104*(24), 241908. DOI: 10.1063/1.4884278

Zhang, J. (2020). Deploying blockchain technology in the supply chain. In *Computer security threats*. IntechOpen., DOI: 10.5772/intechopen.86530

Zhang, Y. F. (2022). A review of arc-discharge method towards large-scale preparation of long linear carbon chains. *Chinese Physics B*, *31*(12), 125201. DOI: 10.1088/1674-1056/ac833f

Zhan, Z., Mao, H., Xue, M., Han, G., Zhou, G., & Zhang, Y. (2024). Ratiometric fluorescence detection of the angiotensin-converting enzyme via single-excitation and double-emission biomass-derived carbon quantum dots. *Methods and Applications in Fluorescence*, *12*(1), 015004. DOI: 10.1088/2050-6120/ad02dd PMID: 37827162

Zhao, H., Kamp, L. M., & Lukszo, Z. (2022). The potential of green ammonia production to reduce renewable power curtailment and encourage the energy transition in China. *International Journal of Hydrogen Energy*, *47*(44), 18935–18954. Advance online publication. DOI: 10.1016/j.ijhydene.2022.04.088

Zhao, P., Tang, D., Xie, J., & Zhang, C. (2024). Magnetism and Magnetic Materials. In *Magnetic Levitation: Innovation of Density-Based Applications* (pp. 1–15). Springer Nature Singapore. DOI: 10.1007/978-981-99-8314-8_1

Zhao, Y., Yu, M., Xiang, Y., Kong, F., & Li, L. (2020). A sustainability comparison between green concretes and traditional concrete using an emergy ternary diagram. *Journal of Cleaner Production*, *256*, 120421. DOI: 10.1016/j.jclepro.2020.120421

Zheng, W., Guo, Z., Wang, Z., Boay, C. G., Wang, X., & Li, Y. (2022). Modeling analysis of elastic properties of graphene-carbon nanotube (G-C) reinforced composites. *Polymer Composites*, *43*(5), 3136–3149. DOI: 10.1002/pc.26606

Zhou, C., Senegor, R., Baron, Z., Chen, Y., Raju, S., Vyas, A. A., Chan, M., Chai, Y., & Yang, C. Y. (2016). Synthesis and interface characterization of CNTs on graphene. *Nanotechnology*, *28*(5), 054007. DOI: 10.1088/1361-6528/28/5/054007 PMID: 28029110

Zhou, L., Boyd, E. P., Dengb, H., Wigen, P. E., & Shoreb, S. G. (1993). Synthesis and evaluation of magnetic thin films. *IEEE Transactions on Magnetics*, *29*(6), 3052–3054. DOI: 10.1109/20.281114

Zhou, Y., He, J., Chen, R., & Li, X. (2022). *Recent Advances in Biomass-derived Graphene and Carbon Nanotubes*. Materials Today Sustainability. DOI: 10.1016/j.mtsust.2022.100138

Zhou, Y., Panteli, M., Moreno, R., & Mancarella, P. (2018). System-level assessment of reliability and resilience provision from microgrids. *Applied Energy*, *230*, 374–392. Advance online publication. DOI: 10.1016/j.apenergy.2018.08.054

Zimmermann, N. E., Vorselaars, B., Quigley, D., & Peters, B. (2015). Nucleation of NaCl from aqueous solution: Critical sizes, ion-attachment kinetics, and rates. *Journal of the American Chemical Society*, *137*(41), 13352–13361. DOI: 10.1021/jacs.5b08098 PMID: 26371630

Zitouni, I., Rhimini, H., & Chouaf, A. (2024). Modeling the Propagation of Ultrasonic Guided Waves in Four-Layer Laminated Composites. In Azari, Z., El Had, K., Ait Ali, M. E., El Mahi, A., Chaari, F., & Haddar, M. (Eds.), *Advances in Applied Mechanics* (pp. 25–33). Springer Nature Switzerland., DOI: 10.1007/978-3-031-49727-8_3

Zlobin, A. V. (2014). *Superconducting Magnets - Principles*. Operation, and Applications.

About the Contributors

Mohamed El Khaili received his PhD from the University of Casablanca (Morocco) in 1998 in image processing and his MBA from Toulouse Business school (France). He is currently a Associate Professor in the Department of Electrical Engineering at the ENSET School. He has more than 50 scientific publications in the areas of Mobile Robotics, Image Processing, Telecoms, Supply Chain and Embedded Systems. He is a member of the Electrical Engineering and Intelligent Systems Laboratory at ENSET. He contributes mainly by developing hardware and software required to make our technologies optimally for the implementation of physical systems and the use of IoT in several context such as smart farm, smart city, predictive maintenance, supply chain, etc. M. EL KHAILI teaches both undergraduate and postgraduate courses including Digital and Analog Electronic, VHDL Design, Embedded Technology, Real Time System, Computer Vision, Algorithms and Data Structures and Intelligent Agents, Electrical Energy Production and Management. - IEEE Member - Organization committee Member of international conferences: IRASET (2021, 2022, 2023, 2024), ICEGC (2021, 2023), ICOA (2018) - TPC member of several international conferences - Reviewer for International Journal of Electrical and Computer Engineering (IJECE)

Naoual Belouaggadia is a distinguished Professor at Hassan II University of Casablanca, Morocco where she contributes significantly to teaching and conducting scientific research at the Ecole Normale Supérieure de l'Enseignement Technique de Mohammedia (ENSET). Her scholarly pursuits primarily revolve around machanics, energy and heat engineering, with a specialized focus on sustainable construction and green buildings. Having undergone rigorous training in both Morocco and France, Mrs. BELOUAGGADIA holds a Bachelor's degree in energetics, a Doctorate in mechanics and energetics, and a Habilitation à diriger la recherche (HDR) in these fields. Her training and research journey is further enriched by her extensive publication record, boasting over 50 scientific articles, and her supervision and co-supervision of 10 doctoral theses. Mrs. BELOUAGGADIA's dedication extends beyond academia, as evidenced by her leadership roles within the university, including heading a research team, serving as a member of the Faculty of Science Council, and fulfilling the duties of deputy director of a research laboratory. Presently, she continues her scholarly pursuits in energy mechanics, simultaneously supporting the next generation of scholars by supervising and mentoring several doctoral students at ENSET and in collaboration with national and international partner institutions. Her expertise spans a wide spectrum, with a particular focus on Energy efficiency in industry and buildings, Thermal engineering, Energetics, Fluid mechanics, and Renewable energies (wind power). Additionally, she plays a pivotal role in the Training of trainers, imparting her wealth of knowledge to shape the future of the field.

Abdellah Zamma was born in Taroudante, Morocco. He obtained his Ph.D. in Materials Engineering from Hassan 2 University in Casablanca in 2012, With a solid academic background, Abdellah ZAMMA quickly entered the academic world as a professor at the "ENSET Mohammedia", a renowned institution affiliated with Hassan 2 University. As a professor, he shares his knowledge and expertise in the fields of materials engineering and industrial engineering with his students, preparing them for the challenges of the modern industry. With 13 years of professional experience, Abdellah ZAMMA has worked in various sectors related to materials engineering, including foundry, material elaboration, and nondestructive testing. His expertise in these areas makes him a sought-after and respected specialist, both nationally and internationally. Outside of his academic and professional activities, Abdellah ZAMMA continues to be passionate about research and innovation in the field of materials engineering by mentoring young doctoral students. His work and publications contribute to enriching the body of knowledge in this constantly evolving field.

***

Shalom Akhai is a distinguished engineering academician, researcher, and innovator with more than eighteen years of professional experience in academia, industry, and consultancy. He is the Founder of IGNIS Technical Solutions, an MSME-registered consultancy specializing in sustainable systems, energy-efficient design, and AI-enabled smart technologies. Recognized among the World's Top 2% Scientists (Stanford–Elsevier Global Ranking 2025), Dr. Akhai has authored over 100 publications in SCI, Scopus, IEEE, CRC, Springer, and IGI Global journals and books, along with more than 70 granted or published patents. He has also written nine academic books and presented his work at numerous national and international conferences. A respected reviewer, mentor, and innovation ambassador, Dr. Akhai continues to contribute significantly to sustainable engineering, advanced technologies, and research-driven development.

Houssam Eddine Chakir was born in Morocco, in 1988. He received the M.S. degree in electrical engineering from Claude Bernard Lyon 1 University, France. He receives the Ph.D. degree in Electrical Systems and Control Engineering from the University Hassan II, Morocco in 2019. He joined ENSET Mohammedia, Morocco, as associate professor of Electrical Engineering (2021). His research interests include renewable energy-based generators and energy management systems for micro grids and future smart grids, electrical system modeling and control and power quality improvement.

Karim Choukri is a lecturer in Electrical Engineering Department at the Hassan II University (UH2), Casablanca, Morocco. He received his B.Eng., and Ph.D. degrees in Electrical Engineering from Hassan 1 st University of Settat in 2011 and 2017 respectively. He is an Associate Professor in UH2, Casablanca, Morocco since 2021. He has made significant contributions to the Renewable Energy sector through his publications of articles and reports, including the REMAP2030 report with IRENA and the Arab Future Energy Index RE Report with RCREEE. As a member of the Renewable Energy University Network (REUNET) in Morocco, he demonstrates his dedication to the advancement of sustainable energy solutions. Furthermore, he serves as an expert reviewer for PRIMA, offering his expertise and insights to the initiative. His research interests include the field of Renewable Energy, Green Hydrogen, Converters design, industrial applications, industrial electronics, industrial informatics, power electronics, motor drives.

Lakshmi D is presently designated as a Senior Associate Professor in the School of Computing Science Engineering and Artificial Intelligence and Assistant Director, at the Centre for Innovation in Teaching & Learning (CITL) at VIT Bhopal. From 2016 to 2021 served as an Educational Research Officer, at Vishnu Educational Development and Innovation Centre (VEDIC) Shri Vishnu Educational Society, Hyderabad She has 26 years of teaching experience. She has addressed innumerable guest lectures, acted as a session chair, and was invited as a keynote speaker at several international conferences. She has conducted FDPs that cover approximately ~1,00,000 plus faculty members including JNTU, TEQIP, SERB, SWAYAM, DST, AICTE, MHRD, ATAL, ISTE, Madhya Pradesh Government-sponsored, MALVIYA MISSION TEACHER TRAINING, Central University of Gujarat and self-financed workshops across India as well as Misamis University, Philipines on Artificial Intelligence in Education: The Future and Futuristic Perspectives both theory and practical sessions. The other titles such as Modern Tools and Software in Teaching Learning Practices, Outcome Based Education, Scientific Educational Practices, Flipped Learning, Blended Learning, Activity Based Learning, Marching towards Quality and Excellence in Higher Educational Institutions, NEP 2022, Emotional Intelligence, Python Essential, Data Analytics, IoT, Machine Learning, Deep Learning, Cyber Security, AR/VR in Medicine, Architecture and Military Applications Instructional Strategies, Assessment techniques, Research Writing, etc. She has 20 international conference presentations, and 70 international journal papers and book chapters inclusive of SCOPUS & SCI. A total of 20 SCOPUS-indexed book chapters. A total of 24 patents are in various stages and 18 patents have been granted at both national and international levels. She was appointed as an external examiner for several Ph.D candidates at reputed universities. Under her supervision, one scholar has completed a Ph.D. at Annamalai University. One scholar is pursuing a Ph.D. right now at VIT Bhopal University. She has won three Best Paper awards at international conferences, one at the IEEE conference and another one at EAMMIS 2021 (First European, Asian, Middle Eastern North African Conference on Management & Information Systems). She received two awards in the year 2022 on the eve of Teacher's Day 1. Kamal Sharma Award of Academic Excellence - 2022 from Lexicon Group of Institutions, Pune for the Engaged Pedagogies. from the virtual ceremony. 2. Bharat Academic Excellence “UttamaAdyapika” Award from Braino Vision. She published a book on “Theory of Computation“ in 2003, “Leading Education in the Age of Disruption” in 2021, and “Smart Cities: Disruptive Technologies Driving the Force”, in 2023. In the last three editions, she acted as a SPOC for KAVACH and the Smart India Hackathon. She has created 15 hours of the high-quality “Python Essentials” online course at VITyarthi Online Learning Platform. She has 11 SCOPUS Indexed book edits with three CRS Press, Taylor, and Francis, four IGI Global, and two IEEE Rivers.

Mehdi Et-Taoussi was born in Morocco in 1990. In 2019 he received the Phd degree in Electrical Energy and Automatic from the Hassan II University of Casablanca Morocco. His fields of interest include: nonlinear control, smart grid and energy management.

Mangaiyarkkarasi J is currently working as an Associate Professor of Physics in NMSS Vellaichamy Nadar College, Nagamalai, Madurai- 625 019, Tamil Nadu, India. She has 23 years of teaching experience. Her major research interest includes synthesis of ceramic materials, powder X-ray analysis and electron density studies. She has published 11 research papers in the refereed International journals and five book chapters. She has prepared study materials (UG& PG Physics) for Madurai Kamaraj University Distance Education Course.

Houssine Laaz graduated with a State Engineering degree in Mechanical and Structural Engineering from the École Nationale Supérieure d'Arts et Métiers (ENSAM), Meknes, Morocco in 2012. Following his graduation, he gained valuable experience in the industrial coatings sector from 2012 to 2019. Currently, Mr. Laaz is a Professor of Industrial Engineering Sciences at the École Royale Navale (ERN) in Casablanca. He is also a doctoral candidate at the Laboratory of Mechanics, Engineering and Innovation at Hassan II University Casablanca, Morocco. His research interests focus on the application of artificial intelligence (AI), structural health monitoring (SHM) and non-destructive testing (NDT).

Mekkaoui Moussa received his engineering degree in Mechanical Engineering from ENSAM – Meknes (École Nationale Supérieure d'Arts et Métiers – Meknes) in 2014. He is currently working as a mechanical engineering instructor for preparatory classes. In 2022, he began his doctoral research in non-destructive testing using ultrasound. His research focuses on modeling the generation and propagation of Lamb waves in composite materials. His latest publication presents an innovative method for tracing dispersion curves in composite structures. Mekkaoui is pursuing his PhD at the Laboratory of Mechanics, Engineering, and Innovation (LMII) at Hassan II University.

Salah Nissabouri received the Eng. degree in industrial engineering from ENSAM enginnering school, Meknes, in 2012 and Ph.D. degrees in mechanical engineering from Science and Technic Faculty, Hassan I university, Settat, Morocco. Currently, he is an Assisstant Professor at the Department of Mechanical Engineering, Hassan II University. His research interests include Mechanical Modeling, Mechanical Simulation, Industrial Engineering, Non-Destructive Testing, Finite Element Analysis, Signal processing, acoustic and vibration control, and Artificial intelligence.

Hassan Rhimini received his PHD degree from Chouaib Doukkali University, EL JADIDA in 2009. He joined the National High School of Electricity and Mechanics in 2011. He currently holds the position of Co-Director of the Mechanics Laboratory of Engineering and Innovation (LM2I) at Hassan 2 University of Casablanca. He has published over 36 publications and communications in acoustics and modeling of ultrasound propagation. He is currently interested in the integration of artificial intelligence tools in the Structural Health Monitoring (SHM) of structures in aeronautics, automobiles and civil engineering. Hassan RHIMINI is a Research Professor at Hassan 2 University of Casablanca, Morocco.

Rita Sharma, an Assistant Professor of Physics at Rayat Bahra University, Punjab, holds an impressive academic background. She completed her M.Sc.,

M.Phil., and Ph.D. in Physics from Guru Nanak Dev University (GNDU), Amritsar. With more than eight years of teaching experience, Dr. Sharma has made significant contributions to the field of Material Science. Her expertise spans various areas, including glass formation, structure-property relations, phase-change memory materials, gamma-irradiated chalcogenide glasses, and the electrical and photovoltaic properties of chalcogenides. Dr. Sharma's current research interests focus on computational and theoretical aspects of experimental results. Throughout her academic journey, Dr. Rita Sharma has authored sixteen research publications, demonstrating her commitment to advancing scientific knowledge. Her teaching roles include serving as an Assistant Professor at the Department of Physics, Rayat Bahra University, Mohali, where she imparts knowledge in areas such as object-oriented programming. Additionally, she has contributed significantly to the academic community during her tenure as an Assistant Professor of Physics at The Oxford College of Science, Bangalore. Her research endeavors have led to valuable insights into the properties of chalcogenide materials, contributing to the field's understanding. Dr. Sharma's dedication to research and teaching underscores her passion for advancing scientific knowledge and fostering the next generation of physicists. Her work in material science and computational analysis continues to shape the field, making her a valuable asset to the academic community.

Jagdeep Walia, Assistant Professor, Chandigarh Engineering College, Chandigarh Group of Colleges, Jhanjeri, Mohali, Punjab, India, brings a wealth of academic and professional experience to his role. Her qualifications include a Master of Science (M.Sc.) in Physics, a Bachelor of Education (B.Ed.), and her current pursuit of a Ph.D. This ongoing doctoral work demonstrates Ms. Walia's dedication to continuous learning. Ms. Walia excels in core competencies like flexibility, adaptability, and classroom management, allowing him to thrive in diverse teaching environments and create engaging learning experiences. She fosters a collaborative spirit, contributing to effective teamwork within the academic community. Her achievements include being awarded "Best Mentor" on Teacher's Day and receiving an appreciation letter for outstanding performance results. Ms. Walia actively participates in professional development opportunities. Her memberships in IAENG and EDEN highlight his dedication to continuous learning. Workshops and conferences she has attended include a Short-Term Course on Effective Teaching, sessions on Physics with Homemade Equipment, and a state workshop on Optical Thin Film Coating and its Industrial Applications. Ms. Walia's research interests, while broad, demonstrate a focus on practical applications. Her publications cover topics like cybersecurity challenges, heavy metal removal from wastewater, and fluorescence spectroscopy applications in medicine. This bridge between theory and practical applications benefits both academia and society. Ms. Walia continues to inspire students, collaborate with colleagues, and contribute to the academic community. Her passion for education and ongoing research makes her a valuable asset to academic and research community.

Index

N

O

P

Q

R

S

www.ingramcontent.com/pod-product-compliance
Lightning Source LLC
LaVergne TN
LVHW060005031026
849161LV00025B/1487